Biology in Space and Life on Earth

Edited by
Enno Brinckmann

1807–2007 Knowledge for Generations

Each generation has its unique needs and aspirations. When Charles Wiley first opened his small printing shop in lower Manhattan in 1807, it was a generation of boundless potential searching for an identity. And we were there, helping to define a new American literary tradition. Over half a century later, in the midst of the Second Industrial Revolution, it was a generation focused on building the future. Once again, we were there, supplying the critical scientific, technical, and engineering knowledge that helped frame the world. Throughout the 20th Century, and into the new millennium, nations began to reach out beyond their own borders and a new international community was born. Wiley was there, expanding its operations around the world to enable a global exchange of ideas, opinions, and know-how.

For 200 years, Wiley has been an integral part of each generation's journey, enabling the flow of information and understanding necessary to meet their needs and fulfill their aspirations. Today, bold new technologies are changing the way we live and learn. Wiley will be there, providing you the must-have knowledge you need to imagine new worlds, new possibilities, and new opportunities.

Generations come and go, but you can always count on Wiley to provide you the knowledge you need, when and where you need it!

William J. Pesce
President and Chief Executive Officer

Peter Booth Wiley
Chairman of the Board

Biology in Space and Life on Earth

Effects of Spaceflight on Biological Systems

Edited by
Enno Brinckmann

WILEY-VCH Verlag GmbH & Co. KGaA

The Editor

Enno Brinckmann
Senior Biologist
ESA/HME-GPL, ESTEC
Postbus 299
2200 AG Noordwijk
The Netherlands

Cover
ESA Astronaut Ulf Merbold floating above the Glove Box of the Biorack facility during the first International Microgravity Laboratory mission (IML-1, STS-42) inside the Spacelab Module. Courtesy NASA (1992)

Library of Congress Card No.:
applied for

British Library Cataloguing-in-Publication Data
A catalogue record for this book is available from the British Library.

Bibliographic information published by the Deutsche Nationalbibliothek
The Deutsche Nationalbibliothek lists this publication in the Deutsche Nationalbibliografie; detailed bibliographic data are available in the Internet at <http://dnb.d-nb.de>.

Composition SNP Best-set Typesetter Ltd., Hong Kong

Printing Betz-Druck GmbH, Darmstadt

Bookbinding Litges & Dopf GmbH, Heppenheim

Wiley Bicentennial Logo Richard J. Pacifico

Printed in the Federal Republic of Germany
Printed on acid-free paper

ISBN 978-3-527-40668-5

Contents

Biology in Space and Life on Earth. Effects of Spaceflight on Biological Systems. Edited by Enno Brinckmann

ISBN: 978-3-527-40668-5

Foreword

Flying into space still is an endeavour very much at the fringes of technical feasibility, but the same time it pushes the horizon of mankind's experience into areas unknown so far. In this respect it continues the efforts of our ancestors to explore the world. Men like Marco Polo, Columbus, Vasco da Gama, Livingston, Cook, Amundson are representative for the human spirit so well described in Johann Wolfgang Goethe's drama "Faust", never to be satisfied with what we know, but rather to be driven to go further and to explore "terra incognita" in order to know more. Whether human curiosity is genetically programmed or there are other reasons, that we persistently are striving to understand the world better and better, might still be an open question, but there is no doubt, that it is the essence of philosophy and science.

At present it is for the first time in human history that we have the tools in hand to leave Earth behind at least temporarily. Considering the many contributions to our knowledge made by our ancestors with much less sophisticated tools I feel strongly that we, the now living generation, have no choice. We must take up the challenge of space exploration. Although there is always room to do more, I feel that essentially we are on the right way to live up to it. NASA's moon flights for example mark unique milestones in human history. Regarding the expansion of knowledge I think we also have been very successful by sending fairly elaborated unmanned space probes to the planets of the solar system. The knowledge we acquired in this way is many times more detailed than the knowledge accumulated during all centuries prior to the short era of spaceflight. Although a lot of work is still to be done, I am certain that within the first half of this century astronauts, respectively cosmonauts will fly to Mars. It is also evident that in addition to the exploration of space, particularly of the solar system, the scientific utilization of the unique space environment will be on mankind's agenda. In fact ESA started already decades ago to perform scientific investigations in space by exploiting microgravity and other unique features of space, e.g. the absence of atmosphere and the global view from orbit.

A visionary decision taken by ESA in seventies of the last century was to contribute to NASA's Space Transportation System by providing Spacelab. This element, in fact a full toolkit designed and built in Europe, converted NASA's Shuttles from transporters into laboratories suitable for a huge variety of scientific

Biology in Space and Life on Earth. Effects of Spaceflight on Biological Systems. Edited by Enno Brinckmann

ISBN: 978-3-527-40668-5

investigations including biology and human physiology. As a European astronaut on Spacelab-1 and on IML-1 with "Columbia" I am happy that Spacelab functioned flawlessly, whenever it was in orbit. The same time I regret that we did not adopt a credible European Spacelab-utilization programme, although many fascinating experiments were performed on several flights. By now the International Space Station is on line. Although for the near future first priority is given to its assembly, it will become the most sophisticated laboratory in orbit, as soon as its assembly is completed. Other than Spacelab it will be in orbit not only for a week or two, but for years. Its novel features will provide many more opportunities to the scientific community to do research in orbit.

Reviewing the scientific investigations performed in the last decades clearly reveals that microgravity turned out as the most relevant asset for gaining new insight. Experiments in human physiology and biology in particular led to many new conclusions and pushed the horizon of our knowledge. A number of studies in these disciplines were based on the astronaut's body as test object. In many cases like on ESA's Euromir'95 mission, we scientist-astronauts had to draw blood samples on each other or place electrodes on our bodies in order to acquire physiological data. Perhaps it was for that reason that most of us developed a special interest in many studies dealing with biological problems. In addition it was also evident, that the insight acquired usually had the potential of direct benefits. If we deepen the knowledge with regard to the metabolism of bones there might be a short way to improve the treatment of osteoporosis. Experiments bringing light into the mechanisms of human immune response also have a potential to finally improve the quality of life by finding new ways to stimulate its power. There are many other areas of research focussing on human physiology as well as on plant physiology, on bacteria, on cells, insects etc. Considering the full spectrum of scientific questions related to space flight biology is one field out of many others, but evidently it is one of the most important areas and definitively one of the most dynamical. The current publication "Biology in Space and Life on Earth" is a comprehensive presentation of the relevant research work performed at present.

Ulf Merbold

Preface

Space and Biology – this combination of terms is not very common in science despite the fact that biological phenomena in the unique environment of weightlessness have been analysed from the early days of space flight up to now. The exploration of Space required investigations in an area beyond the experience of man living on Earth: the border presented by the long-established environment of gravity had to be crossed, and, as in the rising era of steam trains when people were questioning the ability to survive in these high-speed vessels, one could not imagine the impact of zero gravity and Space radiation on any organism during space flight.

After more than 40 years of Space research we have a better understanding now about life in weightlessness. The authors of Chapters 4 and 8 summarize the unique results in plant root physiology and in cells of the immune system collected in their experiments over a period of 12–30 years. This wide timeframe indicates a typical fact of Space research: the frequency of experiments is very low, not days or weeks as in typical ground investigations but rather months or years due to the rare flight opportunities, the complex preparation and the many controls, all of them also having an impact on the costs of experimentation in Space. The transfer of an experimental idea from the common ground-based laboratory to the Space environment is usually combined with many tests to prove that the situation in orbit is comparable to that on the ground. Potential failures and side effects are described in Chapter 1, where the basic terms are also defined, especially the term "microgravity", which is very often used by the science community of gravitational and Space biology to describe the near weightless environment in an orbiting spacecraft. From the perspective of a Project Scientist, however, who was responsible on behalf of the Space Agency for a complete experiment from proposal until the final hand-over of the Space-flown samples to the investigator, these reviews allow one to look upon the subjects of investigation from a different angle – the big picture!

We share this perspective with our readers not only from the historical point of view, although right now experiments in Space biology are entering a new era with the International Space Station (ISS), after the retirement soon of the Space Shuttle as the main carrier for numerous experiment facilities. Most experiment platforms are accumulated now on the ISS, with a few others on satellites. The logistics and the experiment protocols are different on the ISS – more complex,

Biology in Space and Life on Earth. Effects of Spaceflight on Biological Systems. Edited by Enno Brinckmann

ISBN: 978-3-527-40668-5

on the one hand, but larger exposure to near weightlessness and Space radiation is possible on the other hand. These expectations will influence future research and our authors reflect these aspects in their conclusions and outlooks.

The other topic of this book is related to "Life on Earth". It was our aim to demonstrate that experiments in Space are not a solitary research field but are bound into a wide spectrum from ground-based fundamental research to application-oriented medical research. Chapters 2–4 discuss in detail the mechanisms affecting the orientation of plants in the gravitational environment. The progress achieved in this part of plant physiology was not possible without the experiments under reduced gravity in low Earth orbits. The reaction chain between the gravity stimulus and the cell-internal response can be described much better now with recent discoveries achieved in Space experiments – many pieces of the mosaic have been collected and implemented, either by falsifying a previous hypothesis (e.g. Chapter 4, Section 4.2.2) or by adding new evidence from facts previously unknown due to the permanent interference of gravity in ground-based experiments.

Human health research has also gained by space flight: Chapters 5–8 analyse investigations in the field of connective tissues, bone metabolism and immune system.

The widely spread bone loss or osteopenia by ageing or by disease, osteoporosis, is accelerated tremendously in weightlessness and is, therefore, a research objective not only in astronauts but also in cellular models, in which the primary reactions and the potential cure of bone loss can be investigated. Removal of the gravitational force is a perfect way for short-term experimentation with cell cultures, allowing deep insight in the primary processes of tissue formation (Chapter 5), bone formation (Chapters 6 and 7) and immune cell response (Chapter 8) *in vitro*. Whilst it seems obvious that the reduction of mechanical loading leads to bone loss (like on Earth during prolonged bed rest), it is not at all evident that cells of our immune system respond to gravity, which has been present during their entire evolution on Earth. Chapter 8 analyses this mysterious phenomenon that was already observed on astronauts in the very early days of human space flight.

Chapters 9 and 10 concentrate on the other feature of space flight: Space radiation and its impact on organisms and isolated cellular systems. Chapter 9 describes one kind of technological approach for radiation research in Space and on ground, focussing on radiation damage of the DNA in single cells. Chapter 10 extends this aspect of radiation research to general questions, ranging from evolution to the habitability of Mars.

We hope that the reader gets a good overview of past and current achievements in this comprehensive description of biological research in Space. Since most of the Space experiments described in the following chapters were performed in facilities of the European Space Agency (ESA), the Introduction summarises typical mission scenarios and describes ESA's experiment platforms for biological research in Space.

Enno Brinckmann
Leer, July 2007

List of Contributors

Rommel G. Bacabac
ACTA-Vrije Universiteit
Department Oral Cell Biology
Van der Boechorststraat 7
1081 BT Amsterdam
The Netherlands

Philippe Baert
University of Ghent
Department Molecular Biotechnology
Coupure Links 653
9000 Gent
Belgium

František Baluška
Institute of Cellular and Molecular Botany (IZBM)
Plant Cell Biology
University of Bonn
Kirschallee 1
53115 Bonn
Germany

Roger Bouillon
Laboratory for Experimental Medicine and Endocrinology
K.Universiteit Leuven
Gasthuisberg, O&N
Herestraat 49
3000 Leuven
Belgium

Markus Braun
Institute of Plant Molecular Physiology and Biotechnology (IMBIO)
Gravitational Biology Research Group
University of Bonn
Kirschallee 1
53115 Bonn
Germany

Dr. Enno Brinckmann
ESA (retired)
Eschenweg 16
26789 Leer
Germany

Geert Carmeliet
Laboratory of Experimental Medicine & Endocrinology
Katholieke Universiteit Leuven
Herestraat 49
3000 Leuven
Belgium

Lieve Coenegrachts
Laboratory of Experimental Medicine & Endocrinology
Katholieke Universiteit Leuven
Herestraat 49
3000 Leuven
Belgium

Biology in Space and Life on Earth. Effects of Spaceflight on Biological Systems. Edited by Enno Brinckmann

ISBN: 978-3-527-40668-5

Augusto Cogoli
Zero-g Life Tec GmbH
Technoparkstr. 1
8005 Zürich
Switzerland

Marianne Cogoli-Greuter
Zero-g Life Tec GmbH
Technoparkstr. 1
8005 Zürich
Switzerland

Dominique Driss-École
Université P. et M. Curie
Laboratory CEMV, case courier 150
Site d'IVRY-Le Raphaël
4 place Jussieu
75252 Paris Cedex 05
France

Gerda Horneck
DLR, Aerospace Medicine
Linder Höhe
51147 Köln
Germany

Jenneke Klein-Nulend
ACTA-Vrije Universiteit
Department Oral Cell Biology
Van der Boechorststraat 7
1081 BT Amsterdam
The Netherlands

Charles A. Lambert
Laboratory of Connective Tissues Biology
University of Liège
Tour de Pathologie, B23/3
4000 Liège
Belgium

Charles M. Lapière
Laboratory of Connective Tissues Biology
University of Liège
Tour de Pathologie, B23/3
4000 Liège
Belgium

Geert Meesen
University of Ghent
Department Molecular Biotechnology
Coupure Links 653
9000 Gent
Belgium

Betty V. Nusgens
Laboratory of Connective Tissues Biology
University of Liège
Tour de Pathologie, B23/3
4000 Liège
Belgium

Gérald Perbal
Université P. et M. Curie
Laboratory CEMV, case courier 150
Site d'IVRY-Le Raphaël
4 place Jussieu
75252 Paris Cedex 05
France

André Poffijn
University of Ghent
Department Subatomic and Radiation Physics
Proeftuinstraat
9000 Gent
Belgium

Jack J.W.A. van Loon
DESC (Dutch Experiment Support Center)
ACTA-Free University
Department Oral Biology
Van der Boechorststraat 7
1081 BT Amsterdam
The Netherlands

Patrick Van Oostveldt
University of Ghent
Department Molecular Biotechnology
Coupure Links 653
9000 Gent
Belgium

Dieter Volkmann
Institute of Cellular and Molecular Botany (IZBM)
Plant Cell Biology
University of Bonn
Kirschallee 1
53115 Bonn
Germany

Introduction

1
Flight Mission Scenarios

This section overviews the various missions that have provided platforms for the experiments presented in this book. In general, there have been four kinds of flights:

(1) parabolic flights in an aircraft with multiple periods of 20 s weightlessness;
(2) sounding rockets with 5–15 min experiment time;
(3) manned space flight missions with 7–16 days in orbit;
(4) robotic missions in unmanned capsules for 12–15 days in orbit.

Tables 1–3 summarize mission data and give a selected list of experiments. Each mission is identified by its flight number and contained several payloads: only those payloads listed here have been used for the experiments described in the following chapters. The selected experiments are given with their dedicated name, sometimes in connection with an identification number, and their Principle Investigator or team leader. The mission scenarios on the different carriers are not only distinguished by the duration of their free-fall (or microgravity) conditions but also by the required ground logistics before (=late access) and after the flight (=early retrieval); this is an important factor when fresh samples with limited life time have to be transported into Space and when the returning samples cannot be preserved in a stable condition or need immediate treatment after landing, e.g. behavioural studies on live animals.

Table 4 shows the late access time for all carriers, indicating that latest moment when the experiment has to be handed over in its flight configuration to the ground personnel for integration into the spacecraft, varying from a few hours to days. Sounding Rockets have a very late integration time due to their less complex payloads complement. The stowage compartment in the Space Shuttle Middeck cannot be loaded with experiments later than 17 hours before launch due to safety precautions, since the countdown has to continue with liquid fuel tanking of the spacecraft. The situation with the Foton satellite is different: the payload has to be

Biology in Space and Life on Earth. Effects of Spaceflight on Biological Systems. Edited by Enno Brinckmann

ISBN: 978-3-527-40668-5

Table 1 Sounding rocket missions with payloads relevant to some experiments described in this book. TEM: Texus experiment module; CIS: cell-in-space module; BIM: biology in microgravity.

Flight	*Date*	*Payload*	*Investigator (country)*
Maxus-1B	8 NOV 1992		Cogoli (CH)
Maxus 2	28 NOV 1995	TEM 06-5MZ	Cogoli (CH)
Maser-3	10 APR 1989	CIS-1	Cogoli (CH)
Maser-4	29 MAR 1990	CIS-2	Cogoli (CH)
Maser-5	9 APR 1992	CIS-3	
Maser-6	5 NOV 1993	CIS-4	
Maser-7	3 MAY 1996	CIS-5, EMEC	
Maser-9	16 MAR 2002	CIS-6	
Maser-10	2 MAY 2005	BIM-1	
Texus 18	6 MAY 1988	TEM-KT	Volkmann (D)
Texus 19	28 NOV 1988	TEM-KT	Volkmann (D)

integrated into the satellite before the satellite is mated with the launcher rocket, which happens two days before the launch. However, only a few small items without electrical interfaces to the satellite can be placed into the capsule on the launch pad around 12 hours before launch through a special hatch, which gives access to the payload located directly behind the hatch; all other payloads cannot be accessed by that time.

The situation is better for the Soyuz missions, where a late integration of payloads not heavier than 10 kg is possible a few hours before the crew climb into their seats. During past Soyuz missions and Foton-11, however, all items had to be transported to the launch site via Moscow to allow for customs inspection and required additional time for the subsequent transport to the launch site, which extended the late access time for live samples considerably. Only recently has some preparation of experiments been possible at the launch site in Baikonur (Kazakhstan). During the two-day flight period to the International Space Station (ISS), experiments in Soyuz were either inactive or limited to automatic functions: specific interaction by the crew was in general only possible after docking to the ISS, which added, for most samples, two more days to the storage period after handover. During descent, the Space Shuttle was usually equipped with ambient and cold stowage compartments, whilst the temperature for samples returning from ISS in the Soyuz capsule was not actively controlled, resulting in temperature peaks up to 31 °C.

Experiment integration into sounding rockets was possible until one hour before launch; however, the period of weightlessness was limited to 6, 6 and 12 minutes for Texus, Maser, Maxus rockets, respectively [2]. The early retrieval time of the payloads was again beneficial for experiments, as the experiment units were often returned to the scientists within one hour after landing.

Table 2 Space flight missions with crew support in the Space Shuttle (Space Transportation System, STS). Flights 61-A, 42 and 65 were research missions with Spacelab (D-1, German Spacelab Mission; IML-1 and IML-2, International Microgravity Laboratory #1 and #2), whereas flights 76, 81 and 84 were dedicated to activities on the Russian Mir Station with Spacehab in the Shuttle's cargo bay (Shuttle-to-Mir Mission, S/MM-03, 05 and 06). The experiment list is reduced to those described in this book.

Flight	*Date*	*Payload*	*Experiment (investigator)*
STS-9 Spacelab SL-1	28 NOV–8 DEC 1983	Biostack, Portable Incubator	Biostack (Bücker) ES029 (Horneck) AO5/17/LS/CH (Cogoli)
STS-61-A (22) Spacelab D-1	30 OCT–5 NOV 1985	Biorack	19-D DOSIMETR (Bücker) 32-CH BLOOD (Cogoli) 33-CH LYMPHO (Cogoli)
STS-40 Spacelab SLS-1	5–14 JUN 1991		781240 (Cogoli)
STS-42 Spacelab IML-1	22–30 JAN 1992	Biorack, LSLE freezer, Photobox	02-NL BONES (Veldhuijzen) 10-D MOROSUS (Bücker) 12-D DOSIMETR (Reitz) 14.1-CH FRIEND (Cogoli) 14.2-CH HYBRID (Cogoli) 14.3-CH CULTURE (Cogoli) 20-F ROOTS (Perbal)
STS-55 Spacelab D-2	24 APR–6 MAY 1993		RD-BIOS (Reitz) RD-UVRAD (Horneck)
STS-65 Spacelab IML-2	8–23 JUL 1994	Biorack, LSLE freezer, NIZEMI, Biostack	01.1-I ADHESION (Cogoli) 01.1-I MOTION (Cogoli) 08-NL BONES (Veldhuijzen) 12.1-D KINETICS (Horneck) 12.2-D REPAIR (Horneck) 19-D DOSIMETRY (Reitz) 32.1-F LENTIL (Perbal) CRESS (Volkmann) Biostack (Reitz)
STS-76 Spacehab S/MM-03	22–31 MAR 1996	Biorack, LSLE freezer	19-D DOSIMETRY (Reitz) 32.2-F STATOCYTE (Driss-École) 401-D STATOCYTE (Volkmann) 486-D X-RAY (Kiefer)
STS-81 Spacehab S/MM-05	12–22 JAN 1997	Biorack, LSLE freezer, Photobox	19-D DOSIMETRY (Reitz) 23-D CRESS (Volkmann) 89-F GRAVITY (Perbal) 27-D CHARA (Buchen) TEMP (van Loon)
STS-84 Spacehab S/MM-06	15–24 MAY 1997	Biorack, LSLE freezer	22-D BETARAY (Kiefer) 33-D DOSIMETRY (Reitz) 89-F ACTIN (Driss-École)
STS-95 Spacehab	29 OCT–7 NOV 1998	Biobox-4	HUDERM (Lapière) MARROW-4 (Bouillon)

Table 3 Space flight missions in unmanned satellites. The European Retrievable Carrier (EURECA) was a satellite with automated experiments, launched and retrieved by the Space Shuttle [1]. Bion and Foton satellites were launched with Russian carriers.

Flight	*Date*	*Payload*	*Experiment*	*Investigator (country)*
EURECA	31 JUL 1992–8 APR 1993	Exobiology & Radiation Assembly	Exobiological Unit	Horneck (D) Reitz (D)
Bion-10	29 DEC 1992–10 JAN 1993	Biobox-1	Bones	Veldhuijzen (NL) & Rodionova (UA)
			Fibro-1	Tairbekov (RUS)
			Oblast-1	Alexandre (F)
			Marrow-1	Schoeters (B) & Rodionova (UA)
Foton-10	16 FEB–3 MAR 1995	Biobox-2	Fibro-2	Tairbekov (RUS) & Lapière (B)
			Oblast-2	Alexandre (F)
			Marrow-2	Bouillon (B)
Foton-11	9–23 OCT 1997	Biobox-3	Fibro-3	Tairbekov (RUS) & Lapière (B)
			Oblast-3	Alexandre (F)
			Marrow-3	Bouillon (B)
		Biopan-2	Survival	Horneck (D)
Foton-12	9–24 SEP 1999	Biopan-3	Survival	Horneck (D)

The parabolic flight campaigns provided a completely different scenario. After experiment preparation at the airport (Bordeaux, France), where the Airbus 300 ZERO-G took off, the experiments were taken on board and mounted in their flight position. After a flight of about 0.5 h to the airspace where the parabolic flights were permitted, the experiments could be activated for one or more of the 30 parabolas. Each parabola had 20 s of hypergravity ($1.8 \times g$), followed by 22 s of near weightlessness ($10^{-2} \times g$) and again 20 s of hypergravity ($1.8 \times g$). After about 2 min the next parabola started. During the parabolas the scientists could execute their experiments themselves, analyse data in-flight and change parameters or test objects. Immediately after the approximately 3.5 h long flight, the scientists could analyse their experiments at the airport and prepare the next flight day. Normally, three flight days in a row were performed.

2 Sounding Rocket Experiments

A typical launch campaign of a sounding rocket flight started with the accommodation of the technicians and the scientists at ESRANGE close to Kiruna in Northern Sweden [2]. Well-equipped laboratories were set up to prepare live samples and to

Table 4 General conditions for experiment preparation, start, duration and return for manned and robotic space missions. Experiments on the Space Shuttle and on Soyuz flights to the International Space Station (ISS) could use crew interface for operations; experiments in the Foton capsule and in Sounding Rockets (Texus, Maser, Maxus) were performed autonomously.

Flight	*Space Shuttle (Middeck)*	*Soyuz Flight to ISS*	*Bion/Foton*	*Texus, Maser, Maxus*
Late access before launch	17–24 h	8–12 h	48 h	1 h
Launch site	Kennedy Space Center (USA)	Baikonur (Kazakhstan)	Plesetsk (Russia)	Kiruna (Sweden)
Experiment preparation at the launch site	Full laboratory facility	Limited laboratory facility	No laboratory facility	Full laboratory facility
Experiment start time in orbit	Launch + 4 h	Launch + 4 h (2–3 days)	Launch+ 9 min	Launch +70 s (Maxus: 96 s)
Experiment duration (maximum)	16 days	10 days	12–15 days	Texus: 6 min Maser: 5–7 min Maxus: 12.5 min
Temperature control at descent	Ambient, cooler, freezer	Ambient	Cooler (Biobox)	Experiment provided
Early retrieval at landing site	Landing + 6–8 h	Landing + 2 h	Landing + 1–2 h	Launch + 1 h

load them into their automatic experiment hardware. Integration into the payload platform of the rocket was carried out as late as 1 h before launch (Table 4). The flight duration varied with the power of the rocket motor: the smaller Texus and Maser rockets achieved an apogee height of 250 km and allowed experiments in microgravity for about 6 min, whereas Maxus reached 710 km and had, therefore, a useful microgravity period of 12.5 min. Many experiments were run autonomously in a programmed sequence. Those experiments could be observed by the investigator via a real-time video downlink and could be controlled by telecommand during the flight. At the end of the free-fall period, the payload was spun up and re-entered the atmosphere. This caused very short but heavy and randomly distributed accelerations. At 5 km altitude a parachute opened and returned the payload to the ground with a sink velocity of $8\,m\,s^{-1}$. The payload was recovered in one piece by a helicopter, which returned it to the launch site within about 1.5 h after lift-off. On request, a second helicopter was provided for immediate recovery

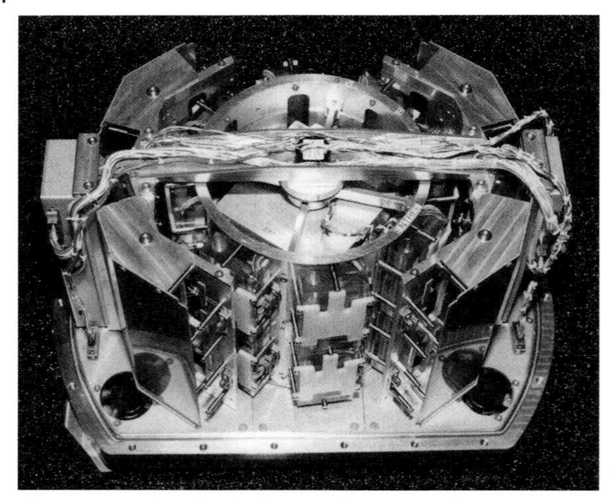

Fig. 1 A typical payload of a sounding rocket: the late access insert (LAI) of the biological incubator module (BIM) as used on the Maser 10 flight for biological experiments. The inner part of the insert consists of a reference centrifuge, which is surrounded by several static racks for accommodation of the experimental units.

of the experiment unit. This allowed a handover of the samples back to the scientists about one hour after launch. Maximum acceleration levels during launch were typically 12×*g* (Maser, Texus) and 13×*g* (Maxus), whilst the microgravity environment was in the order of 10^{-4}×*g*. During re-entry of the payload in the atmosphere, short-term accelerations of up to 50×*g* were possible in all axes, and during impact on the ground shock loads of 50×*g* to 100×*g* could occur.

The experiments were accommodated on circular platforms in experiment-specific modules: Texus Experiment Modules (TEM) were flown on Texus rockets and CIS (cell-in-space) modules on the Maser missions (Table 1). Each module was autonomous; it had its own power supply and electronics unit (Fig. 1). An identical module was often used for biological ground reference experiments. The useful diameter of the experiment deck was 403 mm, the length varied between 160 and 1155 mm, with a mass range of 22–116 kg. These modules provided the desired temperatures for the samples and automated features such as video observation, experiment activation and fixation, control runs on an onboard 1×*g* centrifuge, and data storage or transmission to ground.

3
Biobox on Foton and in the Space Shuttle

The Biobox facility was designed in 1990–1991 for biological experiments on unmanned recoverable capsules of the Bion and Foton type. It was operated in a fully automatic mode, without crew intervention or even telecommanding. After one mission on Bion (Biobox-1) and two on Foton (Biobox-2 and -3), the project was transferred to the US Space Shuttle. After one successful flight on STS-95 (Biobox-4) the career of Biobox on the Shuttle came to a premature end with the STS-107 disaster in 2003 (Biobox-5). A completely re-designed Biobox has been manufactured. Its first flight (Biobox-6) on Foton-M3 is scheduled for launch on 14 SEP 2007 from Baikonur, Kazakhstan. The following information applies to Biobox-1 through to Biobox-4.

Biobox was configured as a programmable single incubator. Experiments sharing that incubator were selected for compatible temperature requirements. On each flight the biological samples consisted of mammalian cell cultures (Table 3), which were accommodated in 30 automatic experiment units, with a standard size of $20 \times 40 \times 80\,mm^3$ (Biorack Type I container and CIS unit). In each unit, one or two 1-mL cultures could be grown. Culture media, biochemical stimulants and fixatives were contained within these units and could be supplied automatically according to a timeline pre-selected before flight.

From the 30 experiment units, six were placed on a centrifuge that generated 1×g acceleration during flight. As an additional reference, a duplicate model of Biobox was operated on the ground almost synchronously with the flight unit. After flight, the results obtained in microgravity were compared with those obtained at 1×g (both from the on-board centrifuge and from those on the ground) to identify biological effects specifically linked to weightlessness.

Before launch, the temperature in Biobox was maintained at 20 °C to suppress the growth and development of the cell cultures before entrance into microgravity. The cells were automatically awakened from their dormancy at 9 min after lift-off, when the in-built micro-accelerometer acknowledged the presence of microgravity. At this moment, the centrifuge kicked into action and the incubator temperature was switched to 37 °C, the optimal value for culturing mammalian cells. Later on in the flight, when all cultures had been stopped by adding fixatives, the centrifuge was switched off and the temperature was lowered to prevent the fixed material decaying. All these events occurred automatically, controlled by internal timers. Full automation was retained on the Space Shuttle (Biobox-4), with the crew operations restricted to the occasional cleaning of the Biobox cooling fans' inlet grid.

Nevertheless, the streamlined simplicity of the automated flight operations was off-set by the complexity and ever-changing demands of the mission operations. Note that Biobox-1 was the very first facility of the European Space Agency (ESA) on any Russian carrier [3]. To provide an impression, the as-flown mission scenarios for Biobox-1–4 are briefly described below, since they were all different.

3.1
Biobox-1

With an inconveniently long late access period of 48 h (Table 4), the decision was made to prepare the experiments, as well as the three Biobox facilities (flight, flight spare and ground), in Moscow. For this purpose a pre-fabricated laboratory (called *Moslab*) was set-up in the Netherlands and, after road transport, re-assembled in

Moscow [3]. Loaded with experiments, the flight model of Biobox was ferried by aircraft from Moscow to the launch site Plesetsk three days before launch. The ground model was retained and operated in *Moslab*.

Nine days into the mission the temperature in the Bion capsule started to drift beyond its nominal upper limit of 28 °C. At this point in time, Biobox was already switched to the cooling mode with all experiments completed and all samples fixed. However, the high satellite temperature and the limited air circulation in the capsule compromised the cooling performance of the Biobox. The temperature problems of the satellite not only affected the Biobox and forced the mission controllers to land the capsule two days earlier than planned.

3.2
Biobox-2

Similar to Biobox-1, all flight preparations were carried out in *Moslab*. The telemetry indicated that Biobox-2 performed flawlessly during this flight. After a successful landing, the Foton capsule was transferred by helicopter to the nearest airbase to remove the scientific hardware. Approaching the airstrip, the helicopter was caught by vehement gusts of wind, making the sling-carried Foton capsule swing like a pendulum. The crew was compelled to release the capsule, which fell to the ground from an altitude of 120 m. The capsule and its payload were ruined. The next morning *Moslab* was informed by phone about the crash. Later that day the remains of Biobox-2 were delivered at *Moslab*. The major part of the Oblast-2 experiment was absent and has never been recovered. The two other experiments (Marrow-2 and Fibro-2) miraculously survived the crash, protected by their sturdy experiment containers.

3.3
Biobox-3

Owing to a changing financial and political climate, *Moslab* could no longer be maintained. An alternative ground operations plan was required. All pre-flight operations, including the experiment preparations, were transferred to the ESA facility (ESTEC) in the Netherlands. To send Biobox as late as possible to the launch site Plesetsk, a special aircraft was chartered. After landing in Moscow for customs clearance and refuelling, Biobox was left on the plane while the ESA personnel disembarked for passport and visa clearance. Six hours later the aircraft was back in the air, destination Plesetsk, leading to a complete transportation time (from experiment handover until launch) of 72 h.

The Biobox ground model was retained at ESTEC. When new telemetry was dumped to ground, ESTEC was informed by phone from the flight control centre in Moscow. The telemetry indicated that the FIBRO-3 experiment was not properly executed. Two days after landing (Fig. 2), when Biobox was returned to ESTEC, the failure of FIBRO-3 was confirmed. This was the first and only time that an experiment was lost due to a technical failure in the Biobox facility.

Fig. 2 Foton-11 satellite after landing in the Russian tundra 180 km East of Orsk. The capsule has been opened and the Biobox is visible inside. ESA's Mission Manager P. Baglioni was at the landing site to retrieve the experiments from Biobox-3.

3.4 Biobox-4

After Biobox-3 a new flight opportunity was offered on the US Space Shuttle. Once more, a brand-new ground operations scenario had to be devised. This time, the experiments were prepared at ESTEC in the Netherlands, while Biobox was simultaneously readied at the Spacehab Payload Processing Facility (SPPF) in Florida. The fully prepared experiments, in thermal boxes at 20 °C, were flown three days before launch from ESTEC to Florida for installation in Biobox, which happened 36 h before launch. Despite the better late-access conditions (36 h for Spacehab, 48 h for Bion/Foton), the lead time for the sample preparations was not improved due to the long transatlantic journey. The Biobox-4 ground model was retained and operated at ESTEC.

4 Biorack in Spacelab and Spacehab

Biorack was the first multi-user facility of the European Space Agency (ESA), designed for biological experiments in Spacelab, the European contribution to NASA's Space Transportation System (STS), better known as the Space Shuttle. It flew three times in Spacelab and three times in Spacehab. Spacelab was the European part of the Space Shuttle science programme, whilst the McDonnell-Douglas-built Spacehab was a kind of cargo-carrier that also offered interfaces to experiments. Both modules provided a pressurized atmosphere with environmental control (oxygen, carbon dioxide, humidity). Table 5 shows the environmental data of a typical Spacehab mission [4]; these data were similar during the Spacelab

Table 5 Spacehab environmental data during the Shuttle-to-Mir Mission #6 (S/MM-06). The data are 12-hourly mean values from the Payload Operations Control Center (POCC) readings and were used for a simulation of the environmental conditions in Spacehab on the ground. Parameters are given in various dimensions for convenience. MET: mission elapsed time; r.h.: relative humidity; MDDK: Shuttle Middeck

MET d/h:min	*Spacehab temperature*		*Dew point temperature*		*Relative humidity*		*Cabin pressure*		*O_2 concentration*			*CO_2 concentration*	
	°F	°C	°F	°C	MDDK (% r.h.)	Spacehab (% r.h.)	psi	hPa	mmHg	psi	%	mmHg	%
0/11:00	76.3	24.6	51.3	10.7	35.3	41	14.8	1023	767	3.0	20.3	1.2	0.1565
0/23:00	79.0	26.1	52.4	11.3	35.0	40	14.9	1027	770	3.0	20.1	2.4	0.3117
1/12:00	80.2	26.8	53.6	12.0	34.9	40	14.8	1023	767	3.0	20.3	1.9	0.2477
1/23:00	80.1	26.7	54.5	12.5	36.5	42	14.8	1023	767	3.0	20.3	1.9	0.2477
2/11:30	80.9	27.2	55.6	13.1	39.9	42	14.6	1007	755	3.1	21.1	2.1	0.2781
3/00:00	80.2	26.8	55.6	13.1	39.2	43	14.9	1027	770	3.1	20.8	4.6	0.5974
3/12:00	75.3	24.1	56.1	13.4	39.5	51	14.9	1027	770	3.1	20.8	2.5	0.3247
4/00:00	74.5	23.6	54.8	12.7	38.4	50	14.9	1027	770	3.1	20.8	4.1	0.5325
4/12:00	76.7	24.8	55.9	13.3	39.7	49	14.9	1027	770	3.1	20.8	2.3	0.2987
5/00:00	73.2	22.9	53.8	12.1	40.3	51	14.9	1027	770	3.1	20.8	3.5	0.4545
5/11:30	75.5	24.2	54.8	12.7	39.6	48	15.1	1041	781	3.5	23.2	2.2	0.2817
5/23:30	72.7	22.6	54.6	12.6	39.3	53	15.1	1041	781	3.5	23.2	3.6	0.4609
6/11:30	74.9	23.8	54.4	12.4	38.1	49	15.3	1055	791	3.7	24.2	1.6	0.2023
6/23:30	72.8	22.7	53.4	11.9	35.9	51	15.1	1041	781	3.8	25.2	1.5	0.1921
7/11:30	75.7	24.3	56.7	13.7	43.9	51	14.8	1020	765	3.5	23.6	2.1	0.2745
8/00:00	73.5	23.1	54.5	12.5	40.4	51	14.7	1014	761	3.3	22.4	3.2	0.4205
8/12:00	74.0	23.0	56.0	13.0	42.2	54	14.7	1014	761	3.1	21.1	2.4	0.3153
9/00:00	73.0	23.0	54.0	12.0	37.8	50	14.8	1020	765	3.0	20.3	3.7	0.4837

flights [5–7]. From Table 5, the environment was obviously similar to ground laboratory conditions, except for the slightly higher and fluctuating temperatures (22.6–27.2 °C) and the intentionally elevated carbon dioxide level to save absorber material: CO_2 levels varied between 0.16% and 0.6%, as compared to about 0.04% in ground laboratories.

In addition to Spacelab and Spacehab, the Shuttle's middeck could also be used for experiment facilities and stowage with the same environmental conditions. The middeck stowage lockers were the latest accessible units for experiment storage (Table 4). In contrast, late access to the Spacelab/Spacehab modules was only possible by climbing down through the (vertical!) tunnel when the Shuttle was on the launch pad. This activity had to be completed about two days before launch. Biorack was always integrated in the modules, but its stowage containers were mainly located in the middeck, providing the shortest possible storage period for the live samples before lift-off. This late access time was also supported by the very late preparation of the samples in fully equipped laboratories in Hangar-L at the Kennedy Space Center close to the launch site.

The specimens were loaded into their experiment-specific hardware inside the standard Experiment Container (EC). Four types of ECs were used [8]: the smaller Type I/0 container with an internal volume of $20 \times 40 \times 82\,mm^3$ (20 mm in direction of the *g*-vector) and the larger Type II/0 container with $87 \times 63 \times 63\,mm^3$ internal volume. Each type was also available with electric connectors (Type I/E and Type II/E), providing power lines for ±12 V and +5 V DC and analogue data transfer to Biorack. The EC was either gas tight or equipped with custom-made holes to allow air exchange. The Type I containers were also used for experiments in sounding rockets and in Biobox. Each experiment used on average about ten ECs, resulting, for example, in a Biorack mission complement of 168 ECs for the IML-1 flight.

The entire Biorack facility was a combination of three incubators, a cooler/freezer, a glovebox and stowage racks (Fig. 3). Each incubator contained two centrifuges (see Fig. 4.8D, p. 83) and static rack positions for the experiment containers, the latter exposing the samples to microgravity conditions. Incubator temperatures were set before flight in the range 22–37 °C. There was no temperature control required in Biorack during launch and landing, since the incubators and cold storage racks were empty. The centrifuges provided 1×g acceleration (centre of EC) for eight Type I containers but had to be stopped for container transfer, leading to accumulated stationary periods of about 0.1–1.7% of the entire running time. When the centrifuge was shared, these stops also affected briefly other experiments. A post-flight evaluation of 22 Biorack experiments revealed that the on-board 1×g reference centrifuge gave in 50% of the experiments the same results as the controls in the identical Biorack ground model [4]. This could be interpreted as an influence of other spaceflight conditions, such as radiation, or as an effect of a delayed, and in several cases interrupted, exposure to the 1×g conditions in orbit (see also Table 1.1). On the Shuttle-to-Mir missions, the centrifuge was also adjustable in-flight to pre-set accelerations of 0.1, 0.2, 0.3 and 0.4×g.

The cooler/freezer unit was flown twice with 4 °C/–15 °C, respectively, but these units were not active during launch and landing. Active cooling during two mis-

Fig. 3 Ulf Merbold floating above the Biorack Glovebox during the Spacelab IML-1 mission inside the Spacelab Module 1992. Two incubators are located above and below the glovebox; the cooler/freezer combination is mounted in the upper tilted part of the rack. (Photo: NASA.)

sions was provided by NASA's LSLE-freezer (Life Sciences Laboratory Equipment, LSLE), set at −22 °C. As an alternative to these power-consuming cooling units, ESA used Passive Thermal Conditioning Units (PTCU), which consist of a large container with a wax, which was solidified before flight at low temperatures, surrounded by a double-walled evacuated Dewar vessel (Ø 21×50 cm) with high insulation capacity. Several kinds of wax served as phase change material and kept the temperature constant for about 20 days until the wax started melting. This way it was possible to store, for example, up to 20 Type I containers refrigerated or frozen, especially during launch and landing, when power supply was limited. Even loading of the PTCU with "warm" containers did not affect the performance of the PTCU considerably. During the last two Shuttle-to-Mir missions (S/MM-05, S/MM-06), Biorack had to rely on the +5 °C PTCU as the only source of refrigeration capacity.

To protect the crew from potential contamination with toxic materials, Biorack was equipped with a sealed glovebox, which had in addition an under-pressure of 7 hPa. Thus, two levels of containment were achieved by the mechanical isolation and the under-pressure in case a leak would occur in the gloves or in the box itself. By this safety precaution, the crew was able to open the Experiment Containers and to perform experimental manipulations, mainly activation or fixation of the samples, with dedicated tools or simple laboratory equipment, like syringes or multi-injectors. A video camcorder or a photo camera above the glovebox window allowed ground controllers to observe the glovebox operations and to document experimental results (Fig. 4).

The stowage racks next to the Biorack facility were used to store Experiment Containers at ambient temperature. In addition, a photo camera, tools, film rolls, video tapes and spare parts were stored in these lockers. A special storage item was the Photobox developed by the French Space Agency (CNES) for experiments

Fig. 4 Payload Commander Rick Hieb working in the Biorack glovebox in Spacelab during the second International Microgravity Laboratory mission (IML-2). A camcorder mounted on the handrail above the glovebox allowed real-time video observation on the ground of the experimental steps performed in the glovebox in space. (Photo: NASA.)

with plant roots (see Chapter 4, Section 4.4.2). The Photobox was $40 \times 17 \times 15$ cm and was battery driven. It allowed time-lapse flash photography of root curvature in microgravity. The crew placed up to six mini-containers into the holder of the Photobox, which had mirrors to reflect the twelve surfaces with plant seedlings into the camera. After crew activation, the Photobox captured automatically the root development on the 35-mm camera during several hours. For the S/MM-06 mission, the Photobox was equipped with a battery driven mini-centrifuge. In this way the root curvature after a gravity stimulus could be followed without crew intervention or any other disturbance of the microgravity environment.

After the retirement of Biorack in 1997 – after six successful flights with 89 experiments performed – a new facility was developed by ESA for biological experiments in the Space Shuttle. It was called Biopack and had the size of two Middeck Lockers. Biopack consisted of an incubator with static racks and three small centrifuges, accommodating the standard Type I and Type II containers. A small built-in refrigerator and freezer as well as a portable glovebox completed the Biopack facility [9]. The experiments could be monitored with telemetry and controlled by telecommanding from ground. However, its maiden flight on STS-107 in 2003 ended with the Columbia disaster.

Since ESA had lost the Biobox flight unit on the failed Foton-M1 launch (2002) and the spare flight model on STS-107 together with Biopack, no facility for biological experiments was left. An immediate alternative was found in the Aquarius incubator, which had been used by the French Space Agency for the Soyuz mission "Andromède" in 2001. Aquarius was basically a thermally controlled transport container with an 8-litre volume. It was launched with Soyuz and then transferred to the Russian complement of the ISS. Aquarius was used also for the Belgium

Soyuz mission "Odissea" in 2002 and the Spanish Soyuz mission "Cervantes" in 2003. For the Dutch Soyuz mission "DELTA" in 2004, a new concept of this – now called Kubik – incubator was developed by ESA: besides a wider temperature range, including refrigeration capacity, an insert with a reference centrifuge for eight Type I containers was added. Since sufficient stowage room is not available during the return of the Soyuz spacecraft, the incubators will stay on board of the ISS and can be used for future experiments, whereas the returning samples are carried in smaller foam-insulated containers to ground. As many experiments demand a fully controlled environment, ESA has also developed two sophisticated incubator systems for future research projects on the ISS [10]: the European Modular Cultivation System (EMCS), which is already performing well in orbit, and Biolab, which resembles a complete small biological laboratory and will be launched in the European Columbus Module end of 2007.

Acknowledgements

We thank the publisher for taking the initiative and inviting us to write this book. The flights of the facilities described were only possible with the help of all teams involved in the development and, finally, in the mission campaigns. A special tribute is paid to the astronauts of the various missions in orbit and to the operators on ground for their skilled experiment performances. I also acknowledge the contributions of my ESA colleagues René Demets and Wolfgang Herfs, who provided all the details about the Biobox experiments and the sounding rocket flights.

Enno Brinckmann
Leer, February 2007

References

1 Innocenti, L., Mesland, D.A.M. (Eds.), Eureca Scientific Results, *Adv. Space Res.* **1997**, *16(8)*.
2 Cogoli, A., Friedrich, U., Mesland, D., Demets, R. *Life Sciences Experiments Performed on Sounding Rockets (1985-1994)*, ESA SP-1206, ESA Publications Division, ESTEC, Noordwijk, **1997**.
3 Demets, R., Jansen, W.H., Simeone, E. *Biological Experiments on the Bion-10 Satellite*, ESA SP-1208, ESA Publications Division, ESTEC, Noordwijk, **2002**.
4 Perry, M. (Ed.), *Biorack on Spacehab: Experiments on Shuttle to Mir Missions 03, 05 & 06*, ESA SP-1222, ESA Publications Division, ESTEC, Noordwijk, **1999**.
5 Longdon, N., David, V. (Eds.), *Biorack on Spacelab D1*, ESA SP-1091, ESA Publications Division, ESTEC, Noordwijk, **1987**.
6 Mattok, C. (Ed.), *Biorack on Spacelab IML-1*, ESA SP-1162, ESA Publications Division, ESTEC, Noordwijk, **1995**.
7 Cogoli, A. (Ed.), Biology under Microgravity Conditions in Spacelab IML-2, *J. Biotechnol.* **1996**, *47*, 65–403.
8 Briarty, L.G. *Biology in Microgravity. A Guide for Experimenters.* ESA TM-02, B. Kaldeich (Ed.), ESA Publications Division, ESTEC, Noordwijk, **1989**; see also "Book Profiles" on web page: http://www.spacebio.net/general/index.html.

9 Van Loon, J.J.W.A. *J. Grav. Phys.* **2004**, *11*(*1*), 57–65.

10 Brinckmann, E., New Facilities and Instruments for Developmental Biology Research in Space, in: *Developmental Biology Research in Space*, Marthy, H.-J. (Ed.), pp. 253–280, Advances in Space Biology and Medicine, Vol. 9, Elsevier, Amsterdam, **2003**.

1
The Gravity Environment in Space Experiments

Jack J. W. A. van Loon

1.1
Introduction to Gravity Research

In 1687, Isaac Newton (1642–1727) produced his famous and important work *Principia*, based on the experiments and analyses of Galileo Galilei (1564–1642). In this work he proposed three main laws. The first was the law of *inertia*. He used the term inertia for the property of matter that causes it to *resist a change* in its state of motion. Inertia of an object must be overcome to set it in motion. The measure of an objects' inertia is its *mass*. The law of inertia states:

> If the net force on an object is zero (i.e. if the vector sum of all forces acting on the object is zero), then the acceleration of the object is zero and the object moves with constant velocity.

$$F_{net} = 0 \tag{1}$$

Newton's second law states:

> The accelerated motion of a body can only be produced by the application of a force to that body. The direction of the acceleration is the same as the direction of the force and the magnitude of the acceleration is proportional to the magnitude of the force.

$$F = ma \tag{2}$$

Commonly, F is referred to as weight, w. The acceleration, a, is referred to as g or gravity, on Earth $9.81\,\mathrm{m\,s^{-2}}$, m=mass (kg). The SI unit for gravity is "g" (lower case), not the often, especially in human gravitational physiology papers, applied "G" that is in SI units the gravitational constant [see Eq. (4)].

Newton's third law tells that single forces cannot occur. Forces always act in pairs. It states:

Biology in Space and Life on Earth. Effects of Spaceflight on Biological Systems. Edited by Enno Brinckmann

ISBN: 978-3-527-40668-5

If object 1 exerts a force on object 2, then object 2 exerts an equal force, oppositely directed, on object 1.

$$F_{12} = -F_{21} \quad (3)$$

All forces have a direction and a magnitude; force is a vector quantity. The SI unit of force (F) is the newton. One newton (N) is the unit mass (kg) times the unit acceleration ($\mathrm{m\,s^{-2}}$). Obviously, these three laws also apply to *microgravity* or, better, "*microweight*" research, or in more general terms to *acceleration* research.

Gravity is one of the four basic forces described in nature today, besides the weak nuclear force, the strong nuclear force and the electrostatic force. Among these forces, gravity is by far the weakest. Compared with a unit gravity or hypergravity environment, various physical phenomena behave differently in the, theoretical, state of zero acceleration, also known as weightlessness. In our universe, zero gravity exists only in theoretical terms, since there is mass in outer space and, hence, gravity fields, as shown by Newton's "universal law of gravitation":

$$F_G = G(mM/r^2) \quad (4)$$

where F_G = the gravitational force, G = the gravitational constant ($6.67 \times 10^{-11}\,\mathrm{N\,m^2\,kg^{-2}}$), m = mass of the object, M = mass of the Earth, r = distance between the centre of the two masses m and M.

As mentioned before, in our Universe, there is no such thing as zero gravity, since masses are present everywhere and, therefore, so are gravitational forces. In an orbiting spacecraft such as the Space Shuttle, Soyuz or the International Space Station there is also a small gravity residuum due to various reasons. A typical spacecraft would experience atmospheric drag (6–$30 \cdot \times 10^{-7} \times g$, at 250 km altitude), solar radiation pressure ($\pm 5 \cdot \times 10^{-9} \times g$), gravity gradient of extended bodies not located in the centre of mass ($\pm 5 \cdot \times 10^{-9} \times g$) and finally self-gravitation due to its own mass ($\pm 10^{-9} \times g$) [1]. Disturbance of the gravity level in space flight experiments is further discussed below.

1.1.1 Principle of Equivalence

An important aspect of Einstein's general relativity has to do with the equivalence of gravitational fields and accelerated motion. If we are in a laboratory on Earth, a mass that is released will fall, or accelerate, downward due to the gravitational attraction of the Earth. Now, let us, theoretically, move this laboratory into space, away from the gravitational influence. We now take the same object and release it within an accelerating rocket. Presume the level of acceleration of the rocket is the same as the Earth gravity, g ($9.81\,\mathrm{m\,s^{-2}}$). In such a situation the rocket will "push" onto the laboratory floor and move this floor towards the "falling" mass. As far as the observations of the two motions of the mass relative to the floor are concerned, the accelerated motion in the two cases will be exactly the same. If the

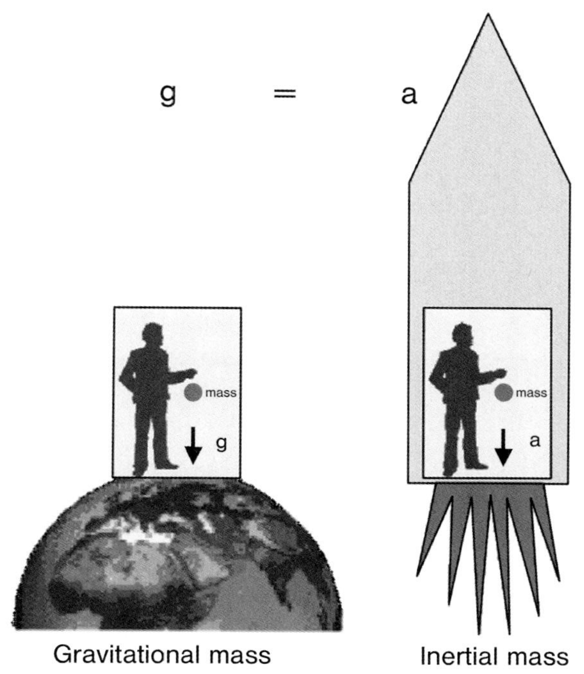

Fig. 1.1 Difference between "gravitational mass" (m_G) and "inertial mass" (m_I). The observer would see no difference between the two accelerations, hence $m_G = m_I$. (Picture used with permission from J. van Loon, DESC, Amsterdam.)

laboratory had no windows, the observer could not distinguish between an acceleration due to gravity (e.g. on Earth) and an acceleration due to a "push" of, for example, a rocket in space (Fig. 1.1). Or, as Einstein described exactly one century ago in his "principle of equivalence":

> In a closed laboratory, no experiment can be performed that will distinguish between the effects of a gravitational field and the effects due to an acceleration with respect to some inertial reference frame.

1.1.2
Microgravity

One of the often made mistakes about microgravity and space flight is to presume that the presence of microgravity in an orbiting spacecraft, such as the Space Shuttle or the International Space Station, ISS, is because "they are in Space" or "outside the Earth's atmosphere", although it might seem that these spacecraft are "far" away from human perspective they are actually quite close to Earth (only 300–500 km above the Earth's surface). So it is not the fact of being up in Space but rather that a spacecraft is orbiting at a particular velocity that provides a near weightlessness environment. These spaceships or satellites actually fall around the Earth with a speed of about 28 000 km h^{-1}. In fact, if the Earth's surface would be as smooth as a snooker ball it would be possible to provide microgravity in a capsule that circulates the Earth's surface at a distance of just 1 mm. For a typical Shuttle mission at an altitude of, for example, 350 km the level of gravity is still 9.04 m s^{-2} that means only 8% less than the gravitational field on the Earth's surface.

It was only at the onset of space exploration and with the possibility of performing experiments in free falling capsules that we became able to perform so-called "microgravity" or better "microweight" experiments. The possibility to compensate the, on Earth, ever present gravitational force provides us with a very useful tool and a unique experiment environment to study the effects of gravity on fluids, dust or plasmas, but also possible effects on living systems such as humans, animals, plants, cells or even smaller structures.

Before we could actually perform spaceflight experiments, there were some reports that stated that, on a cellular level, gravity, either micro- or hypo- or hypergravity, would have no effect on these systems. One of the first papers on the physical background of microgravity studies on a cellular level was written by Pollard over 40 years ago [2]. In this very illustrative work he evaluates phenomena like temperature, Brownian movement, convection, hydrostatic forces, and stresses experienced by a cell membrane. He concluded that cells with a diameter of 10 µm and more would experience gravity. This effect could be due do the redistribution of nucleoli or mitochondria. Although comparing various forces and energies present within living cells may be very illustrative, their magnitude is not necessarily related to their impact. The extent of various cellular physicochemical phenomena has been described in an interesting essay by Albrecht-Buehler [3]. In this paper the author relates the micro-environment of cells with the macro-environment where we, as human beings, relate to. It is calculated that the force of gravity is 400 000× smaller than the force of surface tension, and that the force for moving only one single electron in a typical electrical field of a nerve cell membrane is comparable to the weight of an entire cell. Water appears to be very viscous on a cellular level, while the impact energy, by Brownian movement, of only three water molecules is comparable to the weight of an entire cell. The gravitational potential energy of a cell due to gravity is about 21 kJ mol^{-1}, while the chemical energy of only one hydrogen bond is 17 kJ mol^{-1}. Finally, the contractile force of a single sarcomere is comparable to the weight of 60 cells. In addition, the force of polymerization is far more powerful than the weight of a single cell, while the force needed to add only one sub-unit of, for example, microfilaments, is about ten times larger than the weight of a cell.

All these theoretical studies indicate that, on a (sub-)cellular level, gravity would have no impact whatsoever. However, in recent decades numerous studies, both on orbiting spacecraft and in ground-based facilities, have indicated that gravity does have an effect on small systems (for reviews see Refs. [4–11]). The mechanism of this phenomenon is still a matter of debate. More recent experiments and, especially, future experiments on board of orbiting spacecrafts as well as in ground-based facilities might reveal some of the mechanisms involved.

1.1.3 Artificial Gravity

Since gravity, or acceleration, is a vectorial entity we can produce "artificial gravity" by changing the direction of the vector. This can be done in a centrifuge. In a constantly rotating centrifuge the object moves with a constant velocity. However, since the orientation is constantly changed, the object is accelerated:

$$a_c = \omega^2 r \tag{5}$$

where a_c = centripetal acceleration ($m\,s^{-2}$), ω = angular velocity ($rad\,s^{-1}$), r = radius (m).

The force on this object would be a centripetal force:

$$F_c = ma_c = m\omega^2 r \tag{6}$$

In spaceflight experiments, especially in biology, we often make use of a so-called "on-board" centrifuge. Why would we spend so much effort and money to do a spaceflight/near weightlessness experiment but also provide a 1×g environment?

To be able to draw any conclusions from spaceflight, which requires microgravity experiments, it is important to have a proper control group. To obtain the best control, we have to fully understand all possible influences involved in such experiments. What is that proper control? There are some important differences between a sample exposed to spaceflight microgravity (μg) and a 1×g on-ground control, such as launch vibrations, cosmic radiation and experiment lag time. The better control seems to be an on-board centrifuge. See also Table 1.1.

For biological experiments it was mainly the Biorack facility [4, 12] and later many others that accommodated an on-board 1×g control. However, even though an on-board centrifuge might be the best control, such a configuration also brings about particular artefacts between ground and flight 1×g, one of them being the inertial shear forces [13, 14].

Shear forces can be brought about by inertia (inertial shear) and/or fluid flow (fluid shear). In cell biology, fluid shear is an important physiological phenomenon and most common in blood vessels, where endothelial cells are exposed to blood flow. For endothelial cells the shear stress is in the order of 0.1–0.5 and 0.6–4.0 Pa in venous and arterial vessels, respectively [15]. Not only the cardiovascular system but also the mechano-adaptation of bone is most likely governed by fluid shear forces around osteocytes [16, 17 and Chapter 6]. Inertial shear forces, however, are mostly generated in materials exposed to accelerations. In cells, both fluid shear stress and inertial shear stress will generate cell deformation or strain. In centrifuges, an essential difference between inertial shear force, F_i, and

Table 1.1 Various parameters experienced by samples in-flight (1×g controls and microgravity, μg) as compared with on-ground samples.

Variable	*On-ground 1×g*	*In-flight 1×g*	*In-flight μg*
Launch accelerations	No	Yes	Yes
Launch vibrations[a]	No	Yes	Yes
Cosmic radiation	No	Yes	Yes
Need for centrifuge	No	Yes	NA
Centrifuge vibration[b]	No	Yes	No
Exposure to microgravity before start of experiment[c]	No	Yes	Yes
Temperature gradients/differences[d]	Different	Different	Different
Storage conditions[e]	Different	Different	Different
Flight Crew procedures and manipulations[e]	No	Yes	Yes
Inertial shear forces	No	Yes	No

a Launch vibrations are also accelerations, being at higher frequencies and in random directions.
b Static, non-rotating, samples may still be exposed to remnant vibrations of a centrifuge or other moving parts transmitted through the structure of the facilities or the spacecraft.
c For some facilities, like Biobox, or Kubik, this time is reduced to a minimum.
d Small temperature gradients between, for example, a centrifuge and static sample, or between ground and flight samples.
e Depending on the environment and crew-members involved, there will always be (small) differences in experiment and sample handling.

the force of gravity, F_g, is that inertial shear acts perpendicular to the gravity acceleration vector (Fig. 1.2).

The level of inertial shear experienced by an adherent cell layer depends on the radius of the centrifuge and the location of the cells within the sample surface area. Adherent cells attached to a flat surface will experience a larger inertial shear force, F_i, when located further from the point where the radius is perpendicular to the monolayer surface (Fig. 1.2). This effect is enhanced in centrifuges with a smaller radius. Inertial shear acceleration for adherent cells results in cell deformation that means strain.

For non-adherent and non-actively moving cells the situation is different. Here the cells do not experience inertial shear strain, as they do not attach to the substrate. However, cells on an in-flight 1×g centrifuge arrange themselves differently in the sample volume than they would on Earth. Free moving particles in a rotating system will move to the area of highest acceleration. When we consider a homogeneous suspension of cells and place this, on-ground, in a flat bottomed dish the cells will distribute evenly over the surface area as shown in Fig. 1.3(A). When we apply 1×g to the same dish in an on-board centrifuge the cells will, due to inertial force, move to the experiment container outer edges. In such a configuration the cells will pile up onto each other (Fig. 1.3B). One could imagine that all kind of cells physiological properties might change because of this. Cells might have more cell–cell interactions and medium depletion might become an issue,

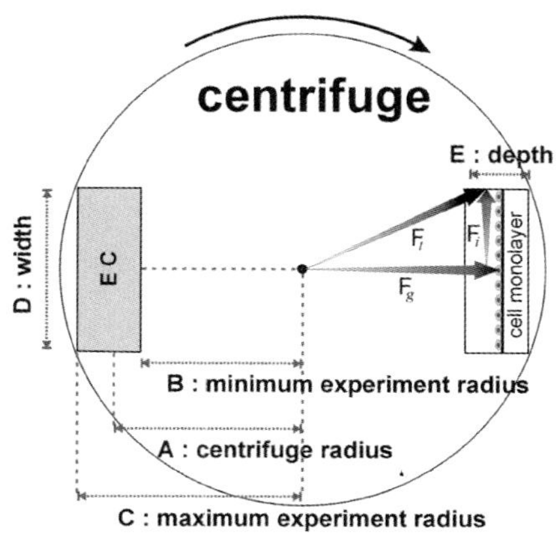

Fig. 1.2 Geometry of an Experiment Container (EC) accommodated on a centrifuge and forces within such a rotating system on board a spacecraft in free fall (weightlessness) condition. The centrifuge radius, *A*, is defined as the distance from the centre of rotation to the centre of the EC. The minimum radius, *B*, is the distance from the centre of rotation to the inner wall of the EC. The maximum radius, *C*, is the distance from the centre of rotation to the outer wall of the EC. Width, *D*, is the maximum lateral width of an EC. The force of gravity, F_g, increases radially from the centre of centrifugation. The inertial shear force, F_i, increases laterally from the centre of the EC, as depicted in the right-hand EC along a plane surface with a schematic monolayer of cells. (Picture adopted from Ref. [13] with kind permission from the *Journal of Biomechanical Engineering*, American Society of Mechanical Engineers, ASME, New York.)

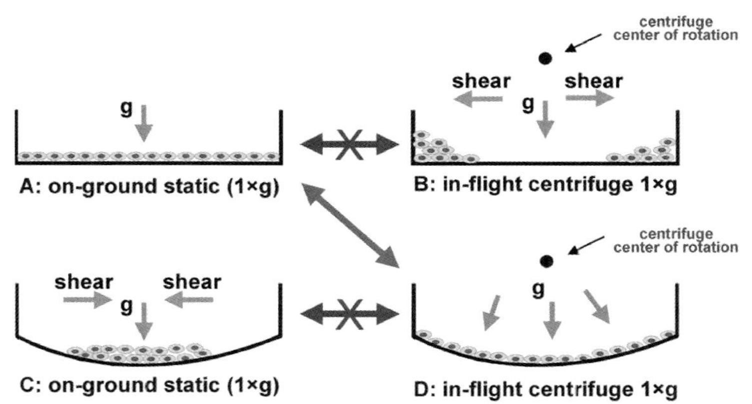

Fig. 1.3 Distribution of non-adherent cells in a 1×g static on-ground environment (A and C) or on a 1×g on-board centrifuge (B and D) in sample chambers of different surface geometry. Note that the mark for the "centre of rotation" and the curvature of the chamber are for clarity of the drawing and are not on the same scale. (Picture adopted from Ref. [13] with kind permission from the *Journal of Biomechanical Engineering*, American Society of Mechanical Engineers, ASME, New York.)

especially in microgravity when there is no convection. Therefore, the design of flight hardware, but also modules used for ground based microgravity simulations such as in a Random Positioning Machine, should ideally reflect these considerations to provide optimal culture conditions to the samples.

1.2 Gravity Phenomena on Small Objects

Several phenomena are expressed differently or have another order of magnitude in an acceleration field (e.g. 1×g Earth's gravity) as compared with a near weightlessness condition. When thinking about any effects of gravity on, e.g. a biological cell, one should distinguish between direct and indirect effects.

Direct effects may be described as effects of acceleration directly onto the cell or its internal components. Indirect effects are effects mediated to the cell by its environment. Some effects reported in cell culture experiments might be interpreted as "direct" while "indirect" effects might not be excluded.

When considering gravitational research we have to be aware of several phenomena that play a role, especially at the level of a biological cell with *in vitro* experiments (i.e. in a "Petri-dish").

1.2.1 Sedimentation

Sedimentation is the downward movement (along the gravity vector g) of an object in such a way that the object moves relative to its surrounding medium. During this process the medium, such as a fluid, exerts a force on the object known as drag force. This drag force is due to the viscosity and also, at high speeds, to turbulence behind the object.

By definition, the sedimentation force (F_S) is the downward force of weight due to linear acceleration (mg) minus buoyancy (F_B) minus frictional forces opposing downward motion (F_F):

$$F_S = mg - F_B - F_F \tag{7}$$

$$F_S = mg - (\rho_f V g) - (6\pi r \eta v) \tag{8}$$

where m=mass of the object (kg), g=the gravity acceleration (m s^{-2}), F_B=buoyant force (see below)=$\rho_f V g$, with ρ_f=specific density of the fluid (g cm^{-3}), V=volume of the displaced fluid (=volume of the object) (m^3).

$$F_F = \text{frictional force, see Stokes equation [Eq. (17)]} = kv \tag{8a}$$

where k from the Stokes equation [Eq. (17)]=$6\pi r \eta$ (for a spherical object), with r=radius of the object (m), η=viscosity constant (N s m^{-2} or Pa s), v=object's velocity relative to the fluid (m s^{-1}).

Sedimentation is one of the main differences between the experimental environment on Earth and in space. From daily experience we know that big (>r) and heavy (>ρ) particles are the quickest settlers.

These relatively big and heavy particles play a role in the positive gravitropism of plant roots. The so-called amyloplasts, localized in specialized cells in the root tip, the statocytes, sediment to the bottom of the cell in response to gravity (for review see Ref. [19]). Also, for the fresh water unicellular ciliate *Loxodes* it is argued that the heavy cell compartments (Mueller organelles), containing barium sulfate crystals, may be a good candidate for the gravisensor in these species (for review see Ref. [9]).

1.2.2
Hydrostatic Pressure

Hydrostatic pressure is also influenced by gravity. In situations of hypergravity part of the pressure that acts on a surface area derives from the weight of the liquid column standing above these surfaces, such as the culture medium above a monolayer of cells. Pressure is defined as force per unit area, where the force, F, is always acting perpendicular to that surface. The SI unit for pressure is the Pascal (Pa), one Pascal being one Newton per square meter ($N\,m^{-2}$).

In liquids or gasses of constant density:

$$P = F/A \tag{9}$$

$$= \rho Agh/A \tag{10}$$

where P=pressure (Pa), F=force (N), A=surface area (m^2), ρ=density of the liquid or gas (kg/m^3), g=unit gravity ($m\,s^{-2}$), h=height of the liquid (or gas) column (m).

Normally, the pressure exerted by a fluid column is additional to atmospheric pressure. The total pressure would be:

$$P_{tot} = \rho gh + P_o \tag{11}$$

where P_o=atmospheric pressure (Pa). Clearly, from this equation, hydrostatic pressure is linearly proportional to g. Under pure weightlessness conditions, hydrostatic pressure due to weight is zero. One should keep in mind, though, that in manned missions the spacecraft interior pressure exerts a hydrostatic pressure in a liquid. During hypergravity experiments, however, in which cells were cultured at 320×g [20] with an assumed liquid column above the cells of 1 mm, the hydrostatic pressure on the surface of the culture plate due to liquid weight would be almost 3.3 kPa (see calculations below), and about 10 kPa with a 3 mm liquid column. In addition to the pressure deriving from the liquid column, the atmospheric pressure should also be taken into account. Atmospheric pressure, also referred to as one atmosphere, is about 100 kPa (=1000 hPa) at sea level.

$$\begin{aligned} P &= \rho g h \\ &= 1.05 \times 10^3 \text{ kg m}^{-3} \times 9.81 \text{ m s}^{-2} \times 320 \times 0.001 \text{ m} \\ &= 3.296 \text{ kPa} \end{aligned} \tag{12}$$

where ρ=specific density of culture medium $\cong 1.05\,\text{kg m}^{-3}$, $g=320\times$unit gravity (m s^{-2}), $h=1\,\text{mm}$ liquid column above the cell layer (m).

In Inoue's experiment in the 320xg group [20], the force due to hydrostatic pressure acting upon a particular cell surface in a monolayer culture with a cell diameter of 10 μm would be around 2.6×10^{-7} N. In comparison, the force of the same cell due to its own weight would be around 5.1×10^{-12} N. This indicates that hydrostatic pressures cannot be neglected when considering cells subjected to hypergravity conditions.

1.2.3 Diffusion

According to kinetic theories, the molecules in a gas or in a liquid are continuously moving around and are colliding with each other. This is the process of diffusion. It can also be visualized microscopically, and is better known as the Brownian movement of small particles in solution. The diffusion process is similar to the conduction of heat through a mass. The conduction of heat takes place in response to a temperature gradient while diffusion takes place in response to a concentration gradient. The physiologist Adolf Fick formulated his equation for diffusion in 1855:

$$\Delta N/\Delta t = -DA \times (\Delta C/\Delta x) \tag{13}$$

where ΔN=number of diffusing molecules (n), Δt=time needed for diffusion (s), A=surface area on which diffusion takes place (m^2), $\Delta C=(C_1-C_2)$ concentration difference at two locations within the surface area, Δx=distance between C_1 and C_2 (m), D=the diffusion coefficient for a particular substance:

$$D = (kT/4\pi\eta r) \tag{14}$$

where k=Boltzmann constant ($1.380658\,\text{J K}^{-1}$), T=temperature (K), η=viscosity constant ($\text{Ns m}^{-2}=\text{Pa s}$) [2], r=radius of diffusing molecule (assumed spherical) (m) – for a spherical particle:

$$r = (3M/4\pi\rho N_A)^{1/3} \tag{15}$$

where M=mass per mole, ρ=density of the molecule (kg L^{-1}), N_A=Avogadro constant (mol^{-1}).

The negative sign in $-DA$ reminds us that the flow of molecules is opposite to the concentration gradient. Diffusion is not altered under microgravity. Therefore, in the absence of sedimentation, diffusion is one of the main remaining *in vitro* "mixing facilities" on the molecular level under weightlessness conditions.

1.2.4 Convection

Convection is the movement of fluid or gas that is based on density variations. When a part of a liquid or gas is heated up it expands, causing its density to decrease. In the same liquid this heated part moves upwards since it is lighter. This phenomenon is only present under conditions of acceleration (that means gravity) and is absent in weightlessness. The impact of density differences within a biological cell due to temperature differences is negligible, based on the high heat conductivity and relatively small temperature differences in cells [2]. The absence of convection during *in vitro* experiments in space could mean that spent culture medium and metabolic waste products, but also autocrine growth factors, accumulate around the cells and tissues under microgravity. In contrast, fresh nutrients from the culture medium are depleted and only supplied to the cell by diffusion processes. This could of course modulate growth and differentiation. The impact would be a secondary effect of microgravity on cell metabolism.

1.2.5 Diffusion/Convection

The effects of diffusion and convection differ considerably in magnitudes. The following gives an idea of diffusion velocity times for non-colliding particles. It is based on Einstein's mean-square or "random-walk", x^2, diffusion distance. The diffusion coefficient of K^+ ions in water is, $D = 1 \times 10^{-5}\,cm^2\,s^{-1}$ [21]. For a one-dimensional displacement over 5 mm by diffusion it would take K^+ ions nearly:

$$t = x^2/D$$
$$t = 25/10^{-3} = 25000\ s = 6.9\ h \qquad (16)$$

where x=displacement distance (mm) and D=diffusion coefficient ($mm^2\,s^{-1}$). Table 1.2 gives some values of D.

Convection currents that result from density differences generate a larger mixing capacity. Assume a spherical particle with a radius r=0.1 mm in water. The density of water $\rho = 1\,g\,cm^{-3}$ and the viscosity $\eta = 0.01$ Poise. Assume the specific gravity of the particle is 1% more than water, $\rho = 1.01\,g\,cm^{-3}$. The Stokes equation for friction, F, would be:

$$F = 6\pi r \eta v = (4/3)\pi r^3 (\rho_{water} - \rho) g$$
$$v = (2/9)\eta (\rho_{water} - \rho) r^2 g$$
$$v = 0.2\ mm\ s^{-1} \qquad (17)$$

Table 1.2 Some examples of diffusion rates.

Particle/molecule	*Diameter*	*Measured in*	*Diffusion coefficient (D) ($cm^2 s^{-1}$)*
Spherical particle	1 mm	Water (25 °C)	5×10^{-12} [a]
Whole cell	~10 µm	–	$<10^{-10}$ [b]
Spherical particle	1 µm	Water (25 °C)	5×10^{-9} [a]
Albumine	–	–	6×10^{-7} [b]
Raffinose	–	Water (15 °C)	3.3×10^{-6} [a]
Spherical particle	1 nm	Water (25 °C)	5×10^{-6} [a]
Sucrose	–	Water (25 °C)	5.2×10^{-6} [a]
1-Butanol	–	Water (25 °C)	5.6×10^{-6} [a]
Glucose	–	Water (25 °C)	6.7×10^{-6} [a]
Ethanol	–	Water (25 °C)	12.4×10^{-6} [a]
Methanol	–	Water (15 °C)	12.8×10^{-6} [a]
Methane	–	Water (25 °C)	14.9×10^{-6} [a]

a Handbook Chemistry & Physics [26].
b Todd, 1989 [27].

where r=radius of particle (m), η=viscosity constant (Pa s), v=velocity ($m s^{-1}$), ρ=density of particle ($g cm^{-3}$)

Or, within 6.9 h this particle would have travelled nearly 5 metres. In this example the convection process is nearly 1000× faster than diffusion.

1.2.6 Buoyancy

The phenomenon of buoyancy-driven convection, or Rayleigh convection, is absent in a weightlessness environment. Density differences that cause this convection are the result of local changes in temperature or composition. This is one of the important differences between ground (1×g) and flight (µg) experiments. The net force acting on an object in relation to its buoyancy is:

$$F_{B,net} = mg + F_1 - F_2 \quad (18)$$

where mg=downwards force due to the object's weight, F_1=downwards force due to the fluid on top of the object:

$$= \rho_f g h_1 A \quad (18a)$$

F_2=upwards force due to buoyancy:

$$= \rho_f g h_2 A \quad (18b)$$

where m=mass of the object (kg), ρ_f=density of the fluid ($g cm^{-3}$), g=the (gravity) acceleration ($m s^{-2}$), h_1=height of the fluid column above the object (m), h_2=height of the fluid column from the surface of the fluid to the bottom of the object (m), A = surface area (of a cylinder or cube) (m^2).

For a hypothetical cylindrical shaped object, $(h_2 - h_1)A = V$ is the volume of this cylinder and $\rho_f \times V = M$ is the mass of the fluid with a volume equal to the volume of the object:

$$\rho_f g (h_2 - h_1) A = Mg = w_f \tag{19}$$

where w_f is the weight of the fluid that has been displaced by the object. Thus:

$$F_{B,net} = mg - w_f \tag{20}$$

The force of buoyancy of an object works opposite to gravity and is proportional to the object's volume and the liquid's weight. Buoyancy is not present under weightlessness conditions, since the term $mg \cong 0$ and the vector of $F_B \cong 0$.

For a hypothetical mammalian cell with a diameter of 10 μm and a density ρ of 1.05 g cm^{-3} in a liquid culture medium of $\rho = 1.01$, the upwards directed buoyant force would be approximately:

$$\begin{aligned} F &= 5.24 \times 10^{-10} \times 1.05 \times 9.81 - (5.24 \times 10^{-10}\, 1.01) \\ &= 4.87 \times 10^{-9}\,\text{N} \end{aligned}$$

Another phenomenon related to buoyancy but actually completely different is the Marangoni convection. It is the effect on mass transfer across a liquid–gas interface. The phenomenon of liquid flowing along an interface from places with low surface tension to places with a higher surface tension is named after the Italian physicist Carlo Giuseppe Matteo Marangoni (1840–1925). The Marangoni number is frequently expressed as:

$$\text{Ma} \cong [(-\delta\gamma/\delta c) c_0 H]/\mu D \tag{21}$$

In this equation, $(\delta\gamma/\delta c)$ is the dependence of surface tension on concentration, c_0 a characteristic concentration, H a characteristic length and D the diffusivity of the solute in the phase of interest. This Marangoni number expresses the ratio of a characteristic diffusion time and a characteristic time for Marangoni driven flow. Usually, the above mentioned buoyancy or Rayleigh effect dominates the flow in liquid layers with dimensions larger than 1 cm. The Marangoni effect usually dominates when the characteristic dimension is smaller than 1 mm. For chemical engineers, the layers of 1 mm are more relevant as these layers are more often encountered in mass transfer equipment. Under normal gravity, however, it is practically impossible to study the flow pattern in such thin layers – not even with the help of a sophisticated Laser-Doppler anemometer. Therefore, the only way to study Marangoni flows in 1 cm layers separately from buoyancy is by experiments in a reduced gravity environment. Although Marangoni convection does not change under microgravity conditions, the near weightlessness environment facilitates a better measurement of this phenomenon. For detailed information on the Marangoni effect see the academic thesis of T. Molenkamp, University of Groningen [22].

1.2.7
Coriolis Acceleration

Coriolis acceleration is a type of g-field that acts upon moving objects within a rotating system such as the Earth or a centrifuge. Coriolis accelerations are so-called cross-coupled responses, owing to the angular motion in two planes. It is an effect of rotation that contributes to the impurity of gravity generated by a rotating system. For gravitational research it is particularly involved in studies of moving objects, such as flagellates in hypergravity fields. In humans or animals, this acceleration is responsible for motion- and Space-sickness while moving in a rotating field, mediated by action on the inner ears' semicircular canals. The definition of this acceleration in gs is:

$$a_{\text{Coriolis}} = (2v \times \omega)/g \tag{22}$$

where a_{Coriolis} = Coriolis acceleration expressed in g, v = velocity of the moving object (m s^{-1}), ω = angular velocity of rotating system (rad s^{-1}), g = acceleration due to gravity (m s^{-2}).

If the angular velocity is expressed in revolutions per minute (rpm), Eq. (22) changes to:

$$a_{\text{Coriolis}} = (2\pi v/30 \times g) \times \text{rpm} \tag{23}$$

where rpm = revolutions per minute = centrifuge speed.

The impact of the Coriolis force upon a sample depends on the axes along which the object moves in respect to the rotation axes. The highest impact is when the motion of the sample occurs in a plane of 90° to the rotation axes.

The impact of this phenomenon is increased in a rapidly changing angular acceleration field such as centrifuges with relatively small diameters for generating a certain g-value compared with the velocity, v, of the object studied. The magnitude, rpm, of the primary acceleration is also important. Two rotating systems with different radii but spinning at similar rpms would generate the same Coriolis accelerations.

For example, for an object like the unicellular *Loxodes* moving at a speed of $200\,\mu\text{m s}^{-1}$ (R. Hemmersbach, DLR, Cologne, Germany, personal communication) in a rotating system of 100 rpm, the Coriolis acceleration would be $4.2 \times 10^{-4} \times g$. If the rotation would generate a g-force of $1.0 \times g$, the perturbation due to the Coriolis acceleration would be less than 0.05%. Coriolis accelerations also play a role in rotating microgravity simulators such as the clinostat [23], the random positioning machine [18, 24] and the rotating wall vessel [25].

References

1 Minster, O., Innocenti, L., Mesland, D. *Looking at Science on Board Eureca*, pp. 3–6, ESA BR-80, ESA Publications Division, ESTEC, Noordwijk, **1993**.

2 Pollard, E.C. *J. Theor. Biol.* **1965**, *8*, 113–123.

3 Albrecht-Buehler, G. *ASGSB Bull.* **1991**, *4*(*2*), 25–34.

4 Mattok, C. (Ed.), *Biorack on Spacelab IML-1*, ESA SP-1162, ESA Publications Division, ESTEC, Noordwijk, **1995**.

5 Perry, M. (Ed.), *Biorack on Spacehab: Experiments on Shuttle to Mir Missions 03, 05 & 06*, ESA SP-1222, ESA Publications Division, ESTEC, Noordwijk, **1999**.

6 Cogoli, A. (Ed.), *Cell Biology and Biotechnology in Space*. Advances Space Biology and Medicine, Elsevier, Amsterdam, **2002**.

7 Burk, W.R. (Ed.), *Biological Experiments on Bion-8 and Bion-9*, ESA SP-1190, ESA Publications Division, ESTEC, Noordwijk, **1996**.

8 Gaubert, F., Schmitt, D., Lapiere, C., Bouillon, R. (Eds.), Cell and molecular biology research in space, *FASEB J.* **1999**, *13*(Suppl.), S1–S178.

9 Häder, D-P., Hemmersbach, R., Lebert, M., *Gravity and the Behavior of Unicellular Organisms*. Chapter 4.2, *Loxodes*, pp. 67–74, Cambridge University Press, Cambridge, **2006**.

10 Marthy, H-J. (Ed.). *Developmental Biology Research in Space*, Elsevier, Amsterdam, **2003**.

11 Moore, D., Cogoli, A., Gravitational and space biology at the cellular level, in: *Biological and Medical Research in Space: An Overview of Life Sciences Research in Microgravity*, Moore, D., Bie, P., Oser, H. (Eds.), pp. 1–106, Springer, Berlin, **1996**.

12 Longdon, N., David, V. (Eds.), *Biorack on Spacelab D-1*, ESA SP-1091, ESA Publications Division, ESTEC, Noordwijk, **1987**.

13 van Loon, J.J.W.A., Folgering, E.H.T.E., Bouten, C.V.C., Veldhuijzen, J.P., Smit, T. H. *ASME J. Biomech. Eng.* **2003**, *125*(*3*), 342–346.

14 van Loon, J.J.W.A., Folgering, E.H.T.E., Bouten, C.V.C., Smit, T.H. *J. Grav. Phys.* **2004**, *11*(*1*), 29–38.

15 Patrick, C.W., Sampath, R., McIntire, L.V. Fluid shear stress effects on cellular function, in: *Biomedical Research Handbook: Tissue Engineering*, Palsson, B., Hubbell, J.A. (Eds.), pp. 1626–1645, CRC Press, Cleveland, Ohio, **1995**.

16 Klein Nulend, J., Van der Plas, A., Semeins, C.M., Ajubi, N.E., Frongos, J.A., Nijweide, P.J., Burger, E.H. *FASEB J.* **1995**, *9*, 441–445.

17 Weinbaum, S., Cowin, S.C., Zeng, Y. *J. Biomech.* **1994**, *27*, 339–360.

18 van Loon, J.J.W.A. Adv. Space Res. **2007**, *39*, 1161–1165.

19 Blancaflor, E.B., Masson, P.H. *Plant Physiol.* **2003**, *133*(*4*), 1677–1690.

20 Inoue, H., Nakamura, O., Duan, Y., Hiraki, Y., Sakuda, M.J. *Dent. Res.* **1993**, *72*(*9*), 1351–1355.

21 Ling, G. *In Search of the Physical Basis of Life*, p. 115, Plenum Press, New York, **1984**.

22 Molenkamp, T. *Marangoni Convection, Mass Transfer and Microgravity*, Academic Thesis, Rijksuniversiteit Groningen, 6 November **1998**.

23 Briegleb, W. *ASGSB Bull.* **1992**, *5*(*2*), 23–30.

24 Mesland, D.A.M. *ESA Microgravity News* **1996**, *9*(*1*), 5–10.

25 Hammond, T.G., Hammond, J.M. *Am. J. Physiol. Renal Physiol.* **2001**, *281*(*1*), F12–F25.

26 Weast, R.C. (Ed.), *Handbook of Chemistry and Physics*, 51st edition, The Chemical Rubber Co., Cleveland, Ohio, **1970**–1971.

27 Todd, P. *ASGSB Bull.* **1989**, *2*, 95–113.

Further Reading

Albrecht-Buehler, G., The simulation of microgravity conditions on the ground, *ASGSB Bull.* **1992**, *5*(*2*), 3–10.

Clément, G., Slenzka, K. (Eds.) *Fundamentals of Space Biology: Research on Cells, Animals, and Plants in Space*, Springer, New York, **2006**.

DeHart, R.L., Davis, J.R. (Eds.), *Fundamentals of Aerospace Medicine*, pp. 346–347, Lea & Febiger, Philadelphia, **1985**.

Mihatov, L. (Ed.), *General Physics*, Prentice-Hall Inc., Englewood Cliffs, New Jersey, **1984**.

Marion, J.B. *General Physics with Bioscience Assays*, John Wiley & Sons Inc., New York, **1979**.

Sober, H.E. (Ed.), *Handbook of Biochemistry Selected Data for Molecular Biology*, 2nd edition, The Chemical Rubber Co., Cleveland, Ohio, **1970**.

Stephenson, R.J. *Mechanics and Properties of Matter*, 3rd edition, pp. 14–17, John Wiley & Sons Inc., New York, **1969**.

2
Primary Responses of Gravity Sensing in Plants

Markus Braun

2.1
Introduction and Historical Background

Gravity is a ubiquitous force that moves celestial bodies, planet systems and galaxies, continuously redesigning the surface of the Earth by rearranging continents (plate tectonics), clouds and ocean currents (weather), and represents the only constant environmental factor that has played a vital role in the origin and evolution of terrestrial life. Gravity has a considerable impact on many biophysical and physiological processes and participates in shaping whole living organisms. It is the main force that mechanically acts on supporting tissues of plants and influences the growth direction of plant organs, enabling them to conquer and explore the space below and above the surface of the Earth.

Mechanisms for sensing the direction of gravity appear to have evolved several times during evolutionary history [1]. Gravity-dependent orientation mechanisms provided the precondition for plants to leave the water and to conquer dry land and landscape almost all habitats on Earth. A germinating seed hidden in the darkness of the soil senses the direction of the gravity vector, guides the primary root downwards to supply water and minerals and the shoot upwards to synthesize sugar in the light. This process, known as gravitropism, is a universal growth response, which almost all plants use to control and correct the optimal orientation of their major growth axes, including roots, shoots, branches, inflorescence stems and even leaves. When the orientation of a plant organ is altered with respect to the gravity vector, e.g. when snow has bent a young tree or wind has flattened a crop in a field, the organ-specific gravitropic response, mostly in the form of differential growth (downwards or upwards), re-establishes the favourable orientation.

Since it was first recognized by Knight [2], gravitropism has been described and studied in a great variety of organs, such as roots, rhizomes, hypocotyls, inflorescence stems, coleoptiles, pulvini, petioles, as well as stalks, styli and stamina of flowers. Knight was also the first to use a device for modifying the gravitational stimulus. He showed that plants do not differentiate between gravitational and centrifugal forces and concluded that plants respond generally to acceleration

Biology in Space and Life on Earth. Effects of Spaceflight on Biological Systems. Edited by Enno Brinckmann

ISBN: 978-3-527-40668-5

forces. By slowly rotating plants on a horizontal axis Sachs [3] eliminated the unilateral effect of gravity and avoided centrifugal accelerations. The so-called clinostat was the first simulator for weightlessness, which has been widely used since then and is still an indispensable tool in gravitational biology research. Sachs and Noll [4] disagreed with the prevailing mechanistic explanation of a passive graviresponse mechanism and favoured a role for gravity as an elicitor of a physiological response. They propagated a pressure effect that relies on gravity-induced settling of intracellular particles that are heavier than the surrounding "sensitive" cytoplasm. Haberlandt [5] and Němec [6] found such sedimenting particles, so-called statoliths, and identified them as starch-filled amyloplasts, e.g. in columella cells of the root cap and in endodermal cells in shoots. When Darwin [7] removed the root cap, he found that he had eliminated the gravitropic response without affecting root growth. It became evident that gravitropism in roots and shoots relies on complex processes that involve responding cell types that are physically separated from the location of specialized cell types where gravity sensing takes place.

The gravitropic signalling pathway can be subdivided into four phases: gravity sensing, signal transduction, signal transmission and the response in the form of differential growth that leads to the reorientation of the organ [8–14]. Gravity sensing involves the initial physical process of gravity-driven statolith sedimentation, named gravity susception, and gravity perception, by which the physical stimulus is transformed into a physiological signal. Experiments on centrifuges, classical and modern fast-rotating and three-dimensional clinostats, various innovative electron microscopic, immunofluorescence, immunocytochemical, cell-biological, and especially molecular techniques have greatly enhanced our understanding of plant gravity sensing mechanisms and gravitropic signalling pathways in the last few decades. However, there are still many unsolved questions and several hypotheses and models are greatly debated.

Today, most researchers accept the starch-statoliths hypothesis. Starch-filled amyloplasts have been found in all gravity-perceiving cells of higher plant organs, which appear to function as susceptors. They are free to move in the direction of gravity and, thereby, transmit the spatial information to competent cellular structures, which trigger gravity perception and initiate the downstream signalling events.

According to the Cholodny-Went theory, the symmetric basipetal transport of the growth hormone auxin regulating root growth is redirected in response to gravistimulation of an organ. A lateral gradient of auxin promotes differential growth of the opposite root flanks in the apical elongation zone. Analyses of several gravitropism-impaired *Arabidopsis* mutants have revealed auxin influx carriers [15] and plasma-membrane associated PIN-formed (PIN) proteins [16] as efflux-carrier components of the auxin transport mechanism. Recently, it has been demonstrated that polar localization of PIN proteins directly regulates the auxin flow and that rapid changes in the PIN polarity are the primary cause for auxin redistribution and gravitropic curvature [17, 18]. PIN3 proteins have been localized in the gravity-sensing columella cells in the root cap and were shown to change their polar localization upon horizontal positioning. Lateral auxin gradients have been confirmed by expressing auxin-responsive reporter genes [19, 20].

Cytokinin is another hormone that may be involved in regulating root growth and gravitropic curvature. The expression pattern of a cytokinin-dependent gene construct [21] and asymmetric cytokinin application on one root flank suggested that cytokinin may be involved in the regulation of differential growth by inhibiting elongation at the lower side and promoting growth at the upper side of the distal elongation zone near the root apex during the early rapid phase of gravity response, which cannot be explained by differences in the auxin concentrations.

With the development of further innovative molecular approaches our knowledge of downstream gravitropic signalling events, and of components of processes involved in the regulation of differential growth, increases on a daily basis. Deciphering the cellular mechanisms and physiological processes plant cells have evolved to sense gravity and to provide useful spatial information, however, is far from complete, is dominated by controversies and still poses a challenge for future research. This chapter reviews the current status of data and the diversity of fascinating experimental approaches, and discusses the models that attempt to explain the early mechanisms and molecular processes involved in gravity sensing of algae and plants. Research on gravity sensing has greatly benefited from experiment platforms of the almost stimulus free environment of microgravity that have been provided during parabolic flights of the airbus A300-Zero-G, of sounding rockets, and during Space Shuttle flights. Data obtained from these experiments are the result of teamwork and the support of many dedicated people, making the results very special and invaluable.

2.2 Evolution of Gravity Sensing Mechanisms under the Earth's Gravity Conditions

The increasing complexity of life forms in the evolution from prokaryotic microorganisms and single-celled eukaryotic protists to multicellular plants and animals was accompanied by the development of sensors for multiple environmental stimuli. Especially, with the evolutionary development of biological structures that amplify the force of gravity, the development of specific gravity sensors required for orientation, balance and movements in a gravity environment was of particular importance. The scaling effect of gravity is well known: the percentage of body mass required for structural support is proportional to the size of a land animal (e.g. 20 g mouse = ~5%, 70 kg human = ~14%). Altered gravitational conditions would require modifications (as can be seen by comparing sea and land animals; for example, sea snakes, land snakes and tree snakes) in the support structures and physiology and may similarly lead to changes in the sensor mechanisms that sense the loads placed upon them [22]. Such sensors could be represented by relatively unspecific plasma-membrane associated mechanoreceptors rather than sensory systems that rely on statolith sedimentation [23, 24]. However, most organisms have developed more sophisticated sensory systems very early in the evolutionary history. Protists exhibit already surprisingly diverse sensing mechanisms that appear to involve mechano- or stretch-sensitive ion channels, cytoskeletal elements and second messengers [9]. Interestingly, in plants and animals light- and gravity-signal transduction pathways are highly conserved and seem to share several genes. The early gravitropic signal transduction pathway in plants seems to have required light and incorporated steps of the earlier evolved light transduction pathway, suggesting that gravitropic sensing may have evolved from the phototropic pathway.

Considering the diversity of sensory systems and the subtle differences in sensing structures and mechanisms, e.g. in gravitaxis, gravimorphism and gravitropism, it seems plausible that gravity sensing may have evolved several times during evolutionary history [1, 25] as it is accepted for light sensing in plants. It will be an interesting point of future research to find out to what extent multiple sensing mechanisms have involved in algae and higher plants, and if independently evolved gravity-sensing mechanisms are an adequate explanation for controversies in this field of research.

In the green alga *Chara*, gravitropic orientation of the tip-growing rhizoids and protonemata is clearly dependent on the sedimentation of $BaSO_4$-based vesicles that function as statoliths, whereas gravimodulation of cytoplasmic streaming and the initiation of a gravitropic curvature response of *Chara* internodal cells do not involve any apparent statoliths [26, 27]. The protist *Loxodes* uses $BaSO_4$-crystals in the so-called "Müllersche" organelles for gravitactic orientation, but seems to be capable of using a further non-statolith based mechanism. These reports suggest that different sensing systems can be realized in one organism.

2.3
Specific Location and Unique Features of Gravity Sensing Cells

In algae and plants, gravity sensing is confined to a few highly specialized cell types. In tube-like rhizoids and protonemata of characean green algae, statolith sedimentation and all subsequent phases of the gravitropic signalling pathway including the response are confined to a specific area, the apical growing region, of a single cell [28, 29]. Positively gravitropic (downward growing) rhizoids have a root-like function and anchor the algal thallus in the sediment. Protonemata are very similar cells but respond negatively gravitropically (upward growing); they are produced by nodal cells in the absence of light, e.g. when the thallus was accidentally buried in the sediment. As soon as protonemata have penetrated the substrate and reach the light, tip growth is arrested and a complex series of cell divisions is initiated that leads to the regeneration of the green thallus [30]. Both cell types are characterized by a strong polar organization of the cytoplasm and exhibit a highly dynamic arrangement of actin microfilaments that are continuously being formed at the tip and rearranged in the different regions of the cells [31]. Statoliths motion is under the control of the actomyosin system, which establishes a dynamically stable resting position of statoliths by counteracting the effect of gravity as long as the cell is in its nominal orientation, i.e. growing upwards (protonemata) or downwards (rhizoids) [32, 33]. Microtubules maintain the polar cytoplasmic organization and a polarized organelle distribution, but they are absent from the apical region and do not interfere with any of the gravitropic signalling processes. The transparency and relatively big size (30 µm diameter) of the cells make it easy to observe the gravity-induced sedimentation of statoliths onto the lower subapical cell flank following any tilting from the vertical. The 50–60 statoliths of about 2 µm in diameters are membrane-bound compartments filled with $BaSO_4$-crystals. Myosin-like motor proteins are associated with the outer statolith membrane and enable these particles to interact with the actin cytoskeleton.

Gravity-sensing cells, named statocytes, in higher plants are found in specific places at distinct developmental stages [25]. They are easily identified by the presence of sedimentable amyloplast-statoliths. The nucleus is the only other organelle that appears to sediment in very few cell types but, in this case, the nucleus always sediments together with amyloplasts [34, 35]. Statocytes constitute cylindrical endodermal sheets mainly in the elongating region of coleoptiles, shoots and inflorescence stems. In roots, the root cap is the sole site where sedimentation can be observed, and only the central columella cells of the root cap function as gravity-sensing statocytes. Neither the meristematic cells that mature into statocytes nor the mucilage-producing peripheral root cap cells allow their amyloplasts any gravity-directed movements [11]. Plant statocytes share the strict polar organization with the gravity-sensing characean cell types. The unique structural polarity of root statocytes was first studied in detail by electron microscopy by Sievers and Volkmann [36, 37]. The amyloplast-statoliths are typically found in the lower distal half of the cells, where in many species ER accumulates in a cushion-like manner. Several features appear to favour unobstructed sedimentation of statoliths: (a) the absence of prominent actin-microfilament strands and microtubule bundles in the cell interior, (b) the peripheral arrangement of endoplasmic-reticulum cisternae, (c) the presence of only small vacuoles, (d) the proximal position of the nucleus, which is stabilized by cytoskeletal elements, and (e) the medium size and starch content of the amyloplasts. Amyloplast-statoliths have been positively labelled for myosins and show actin-dependent random movements. Taking into consideration that the distal accumulation of ER cisternae provides a large sensitive surface for any potential interactions with the highly motile amyloplast-statoliths, it seems attractive to postulate a crucial function in gravitropic sensing. The destruction of individual columella cells by laser ablation revealed that the most decisive information for root curvature comes from the innermost columella cells showing the highest level of polarization and the fastest sedimentation rates [38]. However, these elegant experiments also demonstrated that the lateral statocytes also contribute to the response, albeit with decreasing gravisensitivity.

2.4 Correlation between Statolith Sedimentation and Gravitropic Responses

Stimulated by the work of Nägeli [39] and by the speculations of Berthold [40] and Noll as to the nature of plant starch grains that are freely motile under gravity, in contrast to storage starch showing no such movement, Haberlandt [5] and Němec [6] systematically surveyed plant organs that are sensitive to gravity. They found an almost perfect correlation between the occurrence of statolith tissue (statenchyma) and the localization of gravitropic sensitivity [41–43]. The starch-statolith hypothesis is still persuasive and appears intriguingly plausible. The displacement of starch-filled amyloplasts triggers the initial events in gravity perception that lead to the gravitropic response [10, 11, 25]. But is this theory still adequate and sufficient to explain gravity sensing in plants? Over the past decades numerous elaborate experiments in microgravity and numerous observations on ground have been performed that provide cumulative evidence in support of statolith-based gravity-sensing mechanisms in algae and plants.

Good evidence in favour of a statolith-based sensing mechanism comes from centrifugation experiments. Displacing statoliths into the subapical cytoplasmic region of characean rhizoids and protonemata completely abolished gravitropic responsiveness, which was restored upon the return of statoliths [44, 45]. Cold treatment and starvation was successfully used to reduce the starch content of statoliths reversibly, which was reported to cause a loss or reduction of the gravitropic sensitivity in root statocytes (summarized in Ref. [43]). Treating cress roots with gibberellic acid and kinetin caused a reversible complete destarching of the amyloplast-statoliths and of structural polarity, which inhibited gravitropic curvature [46]. Starchless mutants of *Arabidopsis* and *Nicotiana*, which are defective in plastidic phosphoglucomutase, possess starch-free amyloplast-statoliths. The mutants are still gravitropic, but gravisensitivity is strongly reduced [47–49]. This indicates that plastid-based gravisensing does not require starch, but it also shows that increasing the density of plastids by increasing the starch content strongly improves graviresponsiveness [48, 49]. An additional convincing line of evidence in support for the starch-statolith hypothesis comes from studies that used high-gradient-magnetic fields (HGMF) to displace amyloplasts [50]. When magnetophoretic forces were applied to one site of the root cap of normally downward growing roots, amyloplast-statoliths were laterally displaced due to the diamagnetic susceptibility of starch. The curvature away from the HGMF confirms that a displacement of amyloplasts limited to root statocytes is sufficient for the graviresponse to occur. Amyloplast-statoliths can give sufficient positional information for the initiation of the gravitropic reorientation response. The fact that only amyloplast-statoliths, but not starch-filled amyloplasts in other cell types were displaced, draws our attention to the supportive structures and components, which transform the physical vectorial information originating from the movement of the statoliths into a physiological signal that elicits the signalling cascade leading to the gravitropic growth response.

Many agravitropic *Arabidopsis* mutants have been isolated and several genes have been characterized, which show that endodermal cells are the sole site for shoot gravitropism (*sgr1/scr, sgr7/shr*) [51, 52] and that they can only fulfil their function as gravisensing cells when they contain sedimentable amyloplasts (*eal 1*) [53]. When amyloplasts were displaced in a specific region of *Arabidopsis* inflorescence stems by magnetic forces, curvature was limited to this region [54]. Large vacuoles complicate amyloplast-statolith movements in shoot endodermal statocytes; tonoplast membranes restrict statolith sedimentation and might possibly play a direct role in the gravity sensing process [55].

2.5
Is the Actin Cytoskeleton Involved in Gravity Sensing?

Cytoskeletal elements in plant cells have multiple functions and are of crucial importance for almost all aspects of cellular activities. A great variety of binding proteins, including motor proteins, regulate the arrangement, the dynamics and the different functions of the cytoskeleton. In gravity-sensing cells, the role of the microtubule cytoskeleton appears to be mainly limited to cell cycle and cell progression processes, but microtubules also contribute to the polarized cytoplasmic organization, such as the localization of the nucleus and ER membranes [56]. However, they do not appear to be directly involved in the primary events of gravity sensing [30, 57]. Nevertheless, establishing the structural and functional polarity is an essential requisite for gravity sensing to occur [46].

The actin cytoskeleton has frequently been implicated with all phases of gravitropic signalling pathways (for reviews see Refs. [11, 57]). Statoliths of characean rhizoids and protonemata as well as amyloplast-statoliths in higher-plant roots and shoots were reported to interact with the actin cytoskeleton via myosin motor proteins [58, 59].

The role of the actin cytoskeleton and myosin in the early process of gravity sensing is best analysed and understood in characean rhizoids and protonemata (Fig. 2.1). Several experimental results have accumulated substantial evidence for the critical role of actomyosin forces in the regulation of statoliths positioning and sedimentation and graviperception [32, 33]. Inhibitor treatments and experiments performed on clinostats, centrifuges and in microgravity have shown that the regulation of the statolith position and the control of sedimentation are under the control of actomyosin forces [28, 60]. The resting position of statoliths is regulated by a complex interplay between gravitational and actomyosin forces, which keep statoliths in a dynamically stable position of balance by counteracting each other [33]. To accomplish the resting position, actomyosin forces act oppositely in downward growing rhizoids and upward growing protonemata. Mapping the actomyosin forces has shown that, in all regions of both cell types, actomyosin forces act in both axial directions. To counteract gravity, those actomyosin forces dominate in the subapical region that act on statoliths in the opposite direction of gravity. The highly delicate position of balance is disturbed by any change in the direction and/or the amount of the gravitational force. Actin filaments are mainly axially oriented, running parallel to the longitudinal axis of the cell, which implies that the actomyosin forces act most efficiently on statoliths, when the cell is oriented in parallel to the gravity vector. In any other orientation, actomyosin forces are no longer able to completely compensate the effect of the gravitational force. Thus, upon tilting a rhizoid from the vertical, statoliths do not simply follow the gravity

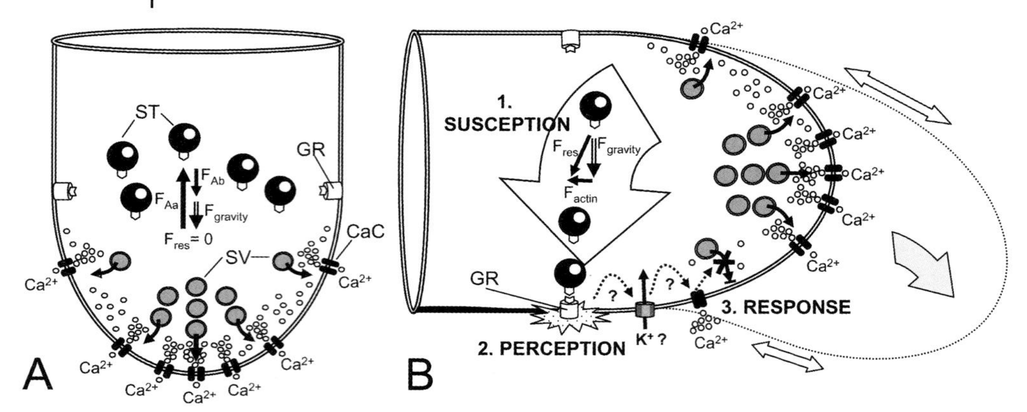

Fig. 2.1 Illustration of the tip-growth and gravity-sensing mechanism in characean rhizoids. (A) In tip-downward growing rhizoids, the statoliths are kept in a dynamically stable resting position by acropetally and basipetally directed actomyosin forces (F_{Aa} + F_{Ab}) compensating gravity ($F_{gravity}$); the resulting force acting on statoliths is zero (F_{res} = 0). Tip growth relies on symmetric exocytosis of secretory vesicles in the apical plasma membrane regulated by calcium channels that establish a tip high gradient of cytosolic free calcium. (B) Upon reorientation, net-basipetally actomyosin forces and gravity guide the statoliths onto the lower subapical cell flank, where statoliths activate the membrane-bound gravireceptors. Gravity perception triggers a short gravisignalling cascade that causes a local inhibition of calcium channels. The resulting local reduction of the calcium concentration impairs exocytosis of secretory vesicles in this area and results in gravitropic curvature response by differential extension of the opposite subapical cell flanks (double-headed arrows). CaC = calcium channel; GR = gravireceptor; ST = statoliths; SV = secretory vesicles.

vector. Instead, guided by the net-acropetally acting actomyosin forces, the sedimenting statoliths are targeted to a specific, belt-like gravisensitive plasma membrane area in the statolith region (10–35 μm from the tip), the only region where statoliths can elicit graviperception [60]. Pushing statoliths onto plasma membrane regions outside the statolith region does not result in a curvature response, which indirectly verifies that gravireceptor molecules are confined to this small subapical membrane area.

Upon tilting protonemata from the vertical, the net-acropetally acting actomyosin forces provide an additional apically directed transport that targets the sedimenting statoliths into the apical dome, where they sediment very close to the tip (a sensitive plasma membrane area 5–10 μm basal to the tip) [60]. Thus, gravity sensing obviously does not work without actin, since actomyosin forces are indispensable for providing the backbone of the gravisensing apparatus and the functional basis of the susception mechanism.

In higher plant statocytes, the role of actin in gravity sensing is less well understood. For a long time all attempts failed to label actin microfilaments in root statocytes, but then Collings et al. [61] succeeded in stabilizing and visualizing the delicate arrangement of actin microfilaments. Disrupting the actin cytoskeleton with specific drugs reduced statolith movement in *Arabidopsis* hypocotyls and inflorescence stems [62, 63] and increased the sedimentation velocity of statoliths in root statocytes [64–66]. Interfering with the actin cytoskeleton appears to increase gravisensitivity and to promote gravitropism rather than to inhibit the response (reviewed in Ref. [57]). Thus, establishing the initial gradient of lateral auxin transport does not seem to require an intact actin cytoskeleton [67]. The promoted graviresponse of roots, which were gravistimulated for only a short time and were then rotated on a clinostat, was speculated to result from the inability of fragmented actin to reset or downregulate the mechanism that relocalizes auxin efflux carriers and terminates the curvature response [68]. These results make a direct role for actin microfilaments as mediating or amplifying structures that transmit the statolith-derived signal to membrane-bound receptor molecules rather unlikely [65, 69, 70]. Although the tensegrity model of gravisensing [71] attractively involves nodal ER structures as possible shielding elements that modulate gravisensing, the hypothesized actin-mediated receptor activation also conflicts with the inhibitor data. In addition, nodal ER structures have only been found in *Nicotiana* root statocytes. It is, in fact, tempting to speculate that actomyosin forces are not essential for gravity sensing *per se*, but are required for controlling and fine-tuning an appropriate and functional resting position and sedimentation of statoliths and, furthermore, actin forces may increase the energetic noise level of the sensing mechanism by randomly moving statoliths, which makes the cells less susceptible for unfavourable fast responses to quickly changing or transient stimulation.

2.6 Gravireceptors

The nature of the gravity receptor molecules that are responsible for gravity perception and initiate the gravitropic signalling cascade in plants is still unknown. Misleadingly, the sedimentable organelles, namely amyloplast-statoliths or $BaSO_4$-crystal-filled statoliths, have frequently been addressed as gravisensors. However, in the first place, statoliths are susceptors – simply particles, whose gravity-induced sedimentation is facilitated and at least partly controlled by the actin cytoskeleton. The receptor or sensor molecule transforms the information of the physical sedimentation process into a physiological signal. Such molecules are commonly discussed as being represented by stretch-activated ion channels, which are localized in the ER or the plasma membrane and open upon statolith-mediated changes in the state of the cells' mechanical stress or tension (reviewed in Refs. [11, 12, 57, 72]). This model has also been challenged by the unimpaired graviresponse of actin-inhibited roots, and there is no clear evidence yet for a mechanosensitive mechanism that activates the gravireceptor.

Again, characean rhizoids were pioneering in terms of the localization and characterization of gravireceptors in plants. Gravireceptors were indirectly pinpointed at a specific subapical area of the plasma membrane, and it was demonstrated that statoliths have to be fully sedimented onto the gravisensitive area of the subapical plasma membrane to trigger gravity perception and to induce the gravitropic response [60]. The sedimentation process itself does not induce a curvature response. But what kind of stimulus activates the gravireceptor? Experiments performed recently during parabolic flights on board of the A300 Zero-G aircraft have contributed significantly to the disclosure of the mode of gravireceptor activation in rhizoids [73]. Statoliths, which were weightless several times for a short period of time but never lost contact with the plasma membrane, were still able to activate the membrane-bound gravireceptor. The experiments ruled out that the pressure exerted by the weight of statoliths is required for gravireceptor activation. This finding was supported by on-ground control experiments that demonstrated that increasing the weight of sedimented statoliths by lateral centrifugation did not enhance the gravitropic response. In contrast, graviperception was terminated within seconds when the contact of statoliths with the plasma membrane was interrupted by inverting gravistimulated cells for only a short period of time. The results demonstrate that graviperception in characean rhizoids relies on direct statolith–gravireceptor contact allowing yet unknown components on the statoliths' surface to interact with membrane-bound receptors rather than on pressure or tension, which is exerted by the weight of statoliths [73].

Parabolic flights experiments are currently under way to address this interesting aspect in higher plant shoots and roots. Preliminary results have shown that abolishing pressure or tensional effects of amyloplast-statoliths on gravisensors in statocytes of *Arabidopsis* roots during several short microgravity periods does not reduce the curvature response and, thus, may not have deactivated the gravireceptor.

Such recent findings prompt a new differentiated look back to the originally proposed hypothesis that amyloplasts trigger gravity perception by directly interacting with membrane-bound gravisensors [74]. Since at least one ER cisterna has been found in all statocytes of roots studied with electron microscopic methods so far, and since statoliths have never been localized in direct contact with the plasma membrane, it appears most likely that statoliths have to interact with gravireceptor molecules, which are located in ER membranes, to trigger gravity perception. The non-uniform shape of the different columella cells of a single orbital segment of the statenchyma, with the often sloped lower and/or lateral cell walls, suggests that in almost all orientations of the roots at least one or a few statoliths come into contact with ER membranes and activate some gravireceptor molecules. Information from those dominates the signals, which are integrated from all statocytes of one orbital segment, presumably in the form of the basipetal auxin flow. The differential information from all segments transmitted to the responding target cells gives rise to the curvature response (Fig. 2.2). Previous experiments have shown that statoliths must not necessarily interact with receptors located in membranes of the distal ER complex. Centrifugation of cress roots in the presence of the actin-disrupting drug cytochalasin D resulted in complete spatial separation of statoliths from the distal ER complex and gravitropic curvature following gravistimulation without any contacts between statoliths and the ER complex [75, 76]. Earlier studies had shown that lateral centrifugation does not affect the kinetics

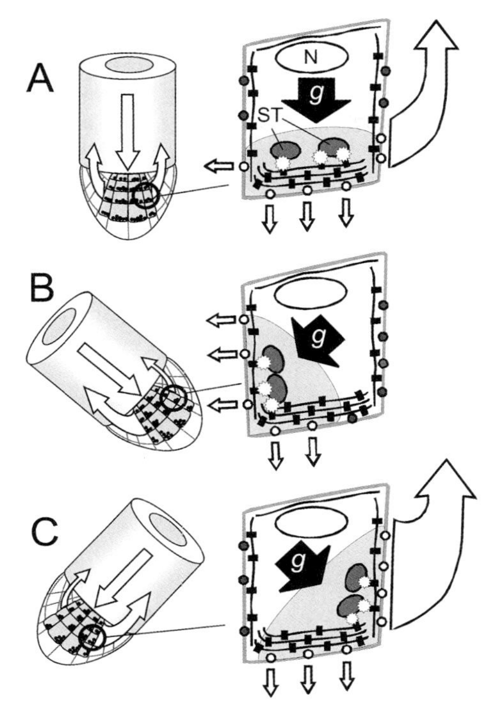

Fig. 2.2 Model for gravity sensing in root statocytes of higher plants, integrating putative membrane-bound gravireceptors, calcium as second messenger and auxin-efflux carriers. (A) In tip-downward growing roots, amyloplast-statoliths activate membrane-bound gravireceptors (white stars) by sedimenting on the distal ER accumulation. Actin-mediated statoliths movements likely avoid continuous static resting on the same membrane area and, rather, trigger continuous dynamic stimulation of gravireceptors. Graviperception results in the activation of calcium channels. The local increase in the distal calcium concentration (shaded area) activates auxin efflux carriers that establish the redirection of the auxin flux (white arrows) mainly towards the lateral root cap cells. The distal ER cisternae might serve to increase gravisensitivity, which is required to detect only small statolith displacements upon minor deviations from the vertical. (B and C) Gravistimulation followed by gravity-directed settlement of statoliths onto gravireceptors in lateral ER membranes. Activation of gravireceptors causes an elevation of the calcium concentration (shaded area) in this cell area, which activates and/or possibly relocalizes and accumulates auxin efflux carriers on the lower cell flank. Auxin is redirected towards the lower cell flank, and the increased basipetal auxin flow inhibits elongation growth of the lower root flank, causing differential growth and, thus, gravitropic curvature. Black squares = gravireceptors; N = nucleus; ST = amyloplast-statoliths; open circles = activated efflux carriers; closed circles = deactivated auxin efflux carriers; shaded area = elevated calcium concentration; white arrows = auxin flux.

of the graviresponse but the mechanism of graviperception, indicating that gravisensitivity increases with the recovering integrity of the distal cell pole. The distal ER complex may represent an amplifying sensory structure, which in cooperation with the actin cytoskeleton fine-tunes gravity sensing, facilitating precise corrections of root growth in response to even small tilts from the vertical. Growing straight is aided by the fact that it is the least energy-requiring mode of growth, which is reflected by the default tendency towards axiality, described as autonomous straightening during growth in microgravity and on clinostats [77]; however, it also requires symmetric basipetal signalling from all orbital segments of the statenchyma.

2.7 Second Messengers in Gravisignalling

Several signalling molecules and second messengers such as cytoplasmic free Ca^{2+}, cytoplasmic pH and $InsP_3$ have regulatory function in several cellular processes and have been associated with early gravitropic signalling in plants (for a review see Ref. [78]). Transient increases of cytoplasmic free Ca^{2+}, cytosolic pH and $InsP_3$ succeed gravitropic stimulation. Cytosolic calcium transients were measured using groups of young seedlings producing aequorin in the cytoplasm. On the cellular level, imaging approaches failed to show gravity-induced changes [79]. Potentially occurring highly localized changes may have escaped the observer due to the limited resolution of the methods used. Such changes have been suggested to activate the gravitropic signal transduction pathways – as is the case in other stimulus response pathways [80]. Using a calcium ionophore and an inhibitor of the ER Ca^{2+}-ATPase to disturb the precise regulation of cytoplasmic free Ca^{2+} in columella cells inhibited gravitropic signalling and the graviresponse [64, 81].

Further reports present indications for gravity-induced changes in cytosolic pH within columella cells of the root cap, which appear to constitute an important element in the early phases of gravitropic signalling; however, their mode of action remains unclear. Apoplastic pH was found to acidify early after gravistimulation, and there are, albeit inconsistent, indications of fast changes in cytosolic pH preceding the gravitropic response [82–84], which may be related to rapid changes in membrane potential found in statocytes of cress roots [85, 86]. Carefully modifying cytosolic pH was shown to reduce graviresponsiveness without affecting general root growth [82, 83]. Only recently it was demonstrated that the enhanced gravitropic response of roots, which had been treated with actin-disrupting inhibitor Latrunculin B, was accompanied by a persistent lateral auxin gradient and unaltered alkalinization of the columella cytoplasm [68]. However, blocking the alkalinization with caged protons reduced the inhibitor-induced promotion of curvature and the lateral auxin gradient. The study indicates that pH changes correlate with early gravitropic processes that do not require an intact actin cytoskeleton.

A contribution of $InsP_3$ in plant gravitropism has recently been found by constitutively expressing an enzyme that specifically hydrolyzes $InsP_3$ in transgenic *Arabidopsis* plants [87]. The $InsP_3$ level was significantly reduced, which did not negatively affect growth and the overall habitus of the plants, but reduced gravitropic curvature by 30%. The $InsP_3$ level, which was reported to increase approx. three-fold following gravistimulation, was not altered in transgenic plants, indicating that $InsP_3$ is involved in early gravitropic signalling. Although there is correlative data for the involvement of signalling molecules and second messengers in gravitropism, the mode of action, especially in the early steps of gravity signalling, remains elusive. More work is needed to clarify the sequence of events that links statolith displacement to the relocalization of PIN proteins and differential auxin transport preceding gravitropic curvature (Fig. 2.2).

2.8
Modifying Gravitational Acceleration Forces – Versatile Tools for Studying Plant Gravity Sensing Mechanisms

As mentioned above, several instruments have been developed to modify the persistent gravitational acceleration. Centrifugation is used to alter the amount and the direction of the mass acceleration that acts on the susceptible masses of gravisensing mechanisms. Centrifugation intensifies weak gravity responses; intracellular processes are more clearly recognizable, e.g. gravity-induced displacement of statoliths, and centrifugal forces have been used to directly interfere with susception and perception processes. The actomyosin forces and the gravisensitive membrane area in characean rhizoids, for instance, have been studied by using centrifugal accelerations in different angles to displace statoliths to different cytoplasmic regions and membrane areas [60, 88]. Statoliths were transported back to the tip of rhizoids by actomyosin forces that worked against centrifugal forces of 70×g, which underscores the efficiency of the actin cytoskeleton in these gravisensing cells. The polar cytoplasmic arrangement of gravisensing cells of cress roots has been studied by centrifugations at 50–2000×g. Disturbing the polar distribution of cell organelles strongly impaired gravity sensing and restoration, and maintenance of cell polarity was found to be actin-dependent [76, 89–91]. A good correlation was found between the increased latent period of the graviresponse and the time needed for restoration of cell polarity [91], suggesting that graviperception requires the specific polar arrangement of intracellular membranes, e.g. the ER cisternae. Subsequent graviresponses of untreated and centrifuged roots were identical.

Furthermore, low-level centrifugation improved graviorientation of starchless gravitropic *Arabidopsis* mutants by increasing the sedimentation rate of the statoliths [92], probably because actomyosin forces act less effectively on statoliths, whose masses have gained weight due to the increased acceleration forces. Most other hypergravity-induced effects on growth and morphogenesis caused by accelerations higher than 50×g are assumed to be independent of gravitropic sensor mechanisms [23].

Attempts to neutralize the unilateral 1-g effect, in other words "to simulate weightlessness", have resulted in the development of the classical clinostat (about 1–10 rpm), fast-rotating clinostats (about 50–120 rpm) and three-dimensional clinostats or random positioning machines, which rotate a sample around two axes.

A new quality of research tools was reached, and multiple methodological options became available with the advent of drop towers, parabolic plane flights, sounding rocket flights, Space Shuttles, satellites and Space Stations, especially the International Space Station. These platforms provide various periods of almost stimulus-free conditions, i.e. microgravity in the range 10^{-3}–10^{-6}×g, which is beyond the susceptibility of biological sensing mechanisms and, therefore, commonly regarded as functional weightlessness. Various studies have been performed in microgravity that address cellular processes involved in plant gravity sensing and graviorientation. Early experiments had demonstrated that gravisensitive organs developed normally (automorphose) in microgravity, indicating that the development of polarity and the mechanisms of gravisensing and graviorientation *per se* are neither induced nor affected by gravity [75, 93, 94]. A reduced starch content of amyloplast-statoliths, an increased amount of ER and increased diameter of lipid bodies were reported in microgravity-grown statocytes [95], which, nevertheless, responded more strongly to short doses of accelerations in microgravity [95, 96].

The absence of gravity also allowed the study of threshold values of gravisensitivity – a good measure for the capacity of the cellular mechanism underlying gravity sensing. Variable acceleration forces can be applied for different periods and, most importantly, the stimulus response can be accomplished unaffected by other stimuli. The minimum acceleration dose (presentation time) of continuous stimulation was estimated to be in a range of 20–30 g×s for microgravity-grown cress roots [95] and were extrapolated from microgravity data to be 27 g×s for lentil roots [75]. Cress roots grown on the 1 g-centrifuge in microgravity exhibited threshold values in a range of 50–60 g×s. Centrifugation experiments on ground had shown that graviresponses of cress roots were observed when an acceleration stimulus of 12 g×s was applied a second time after a period of 118 s.

Experiments performed in microgravity have also shown that the reciprocity rule of plant sensory physiology [97–99], claiming that *dose* = *stimulus* × *time* results in a constant response, is not valid [95]. The application of 0.1×g and 1×g did not result in constant angles even though the applied acceleration doses were the same and were close to the minimum dose. Repeating the experiments with actin-disrupted roots might help to elucidate whether the actomyosin forces increasingly become the limiting factor that non-linearly impairs statolith displacement in untreated statocytes at decreasing hypogravitational conditions.

The perception time defines the minimal stimulation time that is registered by a sensor and that results in a response of clinostatted roots, when the stimulus is repeated. The shortest effective stimulation time of less than 1 s displaces statoliths at 1×g only a fraction of a μm [100–102]. Thus, taking into account (a) that statoliths do not appear to require actin microfilaments as mediators or amplifiers for activating membrane-bound gravireceptor molecules, (b) that graviperception must occur close to statoliths, (c) that most statoliths "free-float" above the distal ER accumulation in random position within weightless statocytes, and (d) in default of other eligible transmitting cellular structures, it is conceivable that due to the diverse geometries of the numerous statocytes there are always some statoliths in contact with or close enough to receptor molecules to trigger graviperception and to provoke differences in the circular flow of former uniform signals strong enough to initiate differential growth and, thus, gravitropic reorientation.

Centrifugation experiments performed during the microgravity periods of TEXUS and MAXUS rockets and in a specially designed slowly-rotating-centrifuge microscope in the Space Shuttle [103] have even offered the possibility to calculate the actomyosin forces that restrict statolith sedimentation in characean rhizoids and protonemata. Acceleration forces between 0.05 and 0.1×g were shown to be required for displacing statoliths laterally, allowing the conclusion that actomyosin forces in the range of 2×10^{-14} N act on statoliths to keep them in place [73]. Gravisensitivity in characean cells, therefore, is mainly determined by actomyosin–statolith interactions. If lateral accelerations exceed the threshold, statoliths are inevitably guided to the graviperception site of the lateral cell flank, where statoliths activate the gravireceptor. The function of actin might be similar, even though possibly less constraining in higher plant statocytes. An active participation of the actin cytoskeleton in statolith sedimentation has been proposed recently for higher plant endodermal statocytes [63].

Under 1×g conditions on Earth, even in normal orientation of the organ, gravisensing cells are continuously stimulated (static gravistimulation), since all gravisensing cells steadily report the direction of gravity or, better, the position of the statoliths to the responding target cells. Clinorotation devices can prevent static stimulation, but largely fail to eliminate dynamic stimulation induced by perturbations and shifting statolith positions [69, 104]. For perfect control with most gravisensing-related research, e.g. in determining the role of cytoplasmic free calcium, $InsP_3$ and pH as second messengers in gravisignalling and the effectiveness of gravitational and actomyosin forces on statoliths, experiments must be conducted in microgravity.

The study of characean rhizoids flown on sounding rockets provides an impressive example for the mode of operation of the gravisusception apparatus. In the short period of microgravity, gravity no longer counteracted the net-basipetally acting actomyosin forces and caused the statoliths to move away from the tip and to adopt a new resting position [32]. Disruption of the actin cytoskeleton by application of specific inhibitors prevented the active movement, indicating that gravity and actomyosin forces precisely balance the delicate resting position of statoliths in normal vertical orientation [32, 33, 69, 105]. Carefully mapping the actomyosin-based forces in further microgravity studies revealed the complexity of the cytoskeleton-based mechanism that regulates and corrects the position of statoliths and facilitates fast and unimpeded sedimentation in the lateral direction on gravireceptor-bearing membrane areas upon gravistimulation [60]. Similar observations in cress root statocytes suggest a related function of actomyosin forces in higher plants [105]. Actomyosin forces restrict random movements of statoliths in microgravity, guide them to new non-random steady state positions [106] and also modulate their sedimentation [96]. However, consistent evidence for the role of actin in graviperception is still missing. Consistent with the fact that graviresponsiveness is promoted rather than inhibited by actin disruption, current experimentation on parabolic plane flights indicates that actomyosin forces neither inhibit nor facilitate gravity sensing but underlie the functional cell polarity and the susception apparatus that modulates and optimizes statolith–gravireceptor interactions to adjust spatial and temporal sensitivity and to calibrate the gravity sensing process for a most beneficial reorientation response.

2.9 Conclusions and Perspectives

Life evolved from the sea where the uniform pressure around the organisms, the only small density differences and, thus, neutral buoyancy avert intensive gravity effects on organisms. However, even aquatic species, e.g. characean algae, have developed sensory systems to use gravity as a directional cue. Increasing the density of intracellular particles, enabling them to sediment freely, and building a receptor system that recognizes the positional information from the sedimenting particles proved to be a successful strategy for orientation in the environment. This capability was especially useful when plants left the aqueous environment and inhabited dry land, where they experienced a heavy gravitational load due to the big density differences between the tissues and air. On land, gravity-sensing mechanisms became crucial for survival by providing at least a reliable orientation and guidance system for roots to reach sources of water and shoots to reach light. Scientists have now left the surface of the Earth, leaving gravitational constraints behind, to unravel the molecular and cellular mechanisms underlying plant gravity sensing in microgravity. Advantage was taken of the absence of the gravitational stimulus in free-fall situations.

Though a comprehensive view of the complex processes involved in gravity sensing is still a distant prospect, current evidence suggests that gravisignalling has evolved from basic signal transduction pathways and largely confirms the starch-statolith theory. The actin cytoskeleton has been identified as a significant element of the susception and perception phase that is proposed to modulate statolith movements and, thus, the statolith-mediated vectorial information to avoid unfavourable and to fine-tune most beneficial gravitropic responses. Current and future microgravity experiments promise to solve the mystery of the nature and localization of gravireceptor molecules in higher plant statocytes. Unicellular algal model systems proved to be especially useful for analysing basic principles of the primary gravity-sensing processes in microgravity. Experiments in microgravity have delivered insights into the complex organization of the actin-based gravisusception apparatus that controls statolith positioning and sedimentation, permitted calculations of molecular actomyosin forces that act on statoliths, and provided evidence for the mechanism by which statoliths interact with and activate gravireceptor molecules to trigger downstream signalling events that lead to gravity-oriented growth responses.

Gravity is essential for life and for the shape of life as we know it. Long duration studies over multiple generations in various species, which have been made available with completion and exploitation of the International Space Station, will provide interesting insights as to whether and how the shapes of life and, in particular, sensor mechanisms will change and adapt in the absence of an environmental cue that has always been present in the past. Conclusions from these studies in Space will provide a valuable basis for unravelling the secrets of life on Earth.

References

1 Barlow, P.W. *Plant Cell Environ.* **1995**, *18*, 951–962.
2 Knight, T. *Philos. Trans. R. Soc. London* **1806**, *99*, 108–120.
3 Sachs, J. *Vorlesungen über Pflanzen-Physiology*, Wilhelm Engelmann, Leipzig, **1887**.
4 Noll, F. *Über heterogene Induktion.* Wilhelm Engelmann, Leipzig, **1892**.
5 Haberlandt, G. *Ber. Dtsch. Bot. Ges.* **1900**, *18*, 261–272.
6 Němec, B. *Ber. Dtsch. Bot. Ges.* **1900**, *18*, 241–245.
7 Darwin, C. *The Power of Movement in Plants*, John Murray, London, **1880**.
8 Chen, R., Rosen, E., Masson, P.H. *Plant Physiol.* **1999**, *120*, 343–350.
9 Hemmersbach, R., Volkmann, D., Häder, D.P. *J. Plant Physiol.* **1999**, *154*, 1–15.
10 Kiss, J.Z. *Crit. Rev. Plant Sci.* **2000**, *19*, 551–573.
11 Sievers, A., Braun, M., Monshausen, G.B., Root cap: structure and function, in: *Plant Roots – the Hidden Half*, 3rd edition, Waisel, Y., Eshel, A., Kafkafi U. (Eds.), pp. 33–47, Marcel Dekker, New York, **2002**.
12 Boonsirichai, K., Guan, C., Chen, R., Masson, P.H. *Annu. Rev. Plant Physiol. Plant Mol. Biol.* **2002**, *53*, 421–447.
13 Blancaflor, E.B., Masson, P.H. *Plant Physiol.* **2003**, *133*, 1677–1690.
14 Morita, M.T., Tasaka, M. *Curr. Opin. Plant Biol.* **2004**, *7*, 712–718.
15 Swarup, R., Friml, J., Marchant, A., Ljung, K., Sandberg, G., Palme, K., Bennett, M. *Genes Dev.* **2001**, *15*, 2648–2653.
16 Friml, J. *Curr. Opin. Plant Biol.* **2003**, *6*, 7–12.
17 Friml, J., Wisniewska, J., Benkova, E., Mendgen, K., Palme, K. *Nature* **2002**, *415*, 806–809.
18 Wiśniewska, J., Xu, J., Seifertová, D., Brewer, P.B., Růžička, K., Blilou, I., Rouquié, D. Benková, E., Scheres, B., Friml, J. *Science* **2006**, *312(12)*, 5775–5883.
19 Rashotte, A.M., DeLong, A., Muday, G.K. *Plant Cell* **2001**, *13*, 1683–1697.
20 Ottenschläger, I., Wolff, P., Wolverton, C., Bhalerao, R.P., Sandberg, G., Ishikawa, H., Evans, M., Palme, K. *Proc. Natl. Acad. Sci. U.S.A.* **2003**, *100*, 2987–2991.
21 Aloni, R., Langhans, M., Aloni, E., Ullrich, C.I. *Planta* **2004**, *220*, 177–182.
22 Morey-Holton, E., The impact of gravity on life, in: *Evolution on Planet Earth: Impact of the Physical Environment*, Rothschild, L. (Ed.), pp. 143–159, Elsevier, **2003**.
23 Soga, K., Wakabayashi, K., Kamisaka, S., Hoson, T. *Planta* **2004**, *218*, 1054–1061.
24 Soga, K., Wakabayashi, K., Kamisaka, S., Hoson, T. *Funct. Plant Biol.* **2005**, *32*, 175–179.
25 Sack, F.D. *Planta* **1997**, *203*, S63–S68.
26 Wayne, R., Staves, M.P., Leopold, A.C. *Protoplasma* **1990**, *155*, 43–57.
27 Ackers, D., Hejnowicz, Z., Sievers, A. *Protoplasma* **1994**, *179*, 61–71.
28 Sievers, A., Buchen, B., Hodick, D. *Trends Plant Sci.* **1996**, *1*, 273–279.
29 Braun, M., Limbach, C., Hauslage, J. Protoplasma **2006**, *229*, 133–142.
30 Braun, M., Wasteneys, G.O. *Planta* **1998**, *205*, 39–50.
31 Braun, M., Hauslage, J., Czogalla, A., Limbach, C. *Planta* **2004**, *219*, 379–388.
32 Buchen, B., Braun, M., Hejnowicz, Z., Sievers, A. *Protoplasma* **1993**, *172*, 38–42.

33 Braun, M., Buchen, B., Sievers, A. *J. Plant Growth Regul.* **2002**, *21*, 137–145.
34 Sack, F.D. *Int. Rev. Cytol.* **1991**, *127*, 193–252.
35 Ridge, R.W., Sack, F.D. *Am. J. Bot.* **1992**, *79*, 328–334.
36 Sievers, A., Volkmann, D. *Planta* **1972**, *102*, 160–172.
37 Sievers, A., Volkmann, D. *Proc. R. Soc. London, Ser. B.* **1977**, *199*, 525–536.
38 Blancaflor, E.B., Fasano, J.M., Gilroy, S. *Plant Physiol.* **1998**, *116*, 213–222.
39 Nägeli, C.W. (Ed.), *Die Stärkekörner. Pflanzenphysiologische Untersuchungen, 2. Heft.* Friedrich Schulthess (Ed.), Zürich, **1858**.
40 Berthold, G.D.W. (Ed.), *Studien über die Protoplasmadynamik*, Arthur Felix, Leipzig, **1886**.
41 Darwin, F. *Proc. Royal Soc London* **1903**, *71*, 362–373.
42 Rawitscher, F. (Ed.), *Der Geotropismus der Pflanzen.* Gustav Fischer, Jena, **1932**.
43 Audus, L.J. *Symp. Soc. Exp. Biol.* **1962**, *16*, 197–226.
44 Sievers, A., Kramer-Fischer, M., Braun, M., Buchen, B. *Bot. Acta* **1991**, *104*, 103–109.
45 Braun, M., Sievers, A. *Protoplasma* **1993**, *174*, 50–61.
46 Busch, M.B. Sievers, A. *Planta* **1990**, *181*, 358–364.
47 Caspar, T., Pickard, B.G. *Planta* **1989**, *177*, 185–197.
48 Kiss, J.Z., Wright, J.B., Caspar, T. *Physiol. Plant.* **1996**, *97*, 237–244.
49 Weise, S.E., Kiss, J.Z. *Int. J. Plant Sci.* **1999**, *160*, 521–527.
50 Kuznetsov, O.A., Hasenstein, K.H. *Planta* **1996**, *198*, 87–94.
51 Fukaki, H., Fujisawa, H., Tasaka, M. *Plant Physiol.* **1996**, *110*, 945–955.
52 Fukaki, H., Wysocka-Diller, J., Kato, T., Fujisawa, H., Benfey, P.N., Tasaka, M. *Plant J.* **1998**, *14*, 425–430.
53 Fujihira, K., Kurata, T., Watahiki, M.K., Karahara, I., Yamamoto, K.T. *Plant Cell Physiol.* **2000**, *41*, 1193–1199.
54 Weise, S.E., Kuznetsov, O.A., Hasenstein, K.H., Kiss, J.Z. *Plant Cell Physiol.* **2000**, *41*, 702–709.
55 Saito, C., Morita, M.T., Kato, T., Tasaka, M. *Plant Cell* **2005**, *17*, 548–558.
56 Hensel, W. *Planta* **1984**, *162*, 404–414.
57 Blancaflor, E.B. *J. Plant Growth Regul.* **2002**, *21*, 120–136.
58 Wunsch, C., Volkmann, D. *Eur. J. Cell. Biol. Suppl.* **1994**, *61*, 46.
59 Braun, M. *Protoplasma* **1996**, *191*, 1–8.
60 Braun, M. *Protoplasma* **2002**, *219*, 150–159.
61 Collings, D.A., Zsuppan, G., Allen, N.S., Blancaflor, E.B. *Planta* **2001**, *212*, 392–403.
62 Yamamoto, K., Kiss, J.Z. *Plant Physiol.* **2002**, *128*, 669–681.
63 Palmieri, M., Kiss, J.Z. *J. Exp. Bot.* **2005**, *56*, 2539–2550.
64 Sievers, A., Kruse, S., Kuo-Huang, L.-L., Wendt, M. *Planta* **1989**, *179*, 275–278.
65 Yoder, T.L., Zheng, H.-Q., Todd, P., Staehelin, L.A. *Plant Physiol.* **2001**, *125*, 1045–1060.
66 Hou, G., Mohamalawari, D.R., Blancaflor, E.B. *Plant Physiol.* **2003**, *113*, 1360–1373.
67 Muday, G.K. *J. Plant Growth Regul.* **2001**, *20*, 226–243.
68 Hou, G., Kramer, V.L., Wang, Y.-S., Chen, R., Perbal, G., Gilroy, S. *Plant J.* **2004**, *39*, 113–125.
69 Sievers, A., Buchen, B., Volkmann, D., Hejnowicz, Z., Role of the cytoskeleton in gravity perception, in: *The Cytoskeletal Basis for Plant Growth and Form*, C.W. Lloyd (Ed.), pp. 169–182, Academic Press, London, **1991**.
70 Volkmann, D., Baluška, F. *Microsc. Res. Tech.* **1999**, *47*, 135–154.
71 Zheng, H.Q., Staehelin, L.A. *Plant Physiol.* **2001**, *125*, 252–265.
72 Perbal, G., Driss-Ecole, D. *Trends Plant Sci.* **2003**, *8*, 498–504.
73 Limbach, C., Hauslage, J., Schaefer, C., Braun, M. *Plant Physiol.* **2005**, *139*, 1–11.
74 Volkmann, D., Sievers, A., Graviperception in multicellular organs, in: *Encyclopedia of Plant Physiology*, New Series, Vol. *7*: *Physiology of Movements*, W. Haupt, M.E. Feinleib (Eds.), pp. 573–600, Springer, Berlin, **1979**.
75 Perbal, G., Driss-Ecole, D., Sallé, D., Raffin, J. *Naturwissenschaften* **1986**, *73*, 444–446.

76 Wendt, M., Kuo-Huang, L.-L., Sievers, A. *Planta* **1987**, *172*, 321–329.

77 Stankovic, B., Antonsen, F., Johnsson, A., Volkmann, D., Sack, F.D. *Adv. Space Res.* **2001**, *27*, 915–919.

78 Fasano, J.M., Massa, G.D., Gilroy, S.J. *Plant Growth Regul.* **2002**, *21*, 71–88.

79 Legué, V., Blancaflor, E.B., Wymer, C., Perbal, G., Fantin, D., Gilroy, S. *Plant Physiol.* **1997**, *114*, 789–800.

80 Sinclair, W., Trewavas, A.J. *Planta* **1997**, *203*, S85–S90.

81 Sievers, A., Busch, M.B. *Planta* **1992**, *188*, 619–622.

82 Scott, A.C., Allen, N.S. *Plant Physiol.* **1999**, *121*, 1291–1298.

83 Fasano, J.M., Swanson, S.J., Blancaflor, E.B., Dowd, P.E., Kao, T., Gilroy, S. *Plant Cell* **2001**, *13*, 907–921.

84 Johannes, E., Collings, D.A., Rink, J.C., Allen, N.S. *Plant Physiol.* **2001**, *127*, 119–130.

85 Sievers, A., Sondag, C., Trebacz, K., Hejnowicz, Z. *Planta* **1995**, *197*, 392–398.

86 Monshausen, G.B., Zieschang, H.E., Sievers, A. *Plant Cell Environ.* **1996**, *19*, 1408–1414.

87 Perera, I.Y., Hung, C.-Y., Brady, S., Muday, G.K., Boss, W.F. *Plant Physiol.* **2006**, *140*, 746–760.

88 Braun, M. *Planta* **1996**, *199*, 443–450.

89 Sievers, A., Heyder-Caspers, L. *Planta* **1983**, *157*, 64–70.

90 Hensel, W. *Protoplasma* **1985**, *129*, 178–187.

91 Wendt, M., Sievers, A. *Plant Cell Environ.* **1986**, *9*, 17–23.

92 Fitzelle, K.J., Kiss, J.Z. *J. Exp. Bot.* **2001**, *52*, 265–275.

93 Volkmann, D., Behrens, H.M., Sievers, A. *Naturwissenschaften* **1986**, *73*, 438–441.

94 Laurinavicius, R., Stockus, A., Buchen, B., Sievers, A. *Adv. Space Res.* **1996**, *17*, 91–94.

95 Volkmann, D., Tewinkel, M. *Plant Cell Environ.* **1996**, *19*, 1195–1202.

96 Perbal, G., Lefrance, A., Jeune, B., Driss-Ecole, D. *Physiol. Plant.* **2004**, *120*, 303–311.

97 Larsen, P., Geotropism: An Introduction, in: *Encyclopedia of Plant Physiology*, Vol. 17: *Physiology of Movements*, W. Ruhland (Eds.), Springer, Berlin-Göttingen-Heidelberg, **1962**.

98 Johnson, A. *Physiol. Plant.* **1965**, *18*, 945–966.

99 Shen-Miller, J., Hinchman, R.R., Gordon, S.A. *Plant Physiol.* **1968**, *43*, 338–344.

100 Fitting, H. (Ed.), *Jb. wiss. Bot.* **1905**, *41*, 221–330, 331–398.

101 Pickard, B.G. *Can. J. Bot.* **1973**, *51*, 1003–1021.

102 Hejnowicz, Z., Sondag, C., Alt, W., Sievers, A. *Plant Cell Environ.* **1998**, *21*, 1293–1300.

103 Friedrich, U.L.D., Joop, O., Pütz, C., Willich, G. *J. Biotechnol.* **1996**, *47*, 225–238.

104 Hoson, T., Kamisaka, S., Masuda, Y., Yamashita, M., Buchen, B. *Planta* **1997**, *203*, S187–S197.

105 Volkmann, D., Buchen, B., Hejnowicz, Z., Tewinkel, M., Sievers, A. *Planta* **1991**, *185*, 153–161.

106 Driss-Ecole, D., Jeune, B., Prouteau, M., Julianus, P., Perbal, G. *Planta* **2000**, *211*, 396–405.

3
Physiological Responses of Higher Plants*

Dieter Volkmann and František Baluška

3.1
Introduction: Historical Overview

Our knowledge concerning plant growth behaviour dates back to at least the 4th century BC, when the Greek philosopher Aristoteles taught his scholars the morphology (μορφη) of plants resulting from different environmental conditions. Charles and Francis Darwin described, in 1880, this behaviour in more detail, using the terms geo- and heliotropism with respect to the Earth's gravitational force and sun light, respectively [1]. At the beginning of the 20th century Němec [2] and Haberlandt [3] discovered in cells of different plant tissues sedimentable particles that they described as statoliths in analogy to mechanosensing systems in animals, ending up with the first theory postulated in plant biology, the statolith theory. However, the never-ending discussion of statoliths as a basis for gravity controlled plant growth documents the weakness of this postulate. The term statolith hypothesis for this, nevertheless, hermeneutic idea would certainly be more realistic.

Subsequently, Cholodny in 1926 [4] and, finally, Went and Thimann in 1937 [5] introduced a totally new aspect into the field of growth under the control of gravity and light, namely the plant growth substance auxin, which was later identified as indole acetic acid (IAA) and is generally accepted as a phytohormone. They established the *Avena* coleoptile test – a methodological approach that finally brought semiquantitative evidence into the field of plant sensor physiology for the first time. A next important step in the field of geotropism was reached when Juniper et al. in 1966 [6] identified the cap hosting sedimentable starch-filled statoliths as the essential prerequisite for gravity controlled growth of plant roots. At that time, research activities aimed to identify the early steps in the process of gravity controlled growth. The research groups of Audus [7], Larsen [8],

* This chapter is dedicated to Dr. Wolfgang Briegleb, one of the nestors of Gravitational and Space Biology, who died tragically by accident on the 16th of September 2006.

Biology in Space and Life on Earth. Effects of Spaceflight on Biological Systems. Edited by Enno Brinckmann

ISBN: 978-3-527-40668-5

Nougarede [9] and Sievers [10] described statoliths-hosting cells, so-called statocytes, in different species by light and electron microscopy. On the basis of correlations between structure and function these authors tried to develop new ideas for the process of sensing the gravitational force, whereas hormone physiologists investigated the effects of different plant hormones on gravity oriented growth [11, 12]. Again some of these ideas had an heuristic character but, ultimately, none of them survived research into the sensing process, which turned out to be more complex.

A qualitative step then occurred concerning the methodological approach for investigations of gravity related processes. The Space Shuttle was born, and also some other smaller systems provided, for the first time, possibilities to investigate plants and other organisms under reduced gravitational force, i.e. under realistic conditions of sensor physiology, as had already been possible for several decades concerning the factor of light. It was at that time that Darwin's term geotropism was logically, and successfully, replaced by the term gravitropism [13]. During the last three decades, space flight experiments with plants have contributed in particular to two aspects of plant biology, firstly to the process of gravisensing and secondly to the field of plant development under microgravity. In addition, experiments with mutants in Earth-bound laboratories, in particular agravitropic mutants [14] and mutants showing deficiencies in supply of phytohormones [15], have enabled huge progress towards a better understanding of the sensing and growth of plants related to the gravity factor (for a recent review see Ref. [16]).

3.2
Terminological Aspects

The classical stimulus response chain of plant sensor physiology is composed of at least three steps: susception, perception, eventually transmission, and response. In this chain the first step of susception is considered to be solely of a physical nature. Taking into account, however, that the earliest measurable signals in the range of milliseconds are triggered by gravi- or mechanostimulation in animals [17], it makes no sense to separate this step from perception. Therefore, there should be agreement in using the term *stimulus transformation* for the first step when an abiotic stimulus – either physical or chemical – is transformed into a biological signal at the cellular level. Finally, the term transmission used in the case of communicating the biological signal from the site of stimulus transformation to the site of response, which means when stimulus transformation and response occur in different cells, should be replaced by the term signal transduction, which is of general meaning for both uni- and multicellular systems (compare also Ref. [18]). As a consequence, the plant stimulus response chain is then generally characterized by the following terminology:

Stimulus → stimulus transformation → signal transduction → response

3.3 Microgravity as a Tool

3.3.1 Equipment

Free falling conditions of any so-called microgravity producing system prevent sedimentation as well as convection. The latter creates problems concerning gas exchange, in particular related to oxygen, carbon dioxide and the plant-specific gas ethylene. Thus, there is a strong demand for proper controls, e.g. on 1×g centrifuges under microgravity conditions in addition to 1×g controls in Earth-bound laboratories as well as on 1×g producing centrifuges. However, 1×g in-flight centrifuges create an additional problem. Owing to several stops of the centrifuge when specific experiments have to be removed for dedicated handling there is always an uncertainty remaining concerning the effects of microgravity on these controls. When interpreting data from space flight experiments all these specific conditions have to be considered. On the basis of Pfeffer's classical clinostat [19], sophisticated equipment was developed for further controls and investigations under simulated weightlessness, e.g. the fast rotating clinostat [20, 21], the two axis clinostat [22, 23], also called the 3D-clinostat, and, finally, the random positioning machine (RPM) developed from the idea of Dick Mesland [24, 25]. Of particular importance concerning clearly defined experimental conditions in Space was the slow rotating centrifuge microscope (NIZEMI = Niedergeschwindigkeits-Zentrifugen-Mikroskop [26]) developed by Wolfgang Briegleb, one of the nestors of life science research in Space.

3.3.2 Testable Hypotheses

Using this specific equipment, experiments under microgravity offered, for the first time, investigations concerning the behaviour of organisms, in particular of plants, under controlled gravitational conditions, i.e. the documentation of graviresponses under stimulus free conditions after defined stimulation on centrifuges. On the basis of these new methodological approaches the new research field of gravitational biology [27] was created, covering the behaviour of all organisms from bacteria through plants and animals to men – as had been realized decades earlier for the factor of light in photobiology. In addition, notably, several hypotheses concerning plant behaviour related to gravity have been developed on the basis of work performed under simulated weightlessness, e.g. on clinostats, on the random positioning machine (RPM), and by magnetic levitation, or under hypergravity conditions on centrifuges. These hypotheses are mainly related to gravisensitivity, stimulus transformation, extracellular matrix as anti-gravitational material and, finally, the existence of gravity related genes. The new experimental approaches under space flight conditions made these hypotheses testable under exact conditions of sensor physiology. They can also prove the quality of results gained under simulated weightlessness provided by different ground-based equipment.

3.3.2.1 Gravisensitivity

Ground-based Data Experiments on clinostats suggested that plants are extremely sensitive towards gravity, as indicated by a minimum dose of 12 g×s, a minimum angle for deviation from vertical of 2–3°, and detecting an absolute threshold of at least 10^{-2}×g (for a review see Ref. [28]).

In-flight Data Sophisticated experiments have demonstrated the high sensitivity of roots from cress [29] and lentil [30] in general (for a review see Ref. [31]). The minimum dose triggering a just measurable graviresponse of cress roots was measured as approximately 80 g×s. Extrapolation by statistical approaches suggested for lentil roots a minimum dose of 27 g×s. These results showed that previous experiments performed on the classical clinostat clearly underestimated the minimum dose of 12 g×s, whereas the threshold for the minimum value of cress roots and shoots was in the range 10^{-2}–10^{-3}×g [32], as suggested by clinostat experiments. For a better understanding of the process of stimulus transformation another result from this series was of particular interest. Cress roots from seedlings cultivated under 1×g conditions on the centrifuge showed about three-times lower sensitivity than those from microgravity culture chambers. Thus, adaptation processes under microgravity culture conditions seem to play a role in sensitivity, stimulus transformation and in graviresponses (compare Fig. 3 in Ref. [29]).

Stimulus summation is another very interesting parameter concerning sensitivity; in this respect it might give some hints for plant memory. Application of two stimuli separated by a pause of two minutes and exceeding in sum the minimum dose resulted in a clear graviresponse [29]. Thus, cress roots show a behaviour corresponding in principle to the response of leaves of the Venus fly trap closing just after consecutive stimulation of two sensory hairs [33]. The idea of a plant memory certainly demands further experimental approaches.

3.3.2.2 Stimulus Transformation: Role of the Actomyosin System

Ground-based Data From several organisms it was already known that the cytosol does not represent a Newtonian fluid, thus thereby probably influencing sedimentation processes (cf. Ref. [28]). On the basis of experiments investigating the role of actomyosin for the establishment of polarity of statocytes [34–36], Sievers' group, in 1989 [37], hypothesized a role of the actomyosin system for the process of stimulus transformation.

In-flight Data This idea was tested during parabolic flights of rockets for roots from cress and rhizoids from the green alga *Chara* [38]. Within six minutes of microgravity statoliths in root cap cells and rhizoid have already moved in the opposite direction of the originally acting gravity vector, indicating that the position of statoliths under 1×g conditions is determined by two counteracting forces, an endogenous force exerted by the actomyosin system and the gravitational force. Application of cytochalasin B in an additional experiment, using again rhizoids from *Chara* [39], confirmed these results and showed a clear role of actin in the positioning process.

Perbal's group [40], using extremely valuable hardware (Chapter 4), demonstrated the positions of statoliths in lentil roots under different experimental conditions, in particular at different times of stimulus-free conditions up to 122 min, which is close to the response time of lentil, and in corresponding experiments under application of cytochalasin D. These results unambiguously show that statoliths reach a stable position under microgravity and that this position depends on a functioning actomyosin system.

3.3.2.3 Extracellular Matrix as Anti-gravitational Material

Ground-based Data Cell wall components (extracellular matrix), in particular the composite of cellulose and lignin (the latter is exclusively found in terrestrial plants), have been characterized by Lewis [41] as the "backbone" of plants living on land under the huge force (1000× larger than in water) caused by the Earth acceleration. Plants growing under hypergravity conditions on centrifuges (300×g) showed increasing amounts of xyloglucans and lignin accompanied by decreasing xyloglucan-degrading enzyme activities. In addition, cell wall rigidity was increased, as indicated by a decreased mechanical extensibility [42, 43]. In contrast, experiments under simulated weightlessness on clinostats indicated a breakdown of cell wall components, in particular the loss of cellulose [44]. Thus, long-term experiments under microgravity conditions were expected to prove the role of cell wall components as anti-gravitational material of plants.

In-flight Data Mainly, the two teams mentioned above had been involved in testing this hypothesis but, however, they reported rather different results. Lewis' group [45] could not find substantial differences in cellulose microfibril organization, cell wall thickening and amount as well as composition of lignins in wheat (*Triticum aestivum*) seedlings after ten days space flight in comparison with ground controls. As a result all plants, regardless of the gravitational field they experienced, were essentially identical in terms of size, height and cell wall thickness. More recently, Stutte et al. [46] did not observe differences in lignin content of leaves from a dwarf mutant of *Triticum aestivum*, too. Contrarily, Hoson's group, using *Arabidopsis thaliana* hypocotyls [47], rice roots [48] and coleoptiles [49] as material, reported for microgravity samples stimulation of elongation growth accompanied by reduced mechanical extensibilities of cell walls, with decreased xyloglucans corresponding to increased xyloglucan-degrading enzyme activities. The cell wall composition and the underlying metabolism are different in plants from different groups like monocots (*Triticum* and *Oriza*) and dicots (*Arabidopsis*) – sometimes even in species of neighbouring families. The question therefore remains as to whether clear differences concerning anti-gravitational material can be explained solely by arguments from systematical groups or even species.

3.3.2.4 Existence of Gravity (Microgravity) Related Genes

Ground-based Data The complexity of the gravitropic stimulus response chains, as well as the response kinetics, demands regulatory processes at the genomic level. The teams of Lewis Feldman [50] and Heike Winter-Sederoff [51] have performed experiments in ground-based laboratories using the method of microarray analysis applied to material coming from the vertical position (control) versus material stimulated horizontally or 135° tilted. Both groups clearly showed differences in expression pattern for up to 200 genes coding for different classes of proteins such as transcription factors, cell wall modifying enzymes, cytoskeletal elements, and signalling molecules. Up and down regulation of several genes occurred even 2 min after stimulation. In the absence of long-term space flights, the results available so far were achieved mainly with experiments under hypergravity [52] and under simulated weightlessness [53]. Both experimental approaches suggest changes in gene expression under varying gravitational conditions, and again different classes of genes, coding for example for cytoskeletal proteins and signalling molecules, are concerned. Using cultures of isolated *Arabidopsis* "non competent cells" in gravisensing, Hampp's group [54, 55] identified changes in gene expression of transcription factors, e.g. members of the MYB and MADS box family, as well as members of signal transduction chains like PIP-4-kinase within 30 min after treatment. These results indicate that hypergravity as well as 2D-clinorotation (60 rpm min^{-1}) might act as stress factor whereas treatment on the random positioning machine and by magnetic levitation suggests a good quality of simulation of weightlessness. In contrast, by comparing the expression patterns of selected genes from *Arabidopsis* seedlings [56] it was shown that genes coding for pectin methyl esterase, Ca^{2+}-ATPase, actin, and auxin transport related proteins are regulated in an opposite manner under hypergravity versus simulated microgravity by clinorotation (1 rpm).

In-flight Data In a most recent publication [46] the authors found no significant differences in gene expression in leaves from *Triticum aestivum* after 24 days cultivation on the International Space Station (ISS). The investigated material, however, came from a mutant showing dwarfism. Taking together these results achieved with different plant material with differences related to the genes selected for investigations and with diverging experimental approaches, there is a high demand for experiments under clearly defined conditions with distinct plant material.

3.3.2.5 Autonomous versus Directed Movements

Ground-based Data Autonomous movements of plants have been known for decades [19]; for reviews, see Johnsson (1979) on circumnutation [57], Kang (1979) on epinasty [58], and Stankovic et al. (1998) for terminology [59]. Experiments under simulated weightlessness demonstrated a high complexity for autonomous movements, eventually interfered with and improved by gravity ([60] for shoots, [61, 62] for roots, [63] for coleoptiles).

In-flight Data When plant organs like coleoptiles [64] or roots [65] are stimulated on a 1×g centrifuge in Space followed by directed movements under stimulus free conditions in microgravity they show clear autonomous (autotropic) straightening. This straightening occurs in different areas of the organ in different quality and kinetics, probably depending on the growth status of the cells involved in this process [65]. Autotropic straightening might be interpreted as another example of the memory of plants that demands further investigations.

3.4 Microgravity as Stress Factor

One important consequence of the reduced gravitational force during space flights is the highly decreased convection. As a result, exchange of gases like oxygen, carbon dioxide and ethylene is incomplete and has to be supported by sophisticated equipment. During early phases of development seedlings, and in particular roots, are distinguished by oxygen consumption. However, in later stages the concentration of carbon dioxide as well as ethylene (acting as a phytohormone) has an extreme affect on plant growth and development. Thus, the conditions of early experiments in microgravity that did not provide adequate possibilities for gas exchange gave rise to a clear stress situation for the samples. Sophisticated gas control systems are the basis for separating the effects of microgravity, convection and even radiation.

3.4.1 Cellular Level

Several papers report differences at the cellular level between in-flight specimens and material from distinct controls, e.g. ground controls with and without centrifugation at 1×g and in-flight controls from 1×g centrifuges. Because most space flights allowed only short-term development of seedlings, these differences are of course related to sedimentable particles, probably functioning as statoliths [66–70], but also related to storage material such as starch grains and starch content, lipid droplets and protein vacuoles, and also to cell division, including chromosome anomalies, cytoskeletal elements and cell wall components. Several authors report disturbances of starch metabolism, documented both by smaller [66] and larger starch grains [67], and/or decreased starch content in gravity perceiving as well as starch storing cells [71–73]. However, both the number of lipid droplets and their size increased dramatically in many root cells of cress [67, 74] that belongs to the Brassicacean family and which is distinguished by a large amount of lipid as storage material. Reports on microgravity effects on cell cycle, i.e. prolongation of the first cycle [75], and chromosome structure such as bridge formation, aneuploidy, breakage and fracture of chromosomes have been published [76]; the results suggest that microgravity conditions might exert stress on fast developing seedlings as well as on cell cultures. Multifactorial effects from low gravity and decreased convection are probably responsible for these results, as has been described for some developmental aspects such as chloroplast structure and hydrocarbon content [77].

3.4.2
Developmental Aspects

Developmental aspects are related to short-term experiments using dry seeds and cultivating seedlings in microgravity for a maximum of two weeks, or to seed-to-seed experiments when a new generation of seeds is developed under microgravity conditions. One of the most convincing results was the observation that cell and organ polarity of young seedlings and mature plants from microgravity did not deviate significantly from control groups (for a review see Ref. [31]). Correspondingly, there are several reports that gravity is absolutely not required for germination, proper growth and development of seedlings, plants and even embryos of the second generation [46, 78, 79].

The quality of seeds from generations developed under microgravity, however, is different [78, 79]. Seeds produced on space station MIR had $<20\%$ of the cotyledon cell number found in seeds harvested from the ground control. Cytochemical localization of storage reserves in mature cotyledons showed that starch was retained in the space flight material, whereas protein and lipid were the primary storage reserves in ground control seeds. Protein bodies in mature cotyledons produced in Space were 44% smaller than those in the ground control seeds. Fifteen days after pollination, cotyledon cells from mature embryos formed in Space had large numbers of starch grains, and protein bodies were absent, while in developing ground control seeds, at the same stage, protein bodies had already formed and fewer starch grains were evident. Investigating in particular the development of pollen-producing tissue under microgravity conditions, it was observed [80] that cells of the anther wall and filaments from the space flight plants contained numerous large starch grains like in developing cotyledon tissues, while such grains were rarely seen in the ground controls. The tapetum remained swollen and persisted to a later developmental stage in the space flight plants than in the ground controls, even though most pollen grains appeared normal. These developmental markers indicate that *Brassica* seeds and pollen produced in microgravity were physiologically younger than those produced in 1×g. From those results the authors hypothesized that microgravity limits mixing of the gaseous microenvironments inside the closed tissues and that the resulting gas composition surrounding the seeds and pollen retards their development.

In addition, several authors have reported on microgravity effects on ion content, metabolites, photosynthetic apparatus and activity, i.e. results that might cause differences in development [81] (for reviews, see Refs. [82, 83]). Growth and photosynthesis of wheat (*Triticum aestivum*) plants have been investigated after cultivation onboard the Space Shuttle Discovery for ten days [84]. As compared with ground control plants, the shoot fresh weight of Space-grown seedlings decreased by 25%. In Space-grown plants, the light compensation point of the leaves increased by 33%, which was probably due to an increase (27%) in leaf dark-respiration rates. Related experiments with thylakoids isolated from Space-grown plants showed that the light-saturated photosynthetic electron transport rate from H_2O through photosystems II and I was reduced by 28%. These results demonstrate that photosynthetic functions are affected by the microgravity environment. Furthermore, changes have been reported [85] in the chlorophyll content, structure and number of chloroplasts in the cell, swelling of thylakoids and a decrease in the number and size of starch grains in the chloroplasts. The thylakoids isolated from Space-grown plants showed lower rates of electron transport through photosystems I and II and in the whole chain. However, no differences were found in photosynthetic activity at the moderate light levels [86]; again, in the most recent paper related to experiments under high quality conditions in Space, no remarkable differences are reported concerning plant morphology and cell structures [46].

3.5
Gravity-related Paradoxes

Different experimental approaches, including not always clearly defined culture and growth conditions and the use of highly different plant material, might be one explanation for the divergent and sometimes even contradictory results described above; the complexity of gravity related processes is certainly an additional point that has to be considered. Going carefully through the literature available, several gravity paradoxes can be identified concerning this topic, i.e. the behaviour of roots versus shoots, the root cap as site of stimulus transformation as well as statoliths and cytoskeletal elements as cell structures involved in stimulus transformation and signal transduction.

Despite the overall similarity of root and shoot graviresponse, a detailed scrutiny reveals profound differences. First, root apices grow downwards while shoot apices grow against the gravity vector. But this is generally considered to be granted as roots are underground and shoots are aboveground organs. However, the case is not so simple if we consider rhizoids and protonemata of Characean algae [87]. These two tip-growing cell types show very similar cytoarchitecture and their high gravisensitivity is related to sedimenting vacuoles filled with crystals of barium sulfate. The significant difference between rhizoids and protonemata of Characean is that these statoliths sediment closer to tips in the protonemata than in the rhizoids, which displaces the "Spitzenkörper", acting as a vesicle supply centre, only in protonemata [88, 89]. Nevertheless, the mechanistic basis linking the sedimentation of statoliths to the downward growth of the rhizoids and the upward movement of the protonemata remains elusive [90]. However, we should keep in mind that other cells from *Chara*, e.g. internodal cells, are inherently excitable, too [91].

Detailed analysis of graviresponding shoots and roots also reveals profound differences. First, although the depolymerization of F-actin with latrunculin B prevents sedimentation of amyloplast-based statoliths, this treatment stimulates the gravitropism of shoots [92]. Moreover, although the sedimenting starch-based amyloplasts are found in most gravisensitive cells of both shoots and roots, recent studies revealed that their sedimentation is not essential for shoot gravitropism. Mutants defective only in shoot gravitropism, but showing normal root gravitropism, still have normal sedimentation of starch-based amyloplasts in shoot statocytes [93, 94]. The *grv2* mutant provides strong genetic evidence of profound differences between shoot and root gravitropism as the GRV2 is a single gene of the *Arabidopsis* genome [94]. Although latrunculin B stimulates also gravitropism of roots, their statocyte amyloplasts sediment normally [95]. All these findings implicate the surprising conclusion that the stimulating effects of latrunculin B treatments on gravitropisms of plant organs are not related to sedimentation of their amyloplast-based statoliths. Depolymerization of F-actin via latrunculin B treatments must target some other processes essential both for stimulus transformation and graviresponse.

Last but not least, there is a dramatic difference in the speed with which shoots and roots accomplish their response. While it takes days to complete gravitropism of maize shoots [96, 97], depending on the developmental stage, it takes only two hours to finish gravitropism of maize roots [98, 99]. The nature of the extremely rapid graviresponse of roots is currently unknown. However, it is somehow related to a root-specific organ, known as root cap, that covers the whole root apex. Importantly, the shoot apex lacks such an organ completely and the shoot apex is rather specialized for the development of sexual organs. All this implies that the root apex represents the "head-like" anterior pole of the plant body while the shoot apex acts as posterior pole [100].

Growing root apices show several other directed growth responses, including hydrotropism, oxytropism, electrotropism. Evidently, roots screen a much wider spectrum of physical parameters and perform integration of the obtained information to achieve their complex and active growth behaviour. This much higher sensitivity of root apices, when compared with shoot apices, is related to their root caps, which cover root apices and are specialized for sensing of physical parameters of the root environment [101]. Intriguingly, root cap statocytes, grouped together and locked within the mechanosensitive root cap, resemble in many aspects the vestibular organs of a lower animal [102]. This fits nicely with the recently introduced plant neurobiological perspective in which the root apex represents the anterior pole of plant body [100] that actively seeks for nutrients and avoids dangerous soil parts, resembling the head of lower animals. In addition, there is, meanwhile, good evidence that back up systems exist for stimulus transformation in other root tissues outside the cap [99, 103]. The root cap, therefore, might act as an enhancer for gravisensing, in particular during the early and most dangerous stages of seed development. Large sedimentable starch-filled plastids (amyloplasts = statoliths) hosted by the central cap cells might have an additional enhancing role in this process.

The situation is similarly complicated with respect to the importance of statoliths in gravisensing. Since Němec [2] and Haberlandt [3] proposed, independently, in 1900 the statolith hypothesis for sedimenting starch-based amyloplasts in roots and shoots, respectively, sedimenting starch grains were postulated as plant statoliths, allowing exquisite gravisensing at growing root apices. After initial enthusiasm, this Němec–Haberlandt hypothesis was slowly abandoned but then resurrected after more than 60 years when it was shown that the surgical ablation of maize root cap did not compromise root growth but such a root lost almost completely its gravisensitivity [6]. Later, the genetic approach finally confirmed that starch-based statoliths act as plant statoliths for both roots and shoots [93, 104–107]. Moreover, magnetophoretic manipulation of statolith positions also provided very strong experimental evidence for the statolith status of the sedimenting starch-filled amyloplasts [108–110]. Nevertheless, the above-mentioned tissue outside the cap, acting eventually as a backup system for gravisensing, does not show sedimentable starch-filled amyloplasts acting as statoliths [99, 103]. Thus, it is still a mystery as to how the signal is transduced from statoliths to the relevant processes driving the differential cell growth.

In the 1990s, Andreas Sievers and coworkers [37] postulated that the sedimented amyloplasts push on actin filaments that are anchored at the plasma membrane, preferably at stretch-activated channels which then would be activated [38, 111]. This hypothesis found support in later observations of root cap cells that showed, after immunolabelling, tiny meshworks of actin filaments [112–114]; for a more general review see Ref. [115]. In contrast, more recent reports [98, 116, 117] state that root cap statocytes are devoid of prominent F-actin elements, and depolymerization of F-actin does not compromise the gravisensitivity but rather increases it. These data support the view that plant statocytes are sensitive to gravity because their actin cytoskeleton is less robust and extremely dynamic [98, 116], which prevents actomyosin-based control over larger organelles like starch-filled amyloplasts. Their surface is associated with a unique population of myosins that also seem to be less efficient in controlling statoliths positioning. In accordance with this view, not only depolymerization of F-actin but also inhibition of myosins stimulates root gravitropism [118]. Intriguingly, even decapped maize roots regained their ability to perform root gravitropisms if devoid of F-actin and myosin activities. This result corresponds well with previous findings that suggest that as well as root.

The cap some other root apex tissues are also graviresponsive and can direct root gravitropism [99, 103].

3.6 Gravity and Evolution

By the production of anti-gravitational material, terrestrial plants [119–121] as well as animals [122, 123] have developed strategies to withstand the extreme mechanical load acting on their bodies. These strategies are mainly based on the structural modification of the extracellular matrix by lignification of cell walls and mineralization of bones, respectively. Lignification has finally evolved to its highest effectiveness in gymnosperm and angiosperm trees as well as grasses, in particular in cereals (compare Fig. 2 in Ref. [119]). Unsurprisingly, therefore, plant biologists speak of the lignified "plant backbone" [41].

As a polymerization product of phenylpropanes, lignins are the second most abundant natural product, forming supra-macromolecular networks spreading throughout the whole plant body. As a result, a cellulose–xyloglucan–pectin–lignin composite is formed that has extremely robust mechanical properties. Depending on the particular amounts of cellulose versus lignins, this cell wall composite has characteristics of extremely elastic glass fibres or it is non-compressible like ferroconcrete, where cellulose fibres play a role analogous to iron in acting against tension and the lignin networks act like concrete against compression. It is, therefore, not surprising that lignification always occurs at sites of highest mechanical load. As with lignification in plants, mineralization of bones occurs mainly at sites of high mechanical load and the density of bones is positively correlated with the mechanical load.

There are several analogies [121] between extracellular matrices of land-living plants [124] and animals [125] concerning their composition, their mechanical characteristics and, probably, also their formation [126–129].

Concerning plants, long- as well as short-term experiments under microgravity contribute highly divergent results to this evolutionary problem. During long-term experiments in Space (16 days) no differences in presence, abundance or distribution of pectin epitopes visualized by antipectin antibodies JIM 5 and 7 have been observed on comparing Space- and Earth-grown tubers of potato [130]. In contrast, short-term experiments (6 min) indicate that endocytosis might be changed in tobacco pollen tubes during tip growth, mainly based on secretion of pectins [131]. For other cell-wall components like cellulose, hemicellulose and lignin, reports are similarly divergent. In Space, elongation growth of roots from rice was stimulated [48]. The levels of both cellulose and the matrix polysaccharides per unit length of roots decreased greatly, whereas the ratio of the high molecular mass polysaccharides in the hemicellulose fraction increased in Space-grown roots. The prominent thinning of the cell wall might overwhelm the disadvantageous changes in the cell-wall mechanical properties, leading to the stimulation of elongation growth in rice roots in Space. In addition, the activity of xyloglucan-degrading enzymes extracted from hypocotyl cell walls of *Arabidopsis* increased under microgravity conditions [47]. These results suggest that microgravity reduces the molecular mass of xyloglucans by increasing xyloglucan-degrading activity. Thus, growth and the cell wall properties of rice roots as well as of *Arabidopsis* hypocotyls were strongly modified under microgravity conditions during space flight. From experiments with protoplasts, which are the most convenient material for investigations of cell wall regeneration, the authors report a 54% decrease in the production of cellulose for protoplasts from rape seedlings, and a 71% decrease for carrot protoplasts. Hemicellulose production was also decreased in the flight samples compared with the ground controls [132, 133].

When, however, wheat plants were examined under microgravity and 1×g conditions on Earth to assess the role of gravity on cellulose microfibril organization and secondary wall thickening patterns, data revealed that the cellulose microfibrils of the Space-grown wheat maintained the same organization as their 1×g-grown counterparts; this means that fibrils were randomly interwoven with each other in the outermost layers, and parallel to each other within the individual strata. The angle between microfibrils of consecutive layers was about 80° for both the Space and Earth-grown plants. Space-grown wheat also developed normal protoxylem and metaxylem vessel elements with secondary thickening patterns ranging from helical to regular pit to reticulate thickenings. Finally, both the Space- and Earth-grown plants were essentially of the same size and height, and their lignin analyses revealed no substantial differences in their amounts and composition, regardless of the gravitational field experienced, meaning that, for the purposes of this study, all plants were essentially identical [45]. These results suggest that the microgravity environment did not affect either cell wall polymer synthesis or the deposition of cellulose microfibrils.

Despite these highly divergent results from space flight experiments, on the basis of many analogies found in plants and animals concerning the extracellular matrix as anti-gravitational material, it is reasonable to look for possible common mechanisms that might have evolved under the constant influence of gravitational force, in particular after leaving water and colonizing land. There are two concepts available that, when combined [121], might serve as the basis for an attractive and testable working hypothesis: Ingber's concept of tensegrity [134, 135] and the concept of actin-based vesicle recycling [127, 128]. In this concept (compare Fig. 4 in ref. [121]), integrating molecular and cellular aspects, a continuum of extracellular matrix via plasma membrane and cytoskeleton [136–138], ending up in actin-based vesicle recycling, provides a complicated signalling network for sensing mechanical stimuli followed by signal transduction and finally development and growth response [121, 139].

3.7 Conclusion and Perspectives

Experiments in reduced gravitational fields have revealed several effects on the molecular and cellular level, ranging from disturbances of cell division via differences in cell wall components and cell wall pattern formation up to changes in photosynthetic products. These effects are sometimes extremely divergent, depending on the plant material, culture conditions and experimental procedures. Nevertheless, one result is common for most experiments: plants develop under these critical conditions [140], finally, more or less normally and they are able to produce seeds over more generations. This demonstrates the high capacity of plants for adaptation to stress situations that never occurred during the process of evolution, e.g. it is apparent that, even in a microgravity environment, woody plants can make appropriate corrections to compensate for stress gradients introduced by mechanical bending, thereby enabling the formation of compression wood [141].

Future experiments should be performed under two general aspects. The first is related to basic research, including:

- sensing and signalling of gravity,
- gravity as driving factor for evolution,
- adaptation to stress situations.

The second is related to future Space explorations, including:
- production of multiple generations,
- maintenance of seed quality and food production,
- oxygen production.

Clearly testable hypotheses must be worked out for any experiment and these experiments have to be performed under exactly defined conditions.

Acknowledgments

The authors' experiments were financially supported by grants from the Bundesministerium für Bildung und Forschung (BMBF) and Bundesministerium für Wirtschaft und Technologie (BMWi) via Deutsches Zentrum für Luft- und Raumfahrt (DLR, Cologne, Germany; projects 50WB 9995 and 0434), from the European Space Agency (ESA-ESTEC Noordwijk, The Netherlands; MAP project AO-99-098), and from the Ente Cassa di Risparmio di Firenze (Italy), which are gratefully acknowledged. F.B. receives partial support from the Slovak Academy of Sciences (Grant Agency VEGA, Bratislava, Slovakia; project 2/5085/25).

References

1 Darwin, C., assisted by Darwin, F. *The Power of Movements in Plants*, John Murray, London, **1880**.
2 Němec, B. *Ber. Deutsch. Bot. Ges.* **1900**, *18*, 241–245.
3 Haberlandt, G. *Ber. Deutsch. Bot. Ges.* **1900**, *18*, 261–272.
4 Cholodny, N. *Jb. wiss. Bot.* **1926**, *65*, 447–459.
5 Went, F.W., Thimann, K.V. *Phytohormones*, MacMillan, New York, **1937**.
6 Juniper, B.E., Groves, S., Landau-Schachar, B., Audus, L.J. *Nature* **1966**, *209*, 93–94.
7 Audus, L.J., Geotropism, in: *Physiology of Plant Growth and Development*, Wilkins, M.B. (Ed.), pp. 205–242, McGraw-Hill, London, **1969**.
8 Iversen, T.H., Larsen, P. *Physiol. Plant* **1973**, *21*, 811–819.
9 Perbal, G. *Planta* **1974**, *116*, 153–171.
10 Sievers, A., Volkmann, D. *Planta* **1972**, *102*, 160–172.
11 Pilet, P.E. *Planta* **1973**, *111*, 275–278.
12 Wilkins, M.B., Growth-Control Mechanisms in Gravitropism, in: *Physiology of Movements. Encyclopedia of Plant Physiology*, new series, Vol. *7*, Haupt, W., Feinleib, M. (Eds.), pp. 601–626, Springer, Berlin-Heidelberg-New York, **1979**.
13 Sievers, A., Volkmann, D., Gravitropism in Single Cells, in: *Physiology of Movements. Encyclopedia of Plant Physiology*, new series, Vol. *7*, Haupt, W., Feinleib, M. (Eds.), pp. 567–572, Springer, Berlin-Heidelberg-New York, **1979**.
14 Boonsirichai, K., Guan, C., Chen, R., Masson, P.H. *Annu. Rev. Plant Biol.* **2002**, *53*, 421–447.

15 Friml, J., Wisniewska, J., Benkova, E., Mendgen, K., Palme, K. *Nature* **2002**, *415*, 806–809.

16 Esmon, C.A., Pedmale, U.V., Liscum, E. *Int. J. Dev. Biol.* **2005**, *49*, 665–674.

17 Kernan, M., Zuker, C. *Curr. Opin. Neurobiol.* **1995**, *5*, 443–448.

18 Shropshire, W., Jr., Stimulus Perception, in: *Physiology of Movements. Encyclopedia of Plant Physiology*, new series, Vol. *7*, Haupt, W., Feinleib, M. (Eds.), pp. 10–41, Springer, Berlin-Heidelberg-New York, **1979**.

19 Pfeffer, W. *Pflanzenphysiologie*, W. Engelmann, Leipzig, **1904**, Vol. 2.

20 Briegleb, W. *ASGSB Bull.* **1992**, *5*, 23–30.

21 Ayed, M., Pironneau, O., Planel, H., Gasset, G., Richoilley, G. *Micrograv. Sci. Technol.* **1992**, *98*, 98–102.

22 Hoson, T., Kamisaka, S., Buchen, B., Sievers, A., Yamashita, M., Masuda, Y. *Adv. Space Res.* **1996**, *17*, 47–53.

23 Hoson, T., Kamisaka, S., Masuda, Y., Yamashita, M., Buchen, B. *Planta* **1997**, *203*, S187–S197.

24 Mesland, D.A.M. *ESA Micrograv. News* **1996**, *9*, 5–10.

25 Huijser, R.H., Desktop RPM: New small size microgravity simulator for the bioscience laboratory. http://www.desc.med.vu.nl

26 Friedrich, U.L., Joop, O., Pütz, C., Willich, G. J. *Biotechnol.* **1996**, *47*, 225–238.

27 Volkmann D., Sievers, A. *Naturwissenschaften* **1992**, *79*, 68–74.

28 Volkmann, D., Sievers, A., Graviperception in Multicellular Organs, in: *Physiology of Movements. Encyclopedia of Plant Physiology*, new series, Vol. 7, Haupt, W., Feinleib, M. (Eds.), pp. 573–600, Springer, Berlin-Heidelberg-New York, **1979**.

29 Volkmann, D., Tewinkel, M. *Plant Cell Environ.* **1991**, *19*, 1195–1202.

30 Perbal, G., Driss-Ecole, D. *Physiol. Plant* **1994**, *90*, 313–318.

31 Perbal, G., Driss-Ecole, D., Tewinkel, M., Volkmann, D. *Planta* **1997**, *203*, S57–S62.

32 Merkys, A.J., Mashinsky, A.L., Laurinavichius, R.S., Nechitailo, G.S., Yaroshius, A.V., Izupak, E.A. *Life Sci. Space Res.* **1975**, *13*, 53–57.

33 Hodick, D., Sievers, A. *Planta* **1988**, *174*, 8–18.

34 Sievers, A., Volkmann, D., Hensel, W., Sobick, V., Briegleb, W. *Naturwissenschaften* **1976**, *63*, 343.

35 Hensel, W. *Protoplasma* **1985**, *129*, 178–187.

36 Hensel, W. *Planta* **1988**, *173*, 142–143.

37 Sievers, A., Kruse, S., Kuo-Huang, L.L., Wendt, M. *Planta* **1989**, *179*, 275–278.

38 Volkmann, D., Buchen, B., Hejnowicz, Z., Tewinkel, M., Sievers, A. *Planta* **1991**, *185*, 153–161.

39 Buchen, B., Braun, M., Hejnowicz, Z., Sievers, A. *Protoplasma* **1993**, *172*, 38–42.

40 Driss-Ecole, D., Jeune, B., Proteau, M., Julianus, P., Perbal, G. *Planta* **2000**, *211*, 396–405.

41 Lewis, N.G. *Curr. Opin. Plant Biol.* **1999**, *2*, 153–162.

42 Hoson, T., Nishitani, K., Miyamoto, K., Ueda, J., Kamisaka, S., Yamamoto, R., Masuda, Y. *J. Exp. Bot.* **1998**, *47*, 513–517.

43 Soga, K., Wakabayashi, K., Kamisaka, S., Hoson, T. *Plant Cell Physiol.* **1999**, *40*, 581–585.

44 Hoson, T., Kamisaka, S., Yamashita, M., Masuda, Y. *Biol. Sci. Space* **1999**, *9*, 337–344.

45 Levine, L.H., Heyenga, A.G., Levine, H.G., Choi, J., Davin, L.B., Krikorian, A. D., Lewis, N.G. *Phytochemistry* **2001**, *57*, 835–846.

46 Stutte, G.W., Monje, O., Hatfield, R.D., Paul, A.L., Ferl, R.J., Simon, C.G. *Planta* **2006**, *224*, 1038–1049.

47 Soga, K., Wakabayashi, K., Kamisaka, S., Hoson, T. *Planta* **2002**, *215*, 1040–1046.

48 Hoson, T., Soga, K., Wakabayashi, K., Kamisaka, S., Tanimoto, E. *Plant Soil* **2003**, *255*, 19–26.

49 Hoson, T., Soga, K., Mori, R., Saiki, M., Nakamura, Y., Wakabayashi, K., Kamisaka, S. *Plant Cell Physiol.* **2002**, *43*, 1067–1071.

50 Moseyko, N., Zhu, T., Chang, H.-S., Wang, X., Feldman, L.J. *Plant Physiol.* **2002**, *130*, 720–728.

51 Kimbrough, J.M., Salinas-Mondragon, R., Boss, W.F., Brown, C.S., Winter-Sederoff,

H. *Plant Physiol.* **2004**, *136*, 2790–2805.

52 Yoshioka, R., Soga, K., Wakabayashi, K., Takeba, G., Hoson, T. *Adv. Space Res.* **2003**, *31*, 2187–2193.

53 Kittang, A.I., van Loon, J.J.W.A., Vorst, O., Hall, R.D., Fossum, K., Iversen, T. H. *J. Gravit. Physiol.* **2005**, *11*, P223–P224.

54 Martzivanou, M., Hampp, R. *Physiol. Plant* **2003**, *118*, 221–231.

55 Martzivanou, M., Babbick, M., Cogoli-Greuter, M., Hampp, R. *Protoplasma* **2006**, *229*, 155–162.

56 Centis-Aubay, S., Gasset, G., Mazars, C., Ranjeva, R., Graziana, A. *Planta* **2003**, *218*, 179–185.

57 Johnsson, A. Circumnutation, in: *Physiology of Movements. Encyclopedia of Plant Physiology*, new series, Vol. 7, Haupt, W., Feinleib, M. (Eds.), pp. 627–646, Springer, Berlin-Heidelberg-New York, **1979**.

58 Kang, B.G., Epinasty, in: *Physiology of Movements. Encyclopedia of Plant Physiology*, new series, Vol. 7, Haupt, W., Feinleib, M. (Eds.), pp. 647–667, Springer, Berlin-Heidelberg-New York, **1979**.

59 Stankovic, B., Volkmann, D., Sack, F.D. *Physiol. Plant.* **1998**, *102*, 328–335.

60 Firn, R.D., Digby, J. *Plant Cell Environ.* **1979**, *2*, 149–154.

61 Hoson, T. *Plant Soil* **1994**, *165*, 309–314.

62 Stankovic, B., Volkmann, D., Sack, F.D. *Plant Physiol.* **1998**, *117*, 893–900.

63 Karlsson, C., Johnsson, A., Chapman, D.K., Brown, A.H. *Physiol. Plant.* **1996**, *98*, 325–332.

64 Chapman, D.K., Johnsson, A., Karlsson, C., Brown, A., Heathcote, D. *Physiol. Plant.* **1994**, *90*, 157–162.

65 Stankovic, B., Antonsen, F., Johnsson, A., Volkmann, D., Sack, F.D. *Adv. Space Res.* **2001**, *27*, 915–919.

66 Volkmann, D., Behrens, H., Sievers, A. *Naturwissenschaften* **1986**, *73*, 438–441.

67 Laurinavicius, R., Stockus, A., Buchen, B., Sievers, A. *Adv. Space Res.* **1996**, *17*, 91–94.

68 Smith J.D., Todd, P., Staehelin, L.A. *Plant J.* **1998**, *12*, 1361–1373.

69 Perbal, G., Driss-Ecole, D., Rutin, J., Salle, G. *Physiol. Plant* **1998**, *70*, 119–126.

70 Kern, V.D., Smith, J.D., Schwuchow, J.M., Sack, F.D. *Plant Physiol.* **2001**, *125*, 2085–2094.

71 Moore, R., Fondren, W.M., McClelen, C.E., Wang, C.L. *Am. J. Bot.* **1987**, *74*, 1006–1012.

72 Kordyum, E., Baranenko, V., Nedukha, E., Samoilov, V. *Plant Cell Physiol.* **1997**, *38*, 1111–1117.

73 Guisinger, M.M., Kiss, J.Z. *Am. J. Bot.* **1999**, *86*, 1357–1366.

74 Volkmann, D., Sievers, A., Gravitational Effects On Subcellular Structures of Plant Cells, in: *Proceedings of the Fourth European Symposium of Life Science Research in Space, Triest*, David, V. (Ed.), pp. 497–501, ESA SP-307, ESA Publications Division, ESTEC, Noordwijk, **1990**.

75 Yu, F., Driss-Ecole, D., Rembur, J., Legue, V., Perbal, G. *Physiol. Plant* **1999**, *105*, 171–178.

76 Krikorian, A.D. *Physiol. Plant* **1996**, *98*, 901–908.

77 Musgrave, M.E., Kuang, A., Brown, C.S., Matthews, S.W. *Ann. Bot.* **1998**, *81*, 503–512.

78 Kuang, A., Xiao, Y., McClure, G., Musgrave, M.E. *Ann. Bot.* **2000**, *85*, 851–859.

79 Musgrave, M.E., Kuang, A., Xiao, Y., Stout, S.C., Bingham, G.E., Briarty, L.G., Levenskikh, M.A., Sychev, V.N., Podolski, I.G. *Planta* **2000**, *210*, 400–406.

80 Kuang, A., Popova, A., McClure, G., Musgrave, M.E. *Int. J. Plant Sci.* **2005**, *166*, 85–96.

81 Hilaire, E., Paulsen, A.Q., Brown, C.S., Guikema, J.A. *Plant Cell Physiol.* **1997**, *36*, 831–837.

82 Halstaead, T.W., Dutcher, F.R. *Annu. Rev. Plant Physiol.* **1987**, *38*, 317–345.

83 Kordyum, E.L. *Int. Rev. Cytol.* **1997**, *171*, 1–78.

84 Tripathy, B.C., Brown, C.S., Levine, H.G., Krikorian, A.D. *Plant Physiol.* **1996**, *110*, 801–806.

85 Volovik, O.I., Kordyum, E.L., Guikema, J.A. *J. Gravit. Physiol.* **1999**, *6*, P127–P128.

86 Stutte, G.W., Monje, O., Goins, G.D., Tripathy, B.C. *Planta* **2005**, *223*, 46–56.

87 Sievers, A., Buchen, B., Hodick, D. *Trends Plant Sci.* **1996**, *1*, 273–279.

88 Hodick, D. *Planta* **1994**, *195*, 43–49.

89 Braun, M. *Planta* **1997**, *203*, S11–S19.

90 Braun, M., Buchen, B., Sievers, A. *J. Plant Growth Regul.* **2002**, *21*, 137–145.

91 Wayne, R. *Bot. Rev.* **1994**, *60*, 265–367.

92 Palmieri, M., Kiss, J.Z. *J. Exp. Bot.* **2005**, *56*, 2539–2550.

93 Yano, D., Sato, M., Saito, C., Sato, M.H., Morita, M.T., Tasaka, M. *Proc. Natl. Acad. Sci. U.S.A.* **2003**, *100*, 8589–8594.

94 Silady, R.A., Kato, T., Lukowitz, W., Sieber, P., Tasaka, M., Somerville, C.R. *Plant Physiol.* **2004**, *136*, 3095–3103.

95 Hou, G., Kramer, V.L., Wang, Y.-S., Chen, R., Perbal, G., Gilroy, S., Blancaflor, E.B. *Plant Cell* **2004**, *39*, 113–125.

96 Collings, D.A., Winter, H., Wyatt, S.E., Allen, N.S. *Planta* **1998**, *207*, 246–258.

97 Long, J.C., Zhao, W., Rashotte, A.M., Muday, G.K., Huber, S.C. *Plant Physiol.* **2002**, *128*, 591–602.

98 Baluška, F., Kreibaum, A., Vitha, S., Parker, J.S., Barlow, P.W., Sievers, A. *Protoplasma* **1997**, *196*, 212–223.

99 Mancuso, S., Barlow, P.W., Volkmann, D., Baluška, F. *Plant Signal Behav.* **2006**, *1*, 52–58.

100 Baluška, F., Hlavacka, A., Mancuso, S., Barlow, P.W., Neurobiological View of Plants and their Body Plan, in: *Communication in Plants. Neuronal Aspects of Plant Life.* Baluška, F. et al. (Eds.), pp. 19–35, Springer, Heidelberg, **2006**.

101 Barlow, P.W. *J. Plant Growth Regul.* **2003**, *21*, 261–286.

102 Day, B.L., Fitzpatrick, R.C. *Curr. Biol.* **2006**, *15*, R583–R586.

103 Wolverton, C., Mullen, J.L., Ishikawa, H., Evans, M.L. *Planta* **2002**, *215*, 153–157.

104 Kiss, J.Z., Wright, J.B., Caspar, T. *Physiol. Plant* **1996**, *97*, 237–244.

105 Tsugeki, R., Fedoroff, N.V. *Proc. Natl. Acad. Sci. U.S.A.* **1999**, *96*, 12941–12946.

106 Weise, S.E., Kiss, J.Z. *Int. J. Plant. Sci.* **1999**, *160*, 21–27.

107 Fujihira, K., Kurata, T., Watahiki, M.K., Karahara, I., Yamamoto, K.T. *Plant Cell Physiol.* **2000**, *41*, 1193–1199.

108 Kuznetsov, O.A., Hasenstein, K.H. *Planta* **1996**, *198*, 87–94.

109 Kuznetsov, O.A., Hasenstein, K.H. *J. Exp. Bot.* **1997**, *48*, 1951–1957.

110 Kuznetsov, O.A., Schwuchow, J., Sack, F.D., Hasenstein, K.H. *Plant Physiol.* **1999**, *119*, 645–650.

111 Sievers, A., Buchen, B., Volkmann, D., Hejnowicz, Z., Role of the Cytoskeleton in Gravity Perception, in: *The Cytoskeletal Basis for Plant Growth and Form*, Lloyd, C.W. (Ed.), pp. 169–182, Academic Press, London, **1991**.

112 Koropp, K., Volkmann, D. *Eur. J. Cell Biol.* **1994**, *64*, 153–162.

113 Driss-Ecole, D., Vassy, J., Rembur, J., Guivarc'h, A., Prouteau, M., Dewitte, W., Perbal, G. *J. Exp. Bot.* **2000**, *51*, 521–528.

114 Volkmann, D., Baluska, F., Lichtscheidl, I., Driss-Ecole, D., Perbal, G. *FASEB J.* **1999**, *13*(Suppl.), S143–S147.

115 Volkmann, D., Baluška, A. *Microsc. Res. Tech.* **1999**, *47*, 135–154.

116 Baluška, F., Hasenstein, K.H. *Planta* **1997**, *203*, S69–S78.

117 Voigt, B., Timmers, T., Šamaj, J., Müller, J., Baluška, F., Menzel, D. *Eur. J. Cell Biol.* **2005**, *84*, 595–608.

118 Mancuso, S., Marras, A.M., Magnus, V., Baluška, F. *Anal. Biochem.* **2005**, *341*, 344–351.

119 Volkmann, D., Landgang der Pflanzen mit Folgen: Schwerkraft als Treiber der Evolution, in: *Mensch-Leben-Schwerkraft-Kosmos*, Rahmann, H., Kirsch, K.A. (Eds.), pp. 162–168, Günther Heimbach, Stuttgart, **2001**.

120 Hoson, T. *Biol. Sci. Space* **2003**, *17*, 54–56.

121 Volkmann, D., Baluška, F. *Protoplasma* **2006**, *229*, 143–148.

122 Rittweger, J., Gunga, H.C., Felsenberg, D., Kirsch, K.A. *J. Gravit. Physiol.* **2000**, *6*, P133–P136.

123 Kirsch, K., Gunga, H.C., Mensch und Schwerkaft, in: *Mensch-Leben-Schwerkraft-Kosmos*, Rahmann, H., Kirsch, K.A.

(Eds.), pp. 210–228, Günther Heimbach, Stuttgart, **2001**.
124 Carpita, N.C., Tierney, M., Campbell, M. *Plant Mol. Biol.* **2001**, *47*, 1–5.
125 Vico, L., Lafage-Proust, M.H., Alexandre, C. *Bones* **1998**, *22*, 95S–100S.
126 Tammi, R., Rilla, K., Pienimaki, J.P., MacCallum, D.K., Hogg, M., Luukkonen, M., Hascall, V.C., Tammi, M. *J. Biol. Chem.* **2001**, *276*, 35111–35122.
127 Baluška, F., Hlavacka, A., Samaj, J., Palme, K., Robinson, D.G., Matoh, T., McCurdy, D.W., Menzel, D., Volkmann, D. *Plant Physiol.* **2002**, *130*, 422–431.
128 Šamaj, J., Baluška, F., Voigt, B., Schlicht, M., Volkmann, D., Menzel, D. *Plant Physiol.* **2004**, *135*, 1150.
129 Šamaj, J., Read, N.D., Volkmann, D., Menzel, D., Baluška, F. *Trends Cell Biol.* **2005**, *15*, 425–433.
130 Cook, M.E., Croxdale, J.G. *J. Exp. Bot.* **2003**, *54*, 2157–2164.
131 Lisboa, Y.S., Scherer, G.E.F., Quader, H. *J. Gravit. Physiol.* **2002**, *9*, P239–P240.
132 Iversen, T.-H., Rasmussen, O., Gmünder, F., Baggerud, C., Kordyum, E.L., Lozovaya, V.V., Tairbekov, M. *Adv. Space Res.* **1992**, *12*, 123–131.
133 Rasmussen, O., Klimchuk, D.A., Kordyum, E.L., Danevich, L.A., Tarnavskaya, E.B., Lozovaya, V.V., Tairbekov, M.G., Baggerud, C., Iversen, T.-H. *Physiol. Plant* **1998**, *84*, 162–170.
134 Ingber, D.E. *Annu. Rev. Physiol.* **1997**, *59*, 575–599.
135 Ingber, D.E. *J. Cell Sci.* **2003**, *116*, 1397–1408.
136 Baluška, F., Samaj, J., Wojtaszek, P., Volkmann, D., Menzel, D. *Plant Physiol.* **2003**, *133*, 482–491.
137 Wojtaszek, P., Anielska-Mazur, A., Gabryś, H., Baluška, F., Volkmann, D. *Funct. Plant Biol.* **2005**, *32*, 721–736.
138 Wojtaszek, P., Baluška, F., Kasprovicz, A., Łuczak, M., Volkmann, D. *Protoplasma* **2007**, *230*, 217–230.
139 Hoson, T., Soga, K. *Int. Rev. Cytol.* **2003**, *229*, 209–244.
140 Stout, S.C., Porterfield, D.M., Briarty, L.G., Kuang, A., Musgrave, M.E. *Int. J. Plant. Sci.* **2001**, *162*, 249–255.
141 Kwon, M., Bedgar, D.L., Piastuch, W., Davin, L.B., Lewis, N.G. *Phytochemistry* **2001**, *57*, 847–857.

4
Development and Gravitropism of Lentil Seedling Roots Grown in Microgravity

Gérald Perbal and Dominique Driss-École

4.1
Introduction

The first Space experiment carried out on lentil seedlings took place in 1985 in the frame of the Spacelab D1 mission. It was the starting point of a series of six experiments dealing with the growth and gravitropism of lentil seedling roots grown in Space. To perform these experiments, the Biorack facility was used, which provided several advantages (such as a 1×g control in Space) but had some constraints, mainly due to the volume of the Type I containers (see Fig. 4.8 below). For this reason it was not possible to study lentil seedlings older than 30h. This was sufficient for the analysis of gravitropism since roots of young seedlings show the strongest sensitivity to gravity. However, it was a little too short for examining the development of the primary root and the possible effect of microgravity on root growth. Thus, the different studies presented here concerned only the beginning of growth of the primary root following the hydration of the seed.

During the twelve years that these experiments lasted, our knowledge on root development has evolved considerably by ground investigations, whilst Space experiments have always been only a part of the study of root growth and gravitropism. The results obtained were sometimes expected and sometimes not expected. Obviously, the latter findings were more fruitful, since they opened up new avenues of research. All these analyses were possible only by carrying out several experiments, which enabled us to utilize the results obtained in the frame of one flight to further in our knowledge and then test new hypotheses in a subsequent flight.

The role of gravity on plant growth is not yet completely known [1–3] and the experiments on lentil seedlings were pioneering in the field of gravimorphogenesis. This work must be followed in the future by studies on the development of whole plants. Some projects dealing with the development of *Arabidopsis* should be flown soon on the International Space Station (ISS).

Biology in Space and Life on Earth. Effects of Spaceflight on Biological Systems. Edited by Enno Brinckmann

ISBN: 978-3-527-40668-5

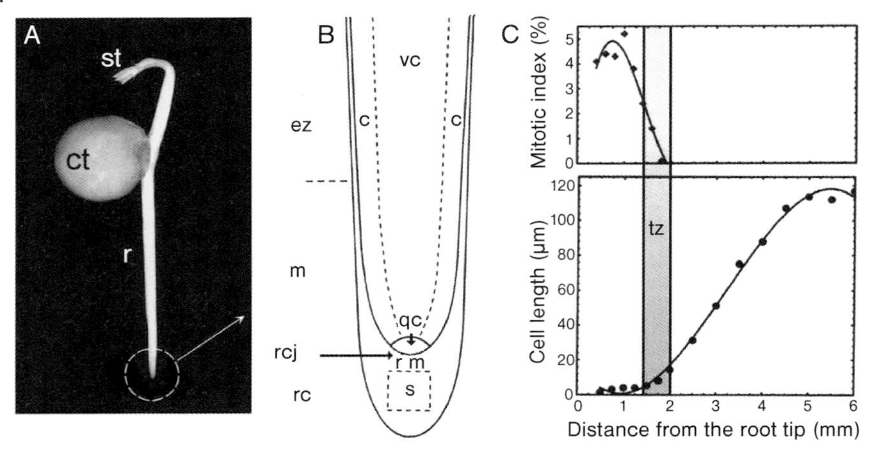

Fig. 4.1 Lentil root structure. (A) Lentil seedling 48-h old; ct, cotyledons without seed coat; r, primary root; st, shoot. (B) Longitudinal axial section of the last 2 mm of the root tip; c, cortex; vc, vascular cylinder; ez, elongation zone; m, meristem; qc, quiescent centre; rc, root cap; rcj, root cap junction; r m, root cap meristem; s, statocytes. (C) Mitotic index (in %) or cell length (in µm) as a function of the distance (in mm) from the root cap junction; tz, transition zone.

4.1.1
Development of Lentil Seedlings on the Ground

Lentil seedlings can start germinating almost immediately after they have been hydrated. Owing to the quantity of reserves in their cotyledons (Fig. 4.1A), the primary root can elongate for several days without any supply of minerals or sugars. This potential made it easy to activate their growth in Space, since only water had to be provided to seeds to start their germination.

4.1.1.1 Functional Zones of the Primary Root

The root tip has a simple structure (Fig. 4.1B), in which one can observe a zone of guidance (the root cap) responsible for fine regulation of root growth, a zone that contains stem cells (quiescent centre) that are normally dividing very slowly, a region where cells are produced intensively in rows (the meristem) and a zone where cells elongate and become specialised. The root is surrounded by an epidermis and is transversely constituted of two major zones (Fig. 4.1B): the cortical cells or cortex (the inner limit of which is the endodermis) and the vascular cylinder (the outermost layer of which is the pericycle).

4.1.1.2 Role of the Root Cap

The root cap may be considered to be a small organ that covers and protects the root meristem [4]. It is formed by the root cap meristem (Fig. 4.1B) which is at the junction between the cap and the quiescent centre. These cells differentiate into statocytes (in the centre of the cap) and into peripheral cells (around the statocytes). It has been shown by different methods (removal of the cap, e.g. Ref. [5]) that the root cap is responsible for perception of gravity. The statocyte contains voluminous organelles (the statoliths) that can sediment under the influence of gravity. These organelles are plastids that contain large starch grains (amyloplasts). The density of starch is greater than that of the surrounding cytoplasm [6] and the statoliths fall within the statocyte because of their weight. It is generally accepted that the statocytes are the gravisensing cells since their destruction by laser beams [7] suppressed the response to gravity, but the role of statoliths is still discussed [8].

4.1.1.3 Meristematic Activity

The meristem corresponds to the zone where cells divide quickly (Fig. 4.1B). To investigate the meristematic activity in the lentil root [9, 10], the cortical cells have been studied since they are more homogeneous than those of the vascular cylinder. These cells differentiate into one tissue (a parenchyma), whereas in the vascular cylinder cells differentiate quickly in many different tissues. Close to the root cap junction, the plane of division of the cortical cells is variable, whereas at some distance from this point the plane of division is almost always perpendicular to the root axis. One simple way to study the meristematic activity that has long been used is to analyse the mitotic index (MI), which represents the percentage of cells in mitosis [11]. Close to the root cap junction, the MI is maximum and then decreases (Fig. 4.1C). It is zero at about 2 mm from the root cap junction. The frequencies of the various phases (prophase, metaphase, anaphase, telophase) of mitosis can also be examined. This study allows us to determine which phase is longer than the others since the phase that is the longest has the highest frequency.

As the MI shows that at a given time only a few percent of the cells are in mitosis, this indicates that the cortical cells, even forming a homogeneous tissue, are not synchronized [10]: they are not dividing at the same time and also not at the same rate. Thus, the study of MI is not sufficient and the cell cycle must be analysed. The cell cycle is composed of four phases: G1, S, G2 phases and the mitosis. The G1, S and G2 phases correspond to the interphase. During the S phase, DNA synthesis occurs, whereas the G2 takes place just before mitosis. There are two major steps in the cell cycle, the first one is a start point that is just before the S phase, the second one is a checkpoint occurring before mitosis. In dry lentil seeds, all cortical cells are in the G1 phase [10]. Following hydration of these seeds, DNA synthesis begins after 13 h (Fig. 4.2A) and there is a peak of mitosis at 25 h (Fig. 4.2B). It has also been demonstrated that all cells are cycling (they have entered the S phase), since 100% of the nuclei of the cortical cells were labelled with IUDR (iododeoxyuridine) after 25 h, which means that all nuclei have synthesised DNA (Fig. 4.2A). Thus, because seedlings not older than 30 h were examined, it can be concluded that, at this time, most of the cortical cells have completed the first cell cycle and entered the second cell cycle. In addition, notably, during the first 30 h there is a partial synchronization of the cell cycle within the cortical cells of lentil seedling roots since there is a peak of mitosis at 25 h.

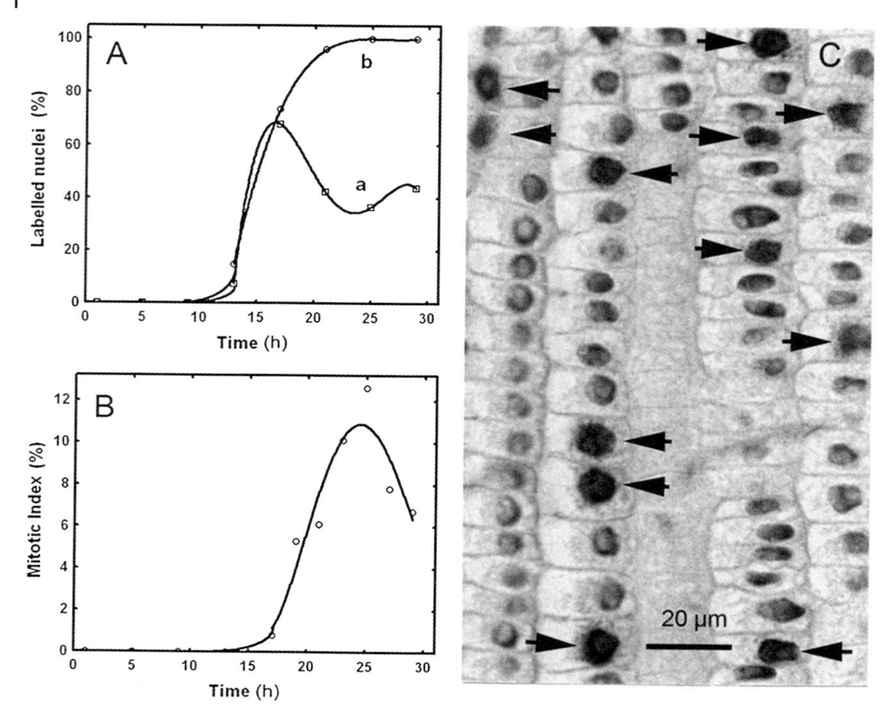

Fig. 4.2 DNA synthesis and mitotic activity in lentil roots. (A) Percentage of nuclei in DNA synthesis in the cortical region (0–6 mm from the root cap junction) of the primary root as a function of time of germination after (a) a 1 h IUdR labelling or (b) a continuous IUdR labelling. Percentages were calculated from one axial section of eight meristems. IUdR, iododeoxyuridine. (B) Mitotic index in the cortical region (0–6 mm from the root cap junction) of the primary root as a function of time of germination. (C) Longitudinal axial section of cortical cells of a lentil primary root after a 1 h incorporation of IUdR at 13 h of hydration. The nuclei marked by arrows were labelled by IUdR and were synthesising DNA. The other nuclei were stained by Gill's N°1 haematoxylin.

4.1.1.4 Cell Elongation

The mitotic activity in the meristem produces rows of cells that are very convenient for following the differentiation of cortical cells. The major features of differentiation for these cells are vacuolization and elongation (Fig. 4.1C) in the direction of the root axis [11]. At the end of this process, a vacuole occupies a large volume in the cell. Cell elongation begins at about 1.5 mm from the root cap junction and the cell length is multiplied by 10 between the 1st and the 5th mm from the root cap junction (Fig. 4.1C). Another functional zone has been distinguished between the meristem and the cell elongation zone. It corresponds to the zone where cells cease to divide and are not yet elongating. This zone was called the "transition zone" by Darbelley et al. [11] and "distal elongation zone" by Ishikawa and Evans [12], whose definition of this zone is slightly different. However, in both cases, it has been shown that this zone is strongly involved in root gravitropism (Section 4.1.2.4).

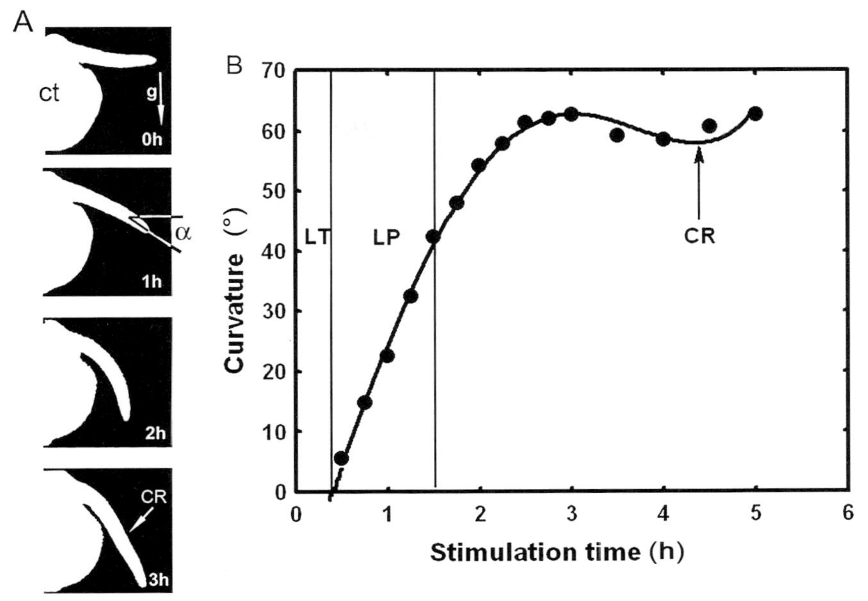

Fig. 4.3 Gravitropic reaction of the lentil root. (A) After 25 h of growth in the vertical position, the seedling was reoriented at 90° with respect to the gravity vector *g* (0 h; gravitropic stimulation). After one hour (1 h) a slight curvature occurred and after 2 h the root tip made a large angle with the gravity vector. A three-hour-stimulation (3 h) led, in most cases, to a counter-reaction (CR): the amplitude of the angle of curvature slightly decreases; ct, cotyledon. (B) Angle of curvature as a function of time; CR, counter-reaction; LP, linear phase during which the rate of curvature is constant; LT, latent time.

4.1.2
Root Gravitropism on Earth

Most plant organs have a gravitational set point angle, GSPA [13], which represents their normal orientation with respect to gravity. Primary roots grow in the direction of the gravity vector and have, therefore, a GSPA of 0°. When a seed germinates, its root tip must enter the soil as fast as possible to find water and minerals necessary for its survival. Root gravitropism is responsible for the downward curvature of the root (Fig. 4.3A and B), which permits the penetration of its tip into the soil. Notably, the cap produces slime to make this possible and to prevent damage of the root meristem.

4.1.2.1 Perception of Gravity

Removal of the root cap suppresses the sensitivity to gravistimulus. The cap is the site of gravisensing, and it has been shown precisely by laser ablation on *Arabidopsis* roots [7] that the statocytes (central root cap cells) are responsible for gravisensing. As these cells (Fig. 4.4) contain voluminous amyloplasts (statoliths) that sediment under the influence of gravity it was hypothesized, at the beginning of the 20th century [14, 15], that these organelles could be the perceptors of gravity.

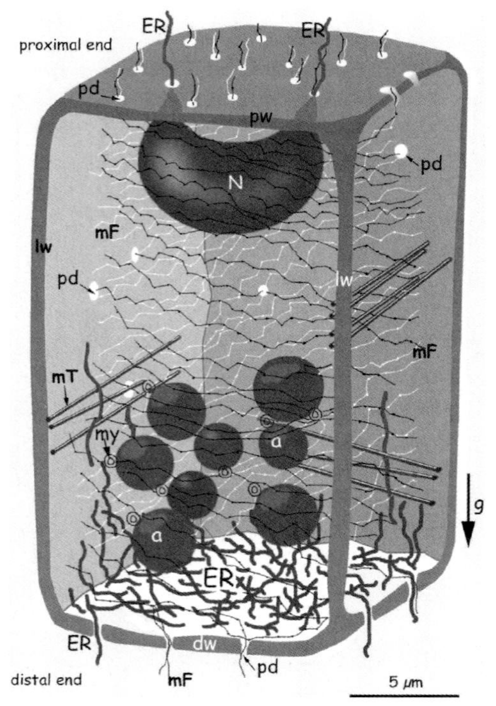

Fig. 4.4 Three-dimensional reconstitution of a lentil root statocyte. Microtubules (mT) are located close to the plasma membrane, perpendicular to the longitudinal axis of the statocyte. The nucleus (N) is maintained near the proximal wall by tubules of endoplasmic reticulum (ER) and by actin microfilaments going through plasmodesmata (pd). Tubules of ER are forming a cup-shaped structure at the distal end. The actin web consists of short filaments (mF, represented in black or white) orientated in different directions and linked by actin binding proteins (small dots) in an overall diagonal orientation. For clarity, the number of amyloplasts (a) was reduced by two-thirds and only a few microtubules are represented. Mitochondria and dictyosomes are not shown. dw, distal wall; g, gravity vector; lw, longitudinal wall; my, myosin; pw, proximal wall.

Many experiments were carried out to test this hypothesis, but none brought very clear cut results. For instance, analysis of the gravitropic response as a function of the natural variations of the volume of the statoliths did not indicate any relationship between these two parameters in lentil seedlings [16]. The most promising studies were done on *Arabidopsis* mutants, the plastids of which were starch depleted [17, 18]. These organelles were not able to move under the effect of gravity [19], but the roots were still able to respond to gravistimulus. Their curvature was much slower, but these roots were still able to perceive gravity. A detailed study with intermediate mutants (mutants containing various quantities of starch) showed that starch was necessary for full sensitivity [20]. Another argument for the involvement of the amyloplasts in gravisensing was given by Kutznetzov and Hasenstein [21]. These authors used intracellular magnetophoresis to move the amyloplasts towards the lateral wall of the statocyte to induce gravitropic curvature of *Arabidopsis* roots. No reaction was observed in starchless mutants subjected to high-gradient magnetic fields. However, root gravitropism could also be induced by a signal originating from outside the cap [22]. All these results were interpreted in two different ways: according to most authors [3, 6, 23, 24] the amyloplasts are the gravisensors, but their movement under the effect of gravity may not be absolutely necessary for gravisensing; according to others [8], the whole cell (protoplast) should be responsible for gravisensing. In our opinion, amyloplasts in higher plants are more efficient than any other cell structure, but the mechanism of perception is such that the whole protoplasm could play this role to a lesser extent [25]. This ability should be considered as an ancestral way of sensing [26].

In all perception chains (whatever the stimulus could be), there is a phase of perception followed by a phase of transduction of the stimulus. In tropisms, the physical stimuli (light for instance) is transformed into a biochemical factor by a receptor. In gravitropism, the nature of the mechanical receptor is not yet known and the only way to discover it is to analyse the statocyte by microscopy and to determine what structure is in contact with amyloplasts and could be involved in the transduction of the gravistimulus.

4.1.2.2 **The Root Statocyte**

The root statocyte shows a vectorial polarity (Fig. 4.4), since their organelles have a specific location along their longitudinal axis, which is parallel to the root axis. For most genera [6, 27], the nucleus is located near the proximal wall of the statocyte (closer to the root meristem), whereas the endoplasmic reticulum is close to the distal wall (Fig. 4.4, ER). Statoliths are the only organelles that do not contribute to the structural polarity because they always sediment in the physically lowest part of the cell. Vacuoles are very small within the statocyte, which is an exception in differentiated plant cells. During the last decade, the mechanism that is at the origin and which maintained the structural polarity has been intensively studied by examining the cytoskeleton of the statocyte [27, 28]. Microtubules encircle the protoplast just under the plasma membrane transversely to the longitudinal axis of the cell. However, actin filaments appear to be elements of the cytoskeleton that maintain the polarity [27]. This has been discovered by using cytochalasin (B or D) treatments, which perturb the polymerization of actin filaments. Under the effect of these drugs, the amyloplasts assume positions very close to the distal wall and the nucleus sediments on the top of the amyloplast bulk (see Fig. 4.19a). The ER tubules remain near their origin of formation, i.e. close to the proximal pole. In non-treated roots, the ER also contributes to the proximal positioning of the nucleus because the nuclear envelope is connected through plasmodesmata to the tubules of ER located at the distal end of the upper neighbour statocyte. This could explain the peculiar shape of the nucleus, which seems to be attached to the proximal wall by one or two regions of its envelope [27].

Studying the statocyte ultrastructure has led to the hypothesis that the ER should be involved in the transduction of the gravistimulus. The contact of the amyloplasts on the ER tubules could induce a Ca^{2+} efflux from these tubules to the cytoplasm in the lower side of the gravisensing cell and activate calcium-dependent proteins that could trigger the transduction of the stimulus [29]. Thus, the mechanoreceptor should be located in the ER tubules.

Another hypothesis is based on the fact that there is a dense cytoplasm (cytogel) along the plasma membrane and that the pressure of the amyloplasts on the cytogel lining the longitudinal walls could be responsible for activating the mechanoreceptors supposed to be located in the plasma membrane [25].

4.1.2.3 Gravisensitivity: The Presentation Time

The presentation time has been used to estimate gravisensitivity [30–32]. This parameter represents the minimal duration of a gravistimulus that is necessary to induce a slight but significant curvature. The gravitropic stimulation can be provoked by gravity, by placing the seedlings in the horizontal position, but the stimulation must be stopped to follow the gravitropic response in absence of any other stimulus. On the ground it is impossible to remove gravity, except by free fall (which could last only a few seconds). Plant physiologists [33] have observed that seedlings rotating (at 1–4 rpm) around a horizontal axis on devices called clinostats (Fig. 4.5A) do not show any gravitropic response. This does not mean that gravity is nullified, but the unilateral direction of this physical factor is replaced by an omnilateral effect. It is generally believed that clinostats simulate to some extent weightlessness, but the quality of this simulation could not be evaluated without analyses in real weightlessness. However, for several decades clinostats were used to estimate gravisensitivity. The simplest way to determine the presentation time was to stimulate seedlings in the horizontal position for various periods (of the order of one to several minutes) and to rotate the plants (for 1–2 h) to follow the gravitropic curvature (Fig. 4.5B).

A dose–response curve was then drawn (Fig. 4.6) and for theoretical reasons (Fechner Law, which states that a response to a stimulus varies linearly as a function of the logarithm of the intensity of this stimulus) it was considered that a logarithmic model best fitted the experimental results (Fig. 4.6) ([Eq. (1)]:

$$R = a + b * \log(t) \tag{1}$$

where R is the curvature in degrees (after 1–2 h), t the stimulation time and a and b are constants.

This was an easy way to estimate gravisensing, since by plotting R as function of the logarithm of the stimulation time and assuming that the model was good, the presentation time was determined by linear extrapolation (Fig. 4.6). However, it appeared very early on that the model did not fit for longer periods of stimulation, such that the authors often did not consider such periods for estimating the presentation time. By this method, the presentation time was estimated to be 30 s [30] to 1 min (Fig. 4.6).

However, plant organs can react to stimuli much shorter than half a minute if the stimulus was repeated several times [34]. This could be demonstrated by stopping a clinostat intermittently for about 1 s. Thus, a continuous stimulation of 30 s

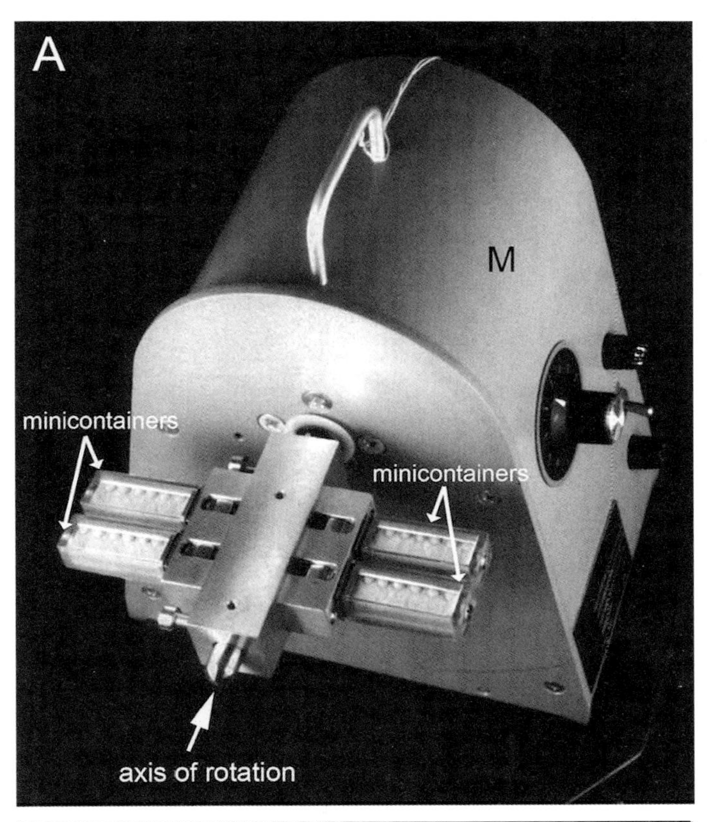

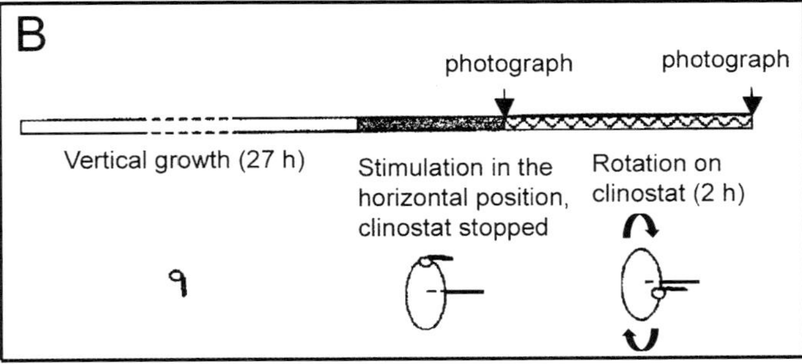

Fig. 4.5 (A) A slowly rotating clinostat. A motor (M) allows the rotation of an axis (1 rotation per min) in the horizontal position. In the minicontainers, the lentil seedling roots are oriented parallel to the axis of rotation. (B) Flow diagram of a typical experiment performed on a slowly rotating clinostat for the determination of the presentation time. After 27 h of growth in the vertical position, the seedlings are placed on the non-rotating clinostat and stimulated in the horizontal position for periods of 1.5 min up to 35 min. At the end of this period, roots are photographed and the rotation of the clinostat starts for 2 h. Another photograph is taken after 2 h of rotation on the clinostat. Figure 4.6 shows the results obtained.

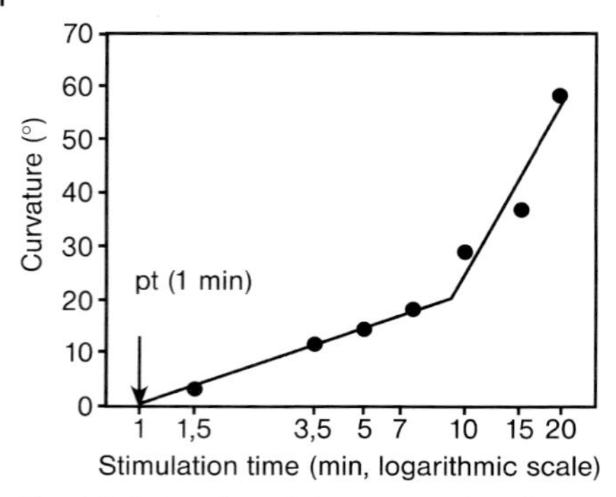

Fig. 4.6 Estimation of the presentation time of lentil roots grown in the vertical position, stimulated in the horizontal position and then placed for 2 h on a clinostat (Fig. 4.5). The amplitude of curvature is plotted as a function of the stimulation time. The presentation time (pt) is estimated by extrapolation down to zero curvature of the regression line fitting the experimental points corresponding to stimulation times of less than 10 min. The model does not fit for longer stimulation times.

can provoke a curvature of the organs, but a repeated stimulus of about 1 s also induces a response. These results indicate that the stimuli could be summed, but it has never been demonstrated that the total period of stimulation with intermittent stimuli must be equal to the presentation time. In addition to stimulation by gravity, stimulation by centrifugal forces has been used [35]. The principle of the experiment remains the same, except that with a centrifuge it is possible to subject plants to various g levels for different periods of times [31]. In this case, the dose–response curve is represented by the mathematical model:

$$R = a + b * \log(t \times \alpha) \tag{2}$$

where R is the curvature in degrees (after 1–2 h), α is the acceleration, t the time of stimulation, $(t \times \alpha)$ is the dose of stimulus expressed in g×s (for instance), a and b are constants.

This study could be performed by using a two axes clinostat that simulate weightlessness by rotating the plants around a horizontal axis and by centrifuging them around a vertical axis. With this device it was technically not possible to subject plants to intermittent stimuli because it is difficult to stop the centrifuge for one second. All these studies assumed that (a) clinostats were able to simulate correctly weightlessness, (b) the method chosen for determining the presentation time was the correct approach.

4.1.2.4 Gravitropic Reaction

The gravitropic curvature of roots is due to a differential growth of the upper and the lower side of these organs, the lower side growing less than the upper side [11, 36, 37]. There have been many studies on this differential growth. Their aim was to compare the growth of both sides of vertically growing roots with the growth

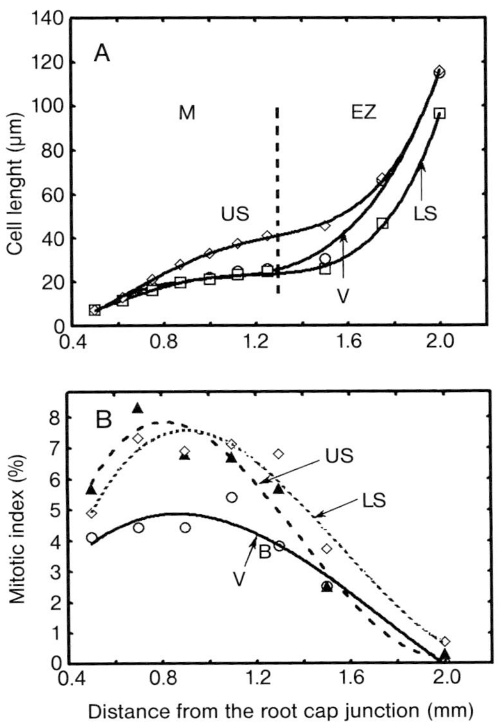

Fig. 4.7 Cell elongation and mitotic activity during gravitropic reaction. Root growth depends upon cell elongation (A) and mitotic activity (B). These parameters are compared in vertical roots (V) and in the lower (LS) and upper (US) sides of horizontally stimulated roots. (A) Cell length versus distance from the root cap junction for vertically and horizontally oriented roots. Similar to the vertically growing roots, the curvature is due to an acceleration of cell elongation in the upper part (US) of the root meristem (M) and to an inhibition of cell elongation in the lower part (LS) of the elongation zone (EZ). (B) A similar analysis was done for mitotic index. The number of mitosis increases in the upper (US) and lower (LS) parts of the meristem (M) for horizontally stimulated roots compared to the vertical ones.

of the lower and upper sides of the stimulated roots. The results obtained were often contradictory and only recently it was understood that several different tissues were involved in the differential growth and can react differently to the gravitropic stimulus. Study of the features of the gravitropic curvature of lentil roots led to the conclusion (Figure 4.7A) that (a) the zone of curvature extends from the distal part of the meristem to the proximal part of the cell elongation zone; (b) the curvature in the distal meristem was due to early elongation of the cells of its upper part; (c) in the proximal part of the cell elongation zone, bending takes place due to inhibition of cell growth in the lower half of the root.

Gravistimulus also has an impact on cell division since the mitotic index (Fig. 4.7B) increases in the upper side and lower side of the meristem of roots placed in a horizontal position for 2 h. The results obtained agree with the hypothesis of a lateral transport of a hormone in gravistimulated roots. This hormone should be an inhibitor of cell growth and should have been present in greater amounts in the lower side of the stimulated root and in lower amount in its upper side than in the vertical controls. The nature of the hormone involved in the gravitropic curvature was studied in the 1960s and 1970s. It was generally accepted that auxin redistribution was responsible for gravitropic curvature [2].

Recently, this hypothesis has been confirmed in roots by analysis of the *Arabidopsis* mutants *aux1* and *pins*. The AUX1 gene family mediates auxin influx [2, 38, 39], whereas members of the PIN family contribute to its efflux [40–42]. This finding led to the so-called "fountain model" of auxin transport in roots [2]. *AUX1* ensures auxin uptake by the statocytes, whereas *PIN3* ensures its efflux. When the root is stimulated in the horizontal position, a lateral movement of auxin occurs [43, 44]. The subcellular localization of *PIN3* is dependent on the root orientation within the gravitational field. Upon gravistimulation PIN3 relocalizes within 2 min, accumulating in the plasma membrane at the bottom side of statocytes [40]. A relocation of PIN3 within the statocytes could therefore be the initial step of the establishment of the lateral auxin gradient upon gravistimulation. Longitudinal transport of auxin towards the zone of gravitropic curvature takes place within the lateral root cap cells and the epidermis and depends on concerted activities of PINs and AUX1 [45]. Recently, it has also been shown that the turnover of PIN2 should modulate the localization of this auxin-efflux facilitator, allowing asymmetrical growth [46]. However, notably, the lateral transport of auxin from the root cap cannot account for some key features of root gravitropism, and there could be a gravitropic signal arising outside of the root cap [22, 47].

4.2 Basic Hardware Used to Perform Space Experiments

Seeds of *Lens culinaris* L. cv. Anicia were chosen because of their small size and their homogeneity of germination. They also had the advantage of being very sensitive to gravity. The seed coat of the dry seeds was removed to facilitate seed germination and root growth. The seeds were placed in small growth chambers (Fig. 4.8A) called minicontainers.

4.2.1 Plant Growth Chambers: The Minicontainers

As mentioned in the introduction, the duration of growth of lentil seedlings could not exceed 30 h, to prevent any contact of the lentil roots with the walls of the plant growth chamber. Apart from this drawback, small growth chambers have the advantage of using a small centrifuge (1×g control in Space), creating a small *g* gradient on this centrifuge (which is recommended for a proper in-flight 1×g control).

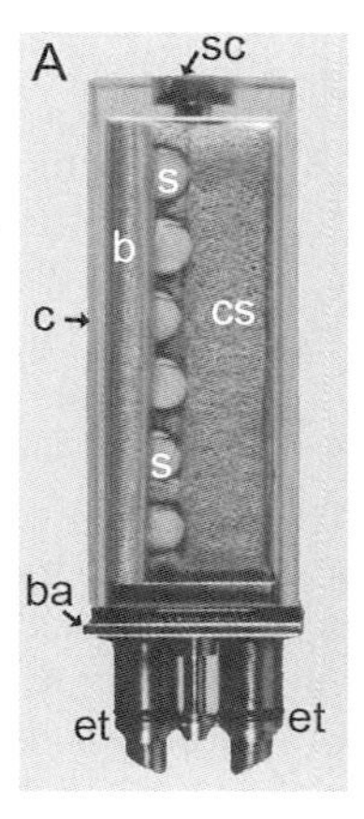

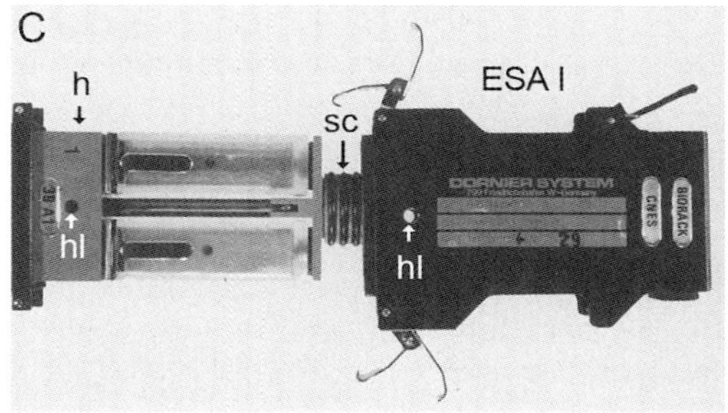

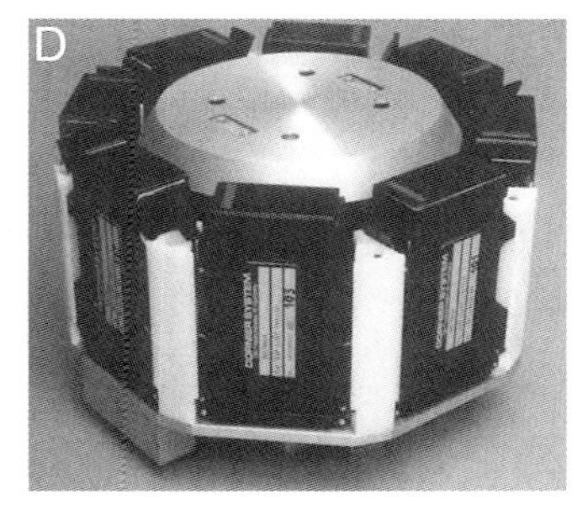

Fig. 4.8 Hardware for Space experiments. (A) The minicontainer: This miniature growth chamber for cultivating lentil seedlings in Space is made of a thick metal partition welded on a base (ba) and of a transparent cover (c) hermetically sealed by a screw (sc). Two entry tubes (et) on the base permit gas exchange and hydration of seeds. A cellulose sponge (cs) is placed on each side of the metal partition. Six dry seeds (s) without their seed coat are held in place on the two cellulose sponges by metal bars (b). (B) Hydration of a minicontainer inside the Biorack glovebox. A syringe (sg) with a special cap is used to inject water into each minicontainer. The experiment is thus activated. This action was performed in Spacelab and Spacehab by the crew. (C) After hydration of the seeds, two minicontainers are placed side by side on a metal holder (h), held in place by a screw (sc) and inserted in an ESA Type I container (ESA I). The holes (hl) permit gas exchange between the minicontainers and the incubator. (D) Eight ESA I containers can be placed on the Biorack centrifuge.

4.2.1.1 Seed Set-up

The minicontainers were made of a transparent cover and a metal base part supporting two sponges. Holes in the metal part allowed hydration of the cellulose sponge (Fig. 4.8A and B). The seeds were held in place on the sponge by metal bars. They were oriented in such way that the root grew either in the direction of the main axis of the minicontainer or more frequently perpendicular to it, depending upon the goal of the experiment. Thus, two configurations of the minicontainers were possible.

4.2.1.2 Hydration of the Seeds

The minicontainers had two entry tubes for fluids (Fig. 4.8A). The first was used to hydrate the sponge by means of a syringe with a special cap coupled to the minicontainer (Fig. 4.8B). Hydration activated the experiment. After the injection of water, two minicontainers were placed together on a metal holder and inserted in an ESA Type I container (Fig. 4.8C). The aperture in the holder and in the Type I container permitted gas exchange between the Biorack incubator atmosphere and the minicontainers. The lentil seedlings germinated in darkness in the incubator (22 °C) for periods varying from 25 to 30 h. The fact that there was gas exchange between the incubator and the inside of the plant growth chambers required a 1×g control in Space on a centrifuge (Fig. 4.8D), since it was impossible on the ground to simulate the cabin atmosphere during the experiment. Nevertheless, a ground control was carried out in parallel in the flight-identical Biorack ground model.

When necessary, photographs were taken by the crew in the Biorack glove box (Fig. 4.8B), by means of a Pentax camera for the first flights or later on by means of a Nikon F5 camera. The dim light of the glove box was sufficient to obtain pictures of good quality.

Importantly, taking pictures did not affect the lentil seedlings grown in microgravity, but it did affect seedlings grown on the 1×g reference centrifuge, since a transfer from 1×g to microgravity was necessary. This transfer did not have a great impact on slow biological processes, but had an effect on some fast phenomena (like the movement of the amyloplasts; see Section 4.4.1). The impact was also apparent when the roots were photographed and chemically fixed or placed back on the 1×g centrifuge. In the latter case the roots were subjected to a microgravity stress (for about 30 min), which could perturb root development. The lentil experiments were carried out in parallel with other experiments but, due to the relatively short period of growth, the 1×g centrifuge was never stopped during the period of growth because of other experiments.

In conclusion, lentil seedlings were sent dry into Space and were hydrated by the crew in microgravity (Fig. 4.8B). The experiment was thus activated in orbit and was, therefore, not affected by the launch. It was possible to run a 1×g control in parallel in Space and on the ground, but taking pictures or fixing seedlings by glutaraldehyde meant stopping the 1×g control in Space. In some cases, this had an impact on the results obtained with the in-flight 1×g reference.

4.2.2 The Glutaraldehyde Fixer

Samples were fixed in 4% (v/v) glutaraldehyde in sodium phosphate buffer (0.1 M) in microgravity. This was carried out by the crew in a special device, into which the minicontainers were inserted (Fig. 4.9). The glutaraldehyde was injected into the minicontainers by a spring-driven piston. Overpressure was avoided by pushing air and water out of the minicontainer and collecting them on the top of the piston. One of the difficulties for the engineers of the manufacturer (COMAT, Toulouse,

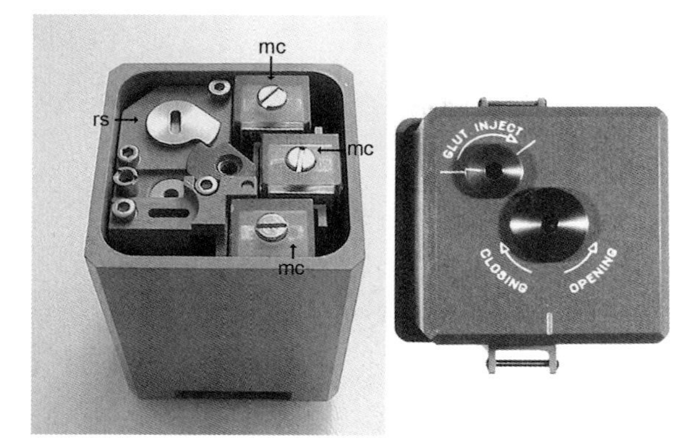

Fig. 4.9 Fixation device and its cover. This device was used at the end of the experiment in Space to chemically fix the samples (4% glutaraldehyde in 0.1 M sodium phosphate buffer). Three minicontainers (mc) can be inserted into this device. The glutaraldehyde, contained in a reservoir (rs), was injected into the minicontainers by a spring-driven piston (not visible). This action was triggered by the crew by turning the screw (glut-inject) on the cover only after the cover was tightly closed. Overpressure was avoided by pushing air and water out of the minicontainers and collecting them on top of the piston.

F) was that the fixative glutaraldehyde was considered to be dangerous for the crew members. The engineers of COMAT were obliged to design a fixer with three levels of containment. It turned out that there was room only for three minicontainers in the fixer due to the volume available in the ESA Type II container, in which the fixers were mounted. This had the drawback that more than one Type I container (with two minicontainers) had to be stopped at a time to fill one fixer.

4.3 Development in Space

For the study of root development, all seeds were set up in the minicontainers with the root tip perpendicular to the main axis, all oriented in the same way (Fig. 4.8A). It was, therefore, easy to analyse the orientation of the root tip in Space with respect to their initial orientation. The experiment carried out during the D1 mission (1985) showed that roots do not grow straight in microgravity.

4.3.1 Root Orientation in Microgravity

Primary roots have a gravitational set point angle, GSPA, of 0° [13], which means they grow in the direction of the gravity vector. In lentil seedling roots grown in microgravity during the first Biorack flight (D1 mission, 1985) [48], as well as in the second flight (IML1 mission, 1992), these organs did not grow straight [9]. The

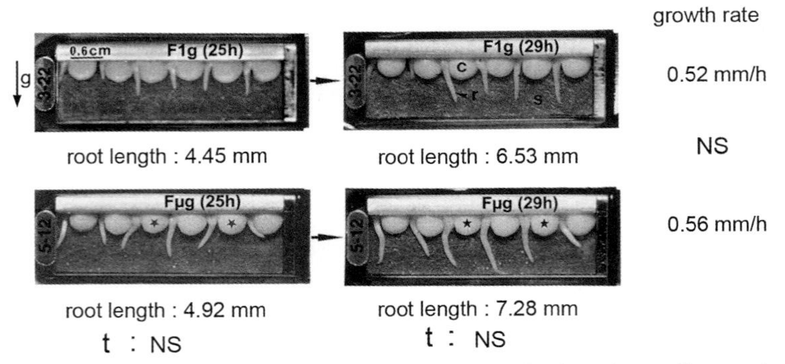

Fig. 4.10 Root length and root orientation of lentil seedlings grown in minicontainers on the Biorack 1×g centrifuge (F1g) or in microgravity (Fμg) for 25 h and 29 h (IML2 Spacelab mission). The arrow (g) indicates the direction of the centrifugal acceleration. Asterisks show the seedlings with a strong change of orientation between the two time points of growth. Differences of root length at 25 h and at 29 h were not significant (t test, NS), neither was the growth rate.

root tip made an average angle of about 30°. Even though the angle of deviation from straight growth was variable, the root tip most often bent away from the cotyledons after 25 h of growth (Fig. 4.10, Fμg 25 h). This deviation was arbitrarily considered as negative since, due to our set up inside the minicontainers, the gravitropic response took place in the opposite direction and was considered as positive. In the 1×g control in Space and on the ground the direction of growth was not significantly different from 0°. This unexpected behaviour was investigated in more detail during the IML2 [49] and the S/MM-06 missions. In the former experiment, photographs were taken after 25 and 29 h (Fig. 4.10).

In microgravity, the average root deviation was –16.7° and –15.5°, respectively, and the orientation was not significantly different from that on the 1×g centrifuge. However, comparison of the photographs corresponding to both times points of growth (25 h and 29 h) showed that the average angle of deviation gave only partial information on root orientation since there was a great variability in the direction of growth of the root tip (Fig. 4.11). Moreover, in some cases (Fig. 4.10), roots could show a strong change of orientation between the two times of growth (see asterisks). They could for instance have a negative deviation at 25 h and a positive deviation after 29 h. In the last experiment (S/MM-06), pictures were taken at 20 min intervals, but owing to technical problems the root orientation was recorded only after 22.5 h. The results showed that after a strong curvature away from the cotyledons (in these cases, 80° and 40°) the roots straightened out for several hours (Fig. 4.12). Thus, all roots reacted in the same way, i.e. they bent strongly away from the cotyledons and then the deviation from their initial orientation diminished progressively until they reached a position close to zero deviation. Notably, such movements were also seen on clinostats and were called "spontaneous curvatures" [30]. In lentil seedling roots grown on a clinostat [50], the deviation from the initial orientation was even stronger but in the same direction (away from the

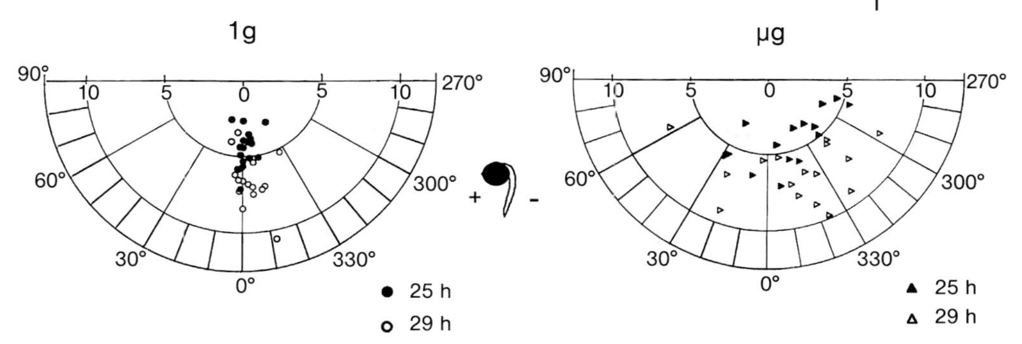

Fig. 4.11 Position, in polar coordinates, of the root tip of lentil seedlings in two samples (1g and μg) after 25 h (filled symbols) and 29 h (open symbols) of growth (IML2 Spacelab mission). The angle of deviation is counted as negative when the root grows away from the cotyledons. 1g, sample grown on the 1×g centrifuge in Space; μg, sample grown in microgravity.

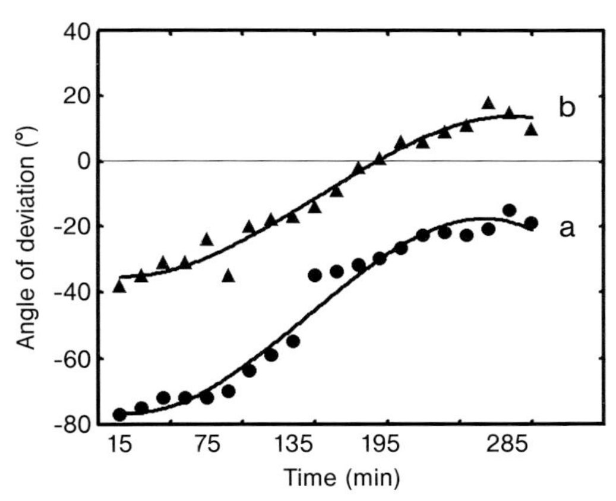

Fig. 4.12 Orientation of two individual roots (a and b) grown in microgravity and measured on photographs taken in the photobox (see Fig. 4.22) (S/MM-06 Spacehab mission). After 22.5 h of growth in microgravity the angle of deviation is strongly negative from their initial orientation (the root grows away from the cotyledons): at –40° and –80°, respectively.

cotyledons). These clinostat experiments could discard the hypothesis of an involvement of hydrotropism (and any other kind of external factors) in the bending, since water distribution in the sponge was homogeneous on this device (due to the constant change of orientation of the minicontainers with respect to gravity). It seems that these movements are due to internal mechanisms following hydration, the effects of which are not visible on the ground because gravity orientates root growth very soon and very strongly. This movement can be considered to be a nastic response, i.e. a natural movement that is not oriented by factors of the environment. This movement should be different from circumnutations, which are rhythmic oscillations of the tip of a plant organ [51]. In any case, it is an unexpected behaviour of the primary root in the unusual environment of microgravity.

4.3.2 Root Growth

The growth of the primary root of lentil seedlings was analysed under three conditions: in microgravity, on a 1×g centrifuge in Space, and on the ground. Notably, the ground control was cultivated in the same instrument (Biorack) as the flight samples and almost at the same time (only with a slight delay of about 2 h). The root growth showed no difference between the 1×g Space and the 1×g ground controls, so in the following text we generally refer to the flight control. After 25 h (Table 4.1, D1 mission), the root length of the controls was exactly the same as in microgravity [48]. However, after 28 h (Table 4.1, IML1 Mission, F) [9], there was a significant difference between the flight control and the microgravity sample (the 1×g control on Earth being the same as the microgravity sample, Table 4.1, G). This slight difference led us to analyse root length later on by taking pictures after 25 h and 29 h, during the IML2 mission, to check if root growth did slow down after 25 h in microgravity. Notably, pictures were taken in microgravity by the crew, which could have had an impact on the 1×g control in Space, which had been subjected to microgravity for about 30 min. The results of this last experiment were clear cut since the root length was found to be the same in microgravity

Table 4.1 Root length (mm) of lentil seedlings grown in space (Spacelab missions D1, IML1, IML2) in microgravity or on the on-board centrifuge (1×g control) or grown on a clinostat (1 rpm) or in the vertical position on the ground (clinostat control). Mean values are given with their interval of confidence (5% level) and compared with t tests. NS, difference not significant or, S, significant at the 5% level. More than 20 roots were studied in each sample. The period of growth is indicated in parenthesis.

	1×g control	*t test*	*Microgravity or simulated microgravity on clinostat*
D1 (25 h)	5.5 ± 0.5	NS	5.5 ± 0.9
IML1 (28 h)			
F (flight)	7.9 ± 0.8	S	} 6.8 ± 0.7
G (ground)	7.3 ± 0.6	NS	
IML2 (25 h)	4.5 ± 0.6	NS	4.9 ± 0.6
IML2 (29 h)	6.5 ± 0.9	NS	7.3 ± 0.8
Clinostat (27 h)	5.9 ± 0.5	NS	5.9 ± 0.4

as on the 1×g centrifuge and the rate of growth was similar for both samples during the period between 25 and 29 h (Table 4.1; Fig. 4.10).

The conclusion of these experiments was that the early growth of the lentil root was not affected by gravity, which did not prove that microgravity has no effect on plant growth since in these studies root length was, at most, 10 mm. Some analyses have shown that root growth should be modified in microgravity only after several days, indicating that microgravity could have a slight but continuous effect on root elongation [52].

4.3.3 Cell Elongation

Cell elongation was first studied out on the D1 mission [53]. The results showed that cell length was the same in the region of the meristem for roots grown in microgravity or on the 1×g centrifuge (Fig. 4.13A). However, in the region where cells elongate strongly, i.e. over 1 mm from the root cap junction, cell growth appeared to be more pronounced in microgravity than on the 1×g centrifuge. A similar analysis was performed during the IML2 mission (Fig. 4.13B) [49], where only a very slight difference was found in cell elongation, which was larger in the 1×g control, in the region 1.5–2.0 mm from the root cap junction. These results must be compared with those obtained with the root tip orientation. In the D1 mission, a deviation of about 30° in microgravity was observed and cell elongation was different in the cell elongation zone (Fig. 4.13A), whereas a not significant deviation of 15.5° and an almost similar cell growth was seen after 29 h in the IML2 mission (Fig. 4.13B). Obviously, a deviation of the root tip is related to a differential cell growth of both sides of the root. This difference observed in microgravity should be due to the change in the bending of the root tip (Fig. 4.10). This result is an example of the need of replicate of experiments. If only the D1 mission would have been performed, it would have been concluded that there was an effect of microgravity on cell elongation, which was indeed true only for a growth period of 25 h.

4.3.4 Meristematic Activity

Several pioneering analyses (see reviews [52, 53]) have led to the conclusion that the meristematic activity could be strongly depressed in plants grown in microgravity. However, these studies were performed on different plants in different facilities. Moreover, the period of growth varied from one or two days to several weeks and most often only the mitotic index (MI, percentage of cells in mitosis) was taken into account. Our Space experiments gave us the opportunity to study MI and cell cycle in roots of young seedlings.

4.3.4.1 Mitotic Activity

The mitotic index, which represents the percentage of mitoses observed on longitudinal sections (stained by Feulgen) of the meristematic zone, was studied in the

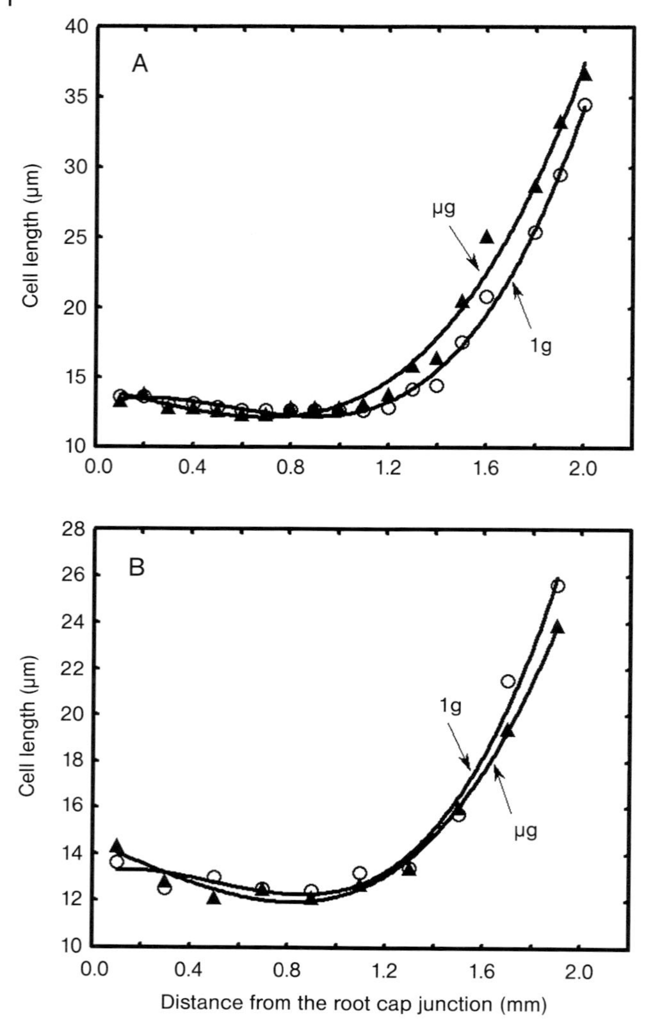

Fig. 4.13 Cell length as a function of the distance from the root cap junction of lentil roots. Seedlings were grown on the 1×g centrifuge (1g) or in microgravity (µg). (A) Lentil seedlings 25-h-old (D1 Spacelab mission). (B) Lentil seedlings 29-h-old (IML2 Spacelab mission).

first three Space experiments (Table 4.2). The mitotic index was found to be higher in microgravity for the D1 experiment [54], but was less for the IML1 [9] and IML2 [10] missions. However, it was always statistically different in microgravity as compared with the 1×g centrifuge. Notably, MI varied as a function of the duration of growth, which is consistent with the fact that cell cycle did not reach a steady state during this phase of development in the two culture conditions.

Table 4.2 Mitotic index (%) in lentil roots grown in microgravity or on the 1×g centrifuge in space (control); ground tests were done on a clinostat and in the vertical position (control). The percentages are compared by χ^2 tests. NS, difference not significant; S, significant at the 5% level.

	Control	*χ^2 test*	*Microgravity*
D1 (25 h)	6.6	S	8.0
IML1 (28 h)	8.9	S	6.8
IML2 (29 h)	6.0	S	3.9
Clinostat (27 h)	6.4	NS	6.1

Table 4.3 IML1 Spacelab mission. Percentage of the various phases of mitosis of the cortical cells of lentil roots. Seedlings were grown in microgravity (F μg), on the on-board 1×g centrifuge (F 1g) or on the ground (G 1g). Results were compared by χ^2 test. Over 1500 nuclei were analysed in each sample. NS, difference not significant at the 5% level. Period of growth: 28 h or 29 h.

Phases	*Percentage (%)*				
	F μg		*F 1g*		*G 1g*
Prophase	50		41		52
Metaphase	18		16		11
Anaphase	10		14		15
Telophase	22		29		22
χ^2 test		2.42 NS		4.32 NS	

MI is a simple and easy to calculate parameter, but its meaning is sometimes not quite clear, because it may vary for several reasons. For instance, it could decrease because cell cycle is lengthened or because mitosis is faster. It may also be larger because the duration of the mitosis itself is extended. Some phases of the mitosis could be eventually more perturbed than others by microgravity. This was why the percentage of the various phases of the mitosis was analysed (Table 4.3). The results obtained showed that there was no statistical difference in the distribution of these percentages in microgravity and on the 1×g control in Space [9, 54]. The 1×g control in Space was also the same as the 1×g control on the ground. This demonstrated that in lentil seedling roots grown for 29 h the mitosis was not perturbed, so that the differences in the mitotic index in microgravity and in 1×g could be due to a change in the duration of the cell cycle.

4.3.4.2 Cell Cycle

As mentioned in the introduction, the first cell cycle after hydration of lentil seeds lasts on average 25 h and the cortical cells are partly synchronized so that after 29 h most of the cells have passed through the first cell cycle and entered the second cell cycle [10]. The DNA content of the nuclei was measured with the PLOIDY program (ALCATEL-TITN) on longitudinal sections stained by Feulgen after fixation by glutaraldehyde in Space. The percentages of the various phases of the cell cycle were determined (Table 4.4) for the IML1 [9] and the IML2 [10] missions. The results obtained in the frame of the IML1 mission were unambiguous, since the distribution of the percentage in the various phases (G1, S, G2, M) was the same in the 1×g Space and ground controls (Table 4.4, G 1g and F 1g; Smirnov test), whereas the microgravity sample was different from the Space control (and the ground control).

For the IML2 experiment, cell cycle in the microgravity sample was also statistically different from that of the 1×g Space control, but the two control data also appeared to be different. This might be because in this experiment photographs were taken after 25 h and 29 h and that the seedlings were chemically fixed after 29 h to study the DNA content within the nuclei. The 1×g Space control was, therefore, perturbed by a period of about 30 min of microgravity occurring at 25 h. This control was, therefore, not an authentic 1×g control. This result indicated that even a short period of microgravity could have modified the cell cycle. Notably, this effect was not found on lentil roots grown on a clinostat [50], which showed that clinorotation should not be a good simulation of microgravity for the study of cell cycle.

Table 4.4 indicates that there were always more cells in the second cell cycle in the 1×g controls than in microgravity. Thus, cell cycle seemed to be slightly delayed

Table 4.4 Percentage of the G2 and M phases of the first cell cycle and of the G1 and S phases of the second cell cycle for roots grown in space [on the on-board 1×g centrifuge (F 1g) or in microgravity (F μg)] or on the ground in the vertical position (G 1g). The Smirnov test was performed to compare the distribution of the DNA content. NS, difference not significant; S, significant at the 5% level.

Mission cycle:		*First*		*Second*		*Smirnov test*
Phase:		*G2*	*M*	*G1*	*S*	
IML1 (28 h)	G 1g	11.2	7.2	55.4	26.2	}NS
	F 1g	11.9	8.9	58.7	20.5	}S
	F μg	27.9	6.8	55.5	9.8	
IML2 (29 h)	G 1g	10.6	4.0	52.8	32.6	}S
	F 1g	19.7	6.0	48.2	26.1	}S
	F μg	17.8	3.9	61.1	17.1	

in microgravity. Because there were numerous cells that have not yet passed the G2 phase of the first cell cycle, it is possible that this phase was perturbed in microgravity or that cells had more difficulty going through the G2/M checkpoint.

In conclusion, the results obtained during the IML1 and IML2 missions clearly showed that the cell cycle of the cortical cells of the root meristem was perturbed in microgravity, even in the early stage of root development. This could be due to a delay in the first cell cycle. However, it is necessary to carry out such analyses on older plants, since this delay could be reduced (because of a kind of adaptation) or amplified after several cell cycles. Clearly, a change in the cell cycle in the root meristem should have an impact on root growth after some days, which could explain why root growth was most often reduced (see reviews [52, 53]) in plants grown for several days in microgravity. New tools to investigate the cell cycle are now available, and the effects of microgravity on cell proliferation will be studied in the future.

4.4 Root Gravitropism in Space

Space experiments gave the unique opportunity to study gravisensing on growing plants, by stimulating them on centrifuges for short periods and by following the gravitropic response without any other stimulus in microgravity (that is in near weightlessness) by time lapse photography. In parallel, it was possible to analyse the movement of the statoliths to better understand their mode of action in the perception of gravity.

4.4.1 Organelle Distribution within the Statocyte

As it was feasible to grow lentil seedlings and to chemically fix them in Space, the distribution of the organelles was studied in statocytes of seedling roots cultivated in microgravity or on a 1×g centrifuge to determine the role of gravity on their structural polarity.

4.4.1.1 Statocyte Polarity

The ultrastructure of the lentil root statocyte was studied on seedlings grown in microgravity, on a 1×g centrifuge in Space or on the ground [48]. The seedlings were fixed with glutaraldehyde in Space or on the ground in the fixers (Fig. 4.9). A morphometrical analysis first carried out on roots grown for 25 h (D1 mission) showed that the structural polarity of the statocyte was not strongly modified in microgravity [55], since only the location of the amyloplasts within these cells was strongly different (Fig. 4.14A and B). This was expected, since on the ground these organelles settled down because of gravity. However, the amyloplasts should have been distributed at random in the statocyte in microgravity, which

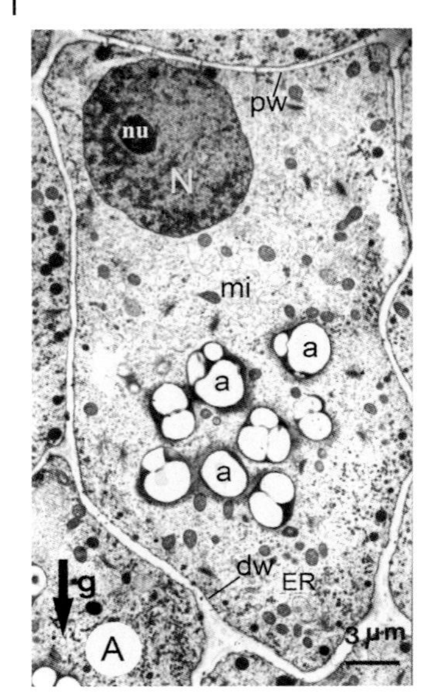

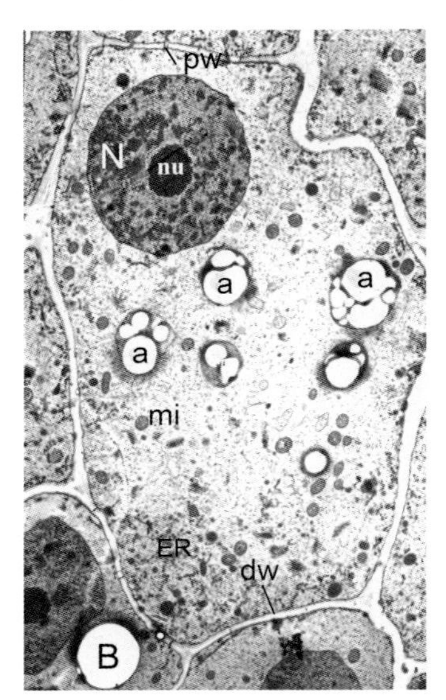

Fig. 4.14 Lentil root statocytes (gravity sensing cells) located in the centre of the root cap. Roots grown on the 1×*g* centrifuge in Space (A) or in microgravity (B). These cells show a structural polarity with the nucleus (N) located near the proximal wall (pw) and the endoplasmic reticulum (ER) near the distal wall (dw). On the 1×*g* centrifuge [in (A) the arrow shows the orientation of the centrifugal acceleration], the amyloplasts (a, or statoliths) sediment in the distal part of the cell. In microgravity (B) the amyloplasts are gathered near the nucleus. mi, mitochondria; nu, nucleolus.

was not the case: they were more numerous in the proximal half of the statocyte (Fig. 4.14B).

A precise analysis of the location of the nucleus also indicated a slight change in microgravity [55]. This organelle was always located close to the proximal cell wall, but in microgravity it was more often found close to the axis of the statocyte and further away from the proximal cell wall (Fig. 4.14B).

Finally, this study also showed that the endoplasmic reticulum, which was forming a kind of cup for the amyloplasts in the controls, was more compact and located closer to the distal wall in microgravity. Figure 4.15(A) and (B) summarizes the results obtained in a 3D reconstruction of statocytes in 1×*g* and in microgravity. Notably, Smith et al. [56] have obtained very similar images of the distribution of the nucleus and the statoliths in the root of white clover.

4.4.1.2 **Positioning of the Nucleus and of the Endoplasmic Reticulum**

As mentioned in the introduction, the nucleus, because of its density, should sediment towards the physical lower part of the 1×*g* controls, the distal wall, whereas the endoplasmic reticulum should not be located close to the distal wall. In fact, the cytoskeleton is involved in the positioning of the nucleus and of the endoplasmic reticulum.

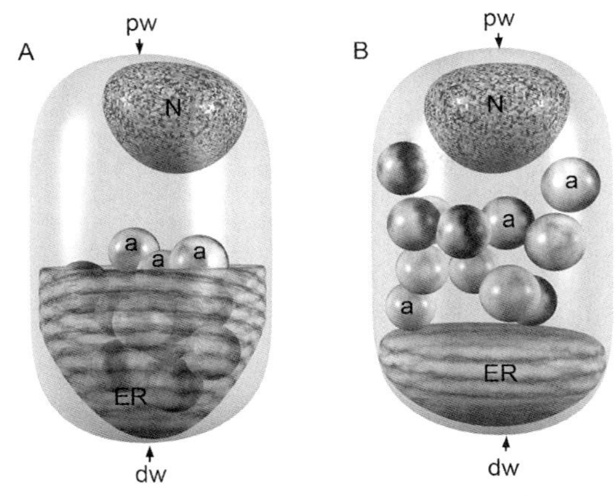

Fig. 4.15 3D reconstruction of lentil statocytes of roots grown in 1×g (A) or in microgravity (B). The comparison shows that in microgravity the nucleus, N, is further from the proximal wall, pw; the endoplasmic reticulum, ER, is closer to the distal wall, dw; the amyloplasts, a, are located in the proximal half of the statocyte, near the nucleus. In 1xg, the ER forms a cup-shaped structure at the distal end.

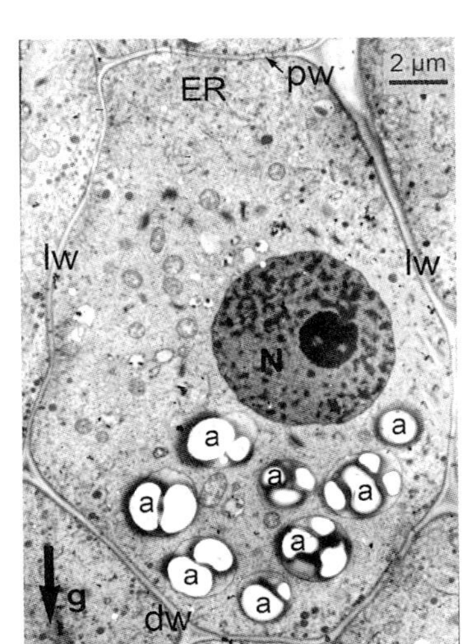

Fig. 4.16 Statocyte of a lentil root treated with cytochalasin B (25 µg mL^{-1} in DMSO 1%) and grown in the vertical position (arrow: direction of g) on the ground. A gravimetric distribution of the organelles is observed: the amyloplasts (a) sediment very close to the distal wall (dw) and the nucleus (N) on the top of the amyloplasts. The endoplasmic reticulum (ER) remains near its origin of formation close to the proximal wall (pw). dw, distal wall; lw, longitudinal wall.

A treatment with cytochalasin B that perturbed the polymerization of actin filaments, provoked on the ground a sedimentation of the nucleus and a gathering of the endoplasmic reticulum at the proximal pole of the statocyte (Fig. 4.16). This experiment proved that actin filaments were responsible for the positioning of these organelles within the statocyte. Centrifugal forces were applied in the direction of the root tip, to move the nucleus experimentally towards the distal

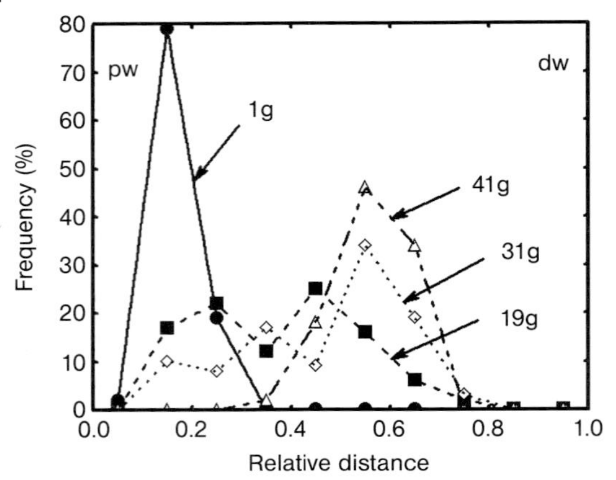

Fig. 4.17 Effect of root tip-directed accelerations on the localization of the nucleus in statocytes of lentil. Roots were grown at 1×*g* for 25.5 h and then either left at 1×*g* or centrifuged at 19×*g*, 31×*g* or 41×*g* for 20 min. The percentage of nuclei is plotted as a function of the relative distance of the nucleus from the proximal cell wall, pw; dw, distal wall.

wall [57]. An acceleration of 19×*g* was necessary to change significantly the position of the nucleus (Fig. 4.17) and an acceleration of 31×*g* could completely detach this organelle from the proximal wall.

These results showed that the nucleus is maintained close to the proximal wall by an actin network. Gravity should exert tension in this network that pulls the nucleus towards the cell wall. In microgravity, this network undergoes a relaxation, since the nucleus is at a greater distance from the proximal cell wall.

The distribution of electron dense chromatin (EDC) of the nucleus has also been studied on micrographs by means of a picture analysis program [58]. This indicated that the texture of EDC was the same in microgravity and in 1×g, but the position of EDC and nucleolus with respect to the longitudinal axis of the statocyte was different in both culture conditions due to the rotation of the nucleus, which occurred in parallel with its displacement towards the cell centre.

The particular shape of the bulk of the endoplasmic reticulum (cup-shaped) at the distal pole (Fig. 4.15A) should be due to the sedimentation of the amyloplasts in 1×g so that the ER tubules can only be positioned close to the distal plasma membrane and along the longitudinal plasma membrane.

4.4.1.3 **Amyloplasts Positioning**

The first Space experiment on lentil seedlings indicated that, in microgravity, most amyloplasts were located in the proximal part of the statocyte near the nucleus [48]. This particular location should have been because the amyloplasts were located around the nucleus in young statocytes and that during the elongation of these cells the amyloplasts did not move and stayed close to the nucleus. The amyloplasts could be maintained near the proximal wall by some mechanism linked to the cytoskeleton. An experiment performed on *Lepidium sativum* roots during a sounding rocket flight has shown that the amyloplasts moved from the distal wall towards the cell centre, when the seedlings where subjected to microgravity for some minutes [59–61]. A similar movement of the statoliths within the *Chara* rhizoid has also been observed [62, 63]. The S/MM-03 mission was carried out to analyse kinetics of the amyloplast movement [64]. Lentil seedlings were grown on the 1×g centrifuge for 28 h and then placed for 13, 29, 46 or 122 min in microgravity (Fig.4.18A–D). They were fixed with glutaraldehyde in Space by the astronauts and the lentil roots were treated for microscopy after retrieval on the ground.

Figure 4.18A, a and B, a shows that, in roots grown in 1×g, the amyloplasts had sedimented close to the distal wall. During the different periods in microgravity, the amyloplasts moved towards the proximal wall and reached a stable position after about 2 h in microgravity, since their distribution was almost the same as that observed in roots grown in microgravity for 28 h (Fig. 4.18C, D, f). This experiment proved that there was a mechanism that positioned the amyloplasts close to the proximal wall in microgravity. This mechanism was not able to counteract gravity on the ground or at 1×g centrifugal accelerations in Space. A distinct role of the cytoskeleton was envisaged, especially for actin filaments [59–61, 65]. This was why, in parallel to the experiments with the amyloplasts movement in microgravity, some seedlings were analysed after cytochalasin D (CD) treatment [64]. The seedlings were hydrated with a solution of 1 μg-CD mL^{-1} in 1% DMSO. The hydration was carried out with a special double syringe (Fig. 4.19). The minisyringes containing the concentrated solution of cytochalasin D were kept at low temperature and were inserted into the water syringes before use. The whole mixture was then injected into the minicontainers. Seedlings were grown on the 1×g centrifuge for 28 h and transferred to microgravity for 2 h and then fixed (Fig. 4.20B).

The movement of the amyloplasts and the nucleus towards the proximal wall occurred also in the CD-treated root statocyte, but at a very low speed. After about 2 h (Fig. 4.20B) the amyloplasts had only moved a little towards the cell centre. This experiment indicated that the amyloplasts that sedimented within the statocytes on the ground or on the 1×g centrifuge were pulled towards the proximal wall of these cells in microgravity by a mechanism depending on actin filaments, since CD treatment slowed down the movement of the amyloplasts considerably. This movement should be due to myosin, a motor protein that can move organelles along actin filaments. This implied that actin filaments were located very close to the amyloplasts and that myosin could be found near these organelles. This was demonstrated by using antibodies [28]. A pre-embedding immunogold silver technique was carried out with a monoclonal antibody raised against actin to study at the electron microscopy level the distribution of the actin filaments in roots grown on the ground. The labelling was scattered in the cytoplasm

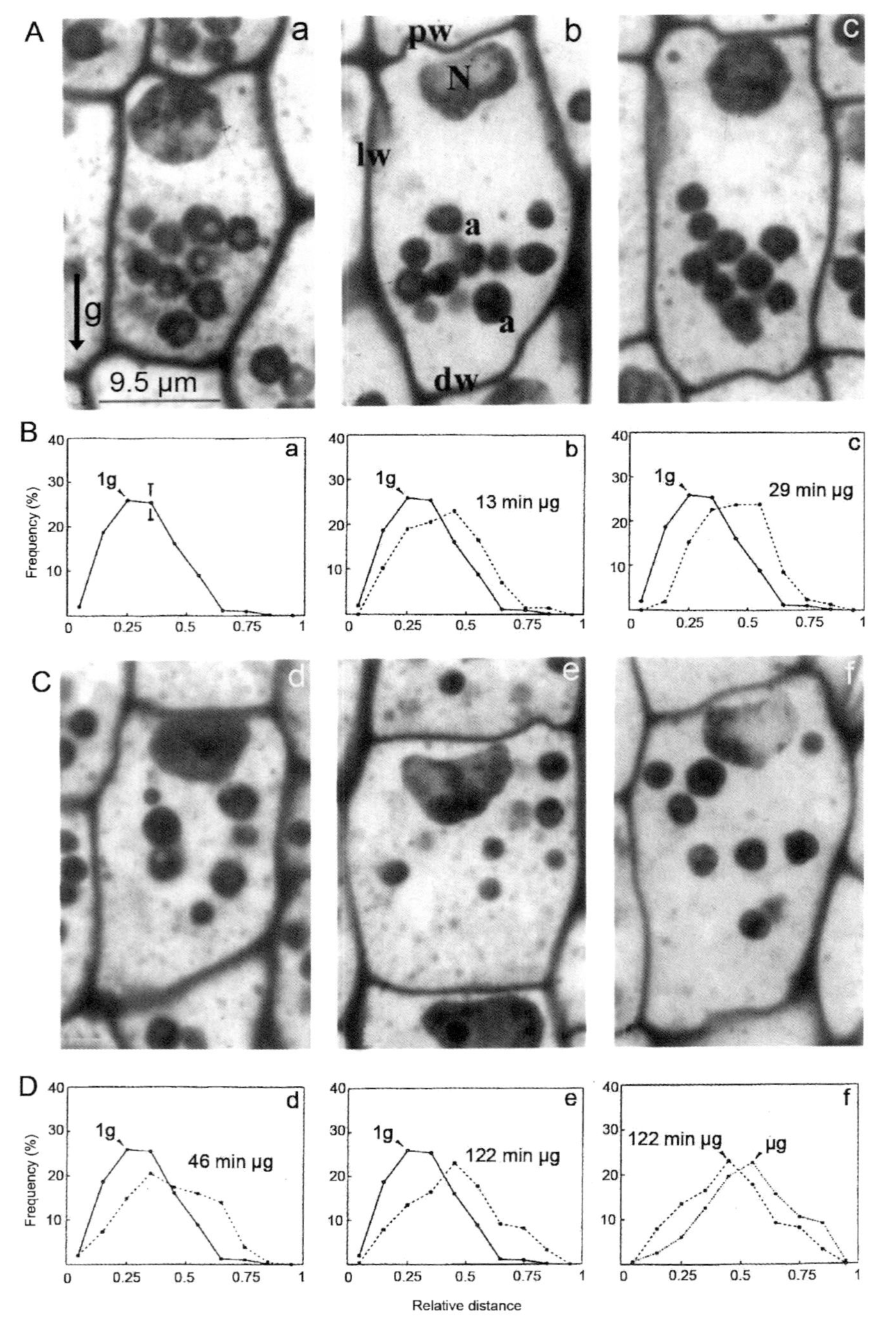

Fig. 4.18 S/MM-03 Spacehab mission. (A and C) Semi-thin sections of lentil root statocytes stained by toluidine blue showing the displacement of the amyloplasts from the distal part to the proximal part of the statocytes in microgravity. (B and D) Percentages of amyloplasts as a function of the relative distance between the distal wall (shown left = 0) and the proximal wall (shown right = 1) of statocytes as those shown in (A) and (C), respectively. (A and B) a: Seedlings grown in 1×g on the ground (solid line); arrow, direction of gravity; (A–D, b–e): seedlings grown first on the Biorack 1×g centrifuge and then placed for 13, 29, 46 or 122 min respectively, in microgravity (dotted lines); (C and D, f): seedlings grown continuously in microgravity (µg) and compared with those grown during 122 min in microgravity. a, amyloplast; dw, distal wall; lw, longitudinal wall; N, nucleus; pw, proximal wall.

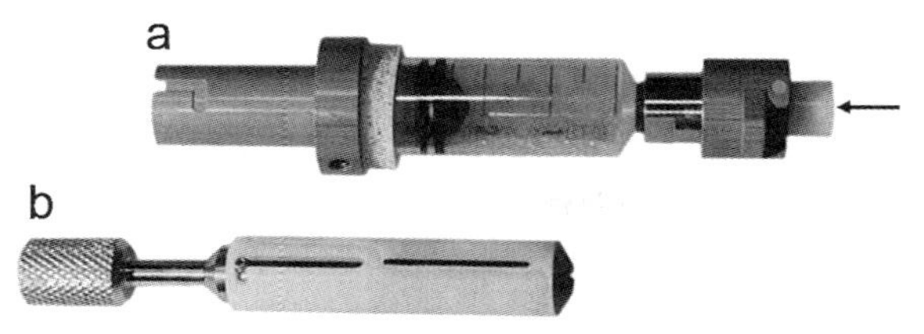

Fig. 4.19 Syringe (a) and mini-syringe (b). The syringe, fitted with a special cap for adaptation to the minicontainer (arrow), was filled with 1.8 mL of water before launch and the mini-syringe (b) filled with a concentrated solution of cytochalasin D (CD) in DMSO (dimethyl sulfoxide). Just before the experiment in Space the crew inserted the mini-syringe into the syringe to inject 20 μL of the concentrated solution of CD (90 μg-CD mL^{-1} in 100% DMSO) in water. The diluted concentration of CD was injected into a minicontainer to start the experiment (final concentration: 1 μg-CD mL^{-1} in 1% DMSO).

(Fig. 4.21A), but frequently the gold-silver particles were associated in files by two to six, which could correspond to isolated microfilaments, some of them abutting the plasma membrane obliquely (Fig. 4.21D). The cell walls, the starch grains in the amyloplasts, the mitochondria and the nucleoplasm did not show any particles (Fig. 4.21A–E). The labelling was always visible close to or on the nuclear and amyloplast envelopes (Fig. 4.21B, C) and these organelles were connected to the cell periphery by microfilaments (Fig. 4.21C). This indicated that each statocyte organelle was enmeshed in an actin web of short filaments arranged in various orientations and never associated into bundles. These results were confirmed by means of a rhodamine-phalloidine staining. The diffuse fluorescence of the cytoplasm could be explained by the fact that the actin network was very thin and the meshes very narrow. Antibodies raised against myosin were also used to analyse the presence of this motor protein and its location in the statocyte. Myosin was found around the amyloplasts (Fig. 4.21F).

4.4.2 Gravisensitivity

4.4.2.1 Presentation Time

Gravisensitivity was, for decades, estimated by measuring the presentation time [33]. This parameter was determined by using a clinostat. However, the mode of action of this tool remains unclear. It removes the unilateral effect of gravity by rotating the plants in the gravitational field, but nobody is able to show that clinostating has no effect on gravisensitivity. The IML1 mission gave the opportunity to measure the presentation time of lentil seedling roots grown for 28 h in microgravity. In this experiment, however, it was not possible to grow lentil seedlings on the 1×g centrifuge and to stimulate them, since the ESA Type I container could be oriented only in one way on the Biorack centrifuge. Very short periods of stimulation on the 1×g centrifuge were also not possible due to the required manual operations. The lentil seedlings grown in microgravity were, therefore, stimulated on the Biorack 1×g centrifuge for periods varying from 5 to 60 min

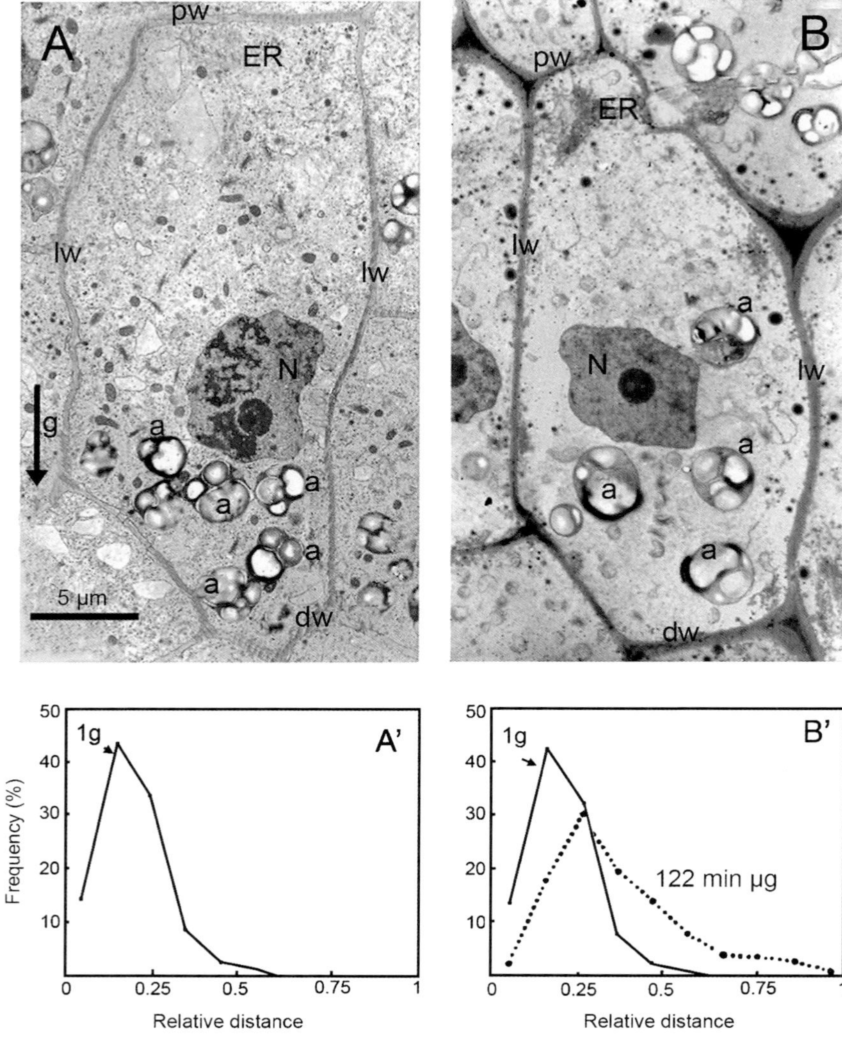

Fig. 4.20 Effect of cytochalasin D on statocyte polarity (S/MM-03 mission of Spacehab). Ultrathin sections of root statocytes of lentil seedlings grown in CD solution (A) in 1×*g* on the ground (arrow, direction of gravity) or (B) first on the Biorack 1×*g* centrifuge and then placed for 122 min in microgravity. a, amyloplast; dw, distal wall; ER, endoplasmic reticulum; lw, longitudinal wall; N, nucleus; pw, proximal wall. (A' and B') Percentages of amyloplasts as a function of the relative distance between the distal wall (shown left = 0) and the proximal wall (shown right = 1) of statocytes such as those shown in (A) and (B). Solid line: seedlings grown in 1×*g* on the ground as in (A); dotted line: seedlings grown first on the Biorack 1×*g* centrifuge and then for 122 min in microgravity as in (B).

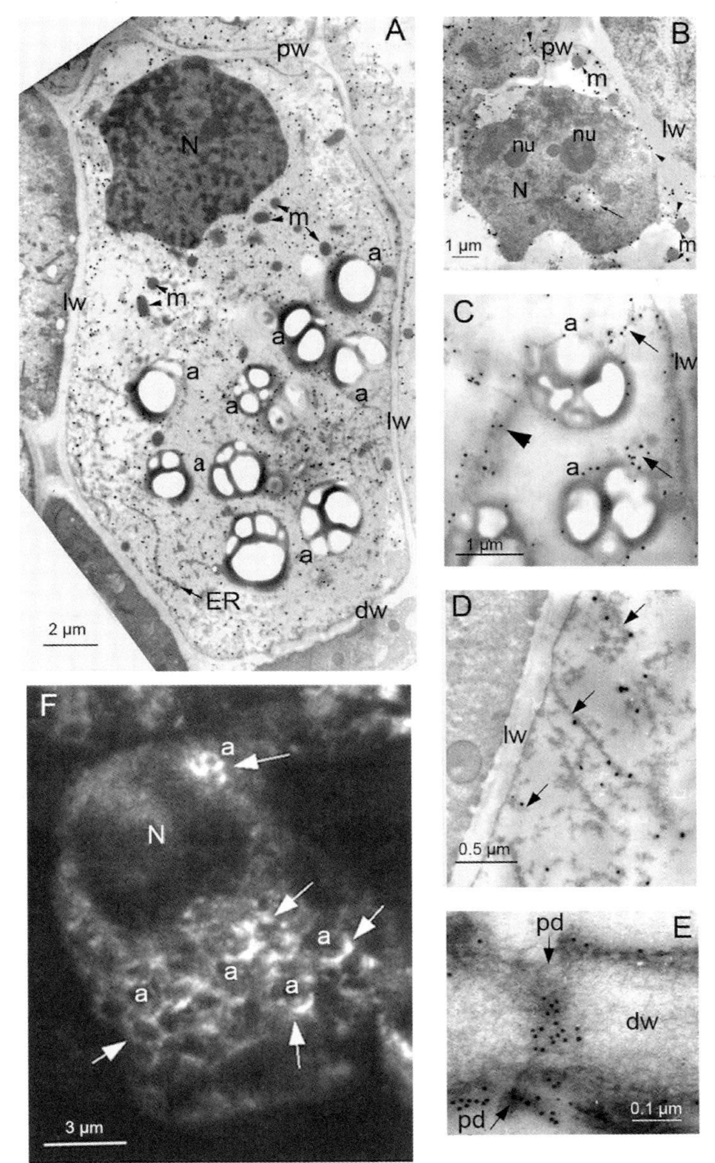

Fig. 4.21 Immunolocalization of actin (A–E, electron microscopy) and myosin (F, confocal microscopy) in statocytes of lentil roots. (A–D) Pre-embedding immunogold silver technique carried out with a monoclonal antiactin antibody. The nucleoplasm N (A and B), the longitudinal wall (lw; A–D), the proximal wall (pw, A) and the distal wall (A) are never labelled. Very few dots are localized in the stroma of the amyloplasts (C). The labelling occurs close to the nucleus envelope (A and B) and the amyloplast envelope (C). The actin is also localized in the remaining strands of cytoplasm (after Triton treatment) linking the amyloplasts to the plasma membrane or between the amyloplasts (C). Some strands of cytoplasm with actin made an angle of about 45° with the plasma membrane (D, arrows). In part (B), the arrow shows a hole in the nucleus, through which the labelling of actin in the cytoplasm can be seen; nu, nucleolus; m, mitochondria. (E) Post-embedding immunogold technique carried out with a polyclonal antiactin antibody. Dots of labelling are present in plasmodesmata, pd, crossing the distal wall, dw (not labelled). (F) Immunofluorescence carried out with a polyclonal antimyosin antibody. The fluorescence is located around the amyloplasts, a (white arrows); N, nucleus.

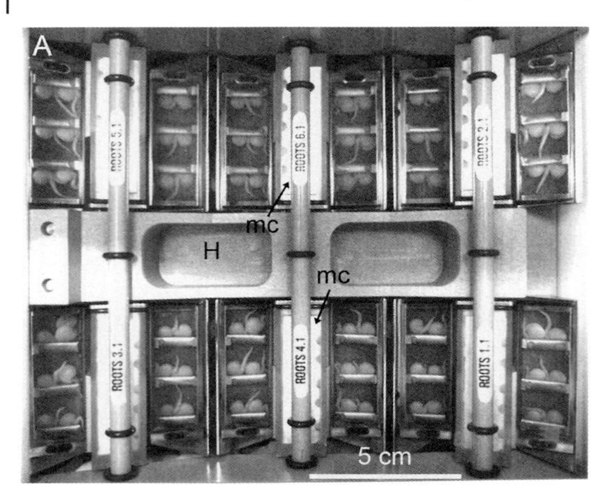

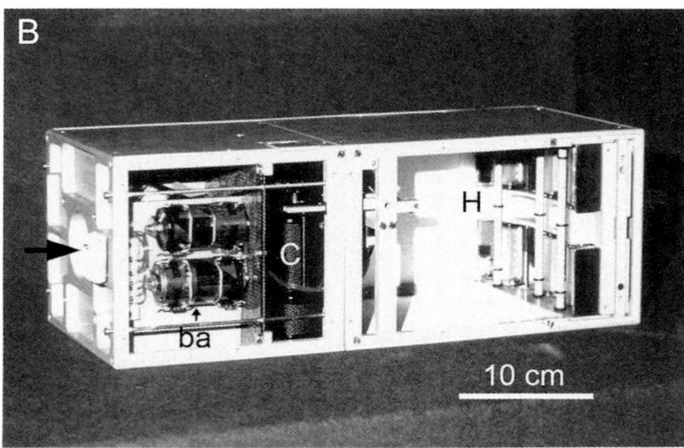

Fig. 4.22 Photobox (IML1 Spacelab mission). (A) Six minicontainers (mc) are placed in a holder (H). Mirrors were mounted at an angle of 45° with respect to the two windows of each minicontainer, allowing photographs of both sides in one shot. (B) The holder is placed in front of a camera (C). Time lapse photography was initiated by the crew, by pushing the external button (arrow). A flash with filtered light (560 nm) allowed photography of the minicontainers every 10 min. The batteries (ba) kept the flash charged during the whole sequence (6 h).

(acceleration perpendicular to the root tip). Then, the minicontainers were placed in a separate photobox (Fig. 4.22A and B), which was built to follow the root curvature in microgravity for several hours by automatic time lapse photography. The presentation time was extrapolated by considering the initial rate of curvature (Fig. 4.3) as a function of the logarithm of the time of stimulation (Fig. 4.23A) and calculating the regression line that best fitted the experimental points [66]. The presentation time was estimated to be 27 s for an acceleration of 1×g. By considering the curvature after 2 h (Fig. 4.23B) the presentation time was estimated to be 26 s, but the fit was not as good as for the previous analysis (Fig. 4.23A).

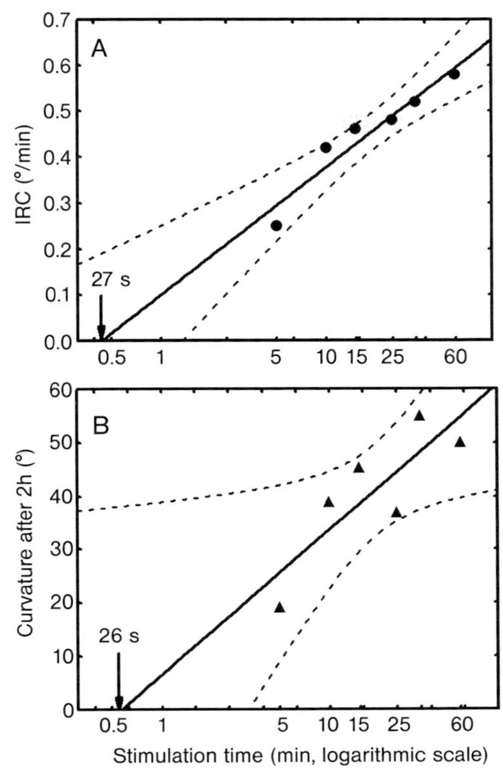

Fig. 4.23 Estimation of the presentation time (pt) of lentil seedling roots grown in Space and stimulated on the Biorack 1×g centrifuge (IML1 Spacelab mission). The initial rate of curvature (A; IRC) or the amplitude of bending after 2 h (B) is plotted as a function of the stimulation time. The pt, calculated by considering the value of the stimulation time for IRC = 0 or curvature after 2 h = 0, is estimated to be 27 and 26 s, respectively. Dotted lines represent the limits of confidence at the 5% level. Clearly, the model fitting the experimental points is much better with IRC than with the curvature after 2 h.

These results permitted comparison of the estimate of the presentation time obtained by means of a clinostat with that obtained in microgravity. On the ground, seedlings were grown for 28 h on a clinostat or in the vertical position (Fig. 4.24) [67]. They were stimulated in the horizontal position for various periods (from 2 to 25 min). The presentation time, estimated in the same way as for the Space experiment, was 25 s for the clinostated seedling and 60 s for the roots grown in the vertical position. However, the dose–response curves obtained were not straight lines so that points corresponding to stimulations longer than 10 min were arbitrarily discarded. These results led to a reconsideration of the method of determination of the presentation time and the carrying out of another Space experiment, in which the gravisensitivity of roots grown in microgravity could be compared with that of roots grown on a 1×g centrifuge.

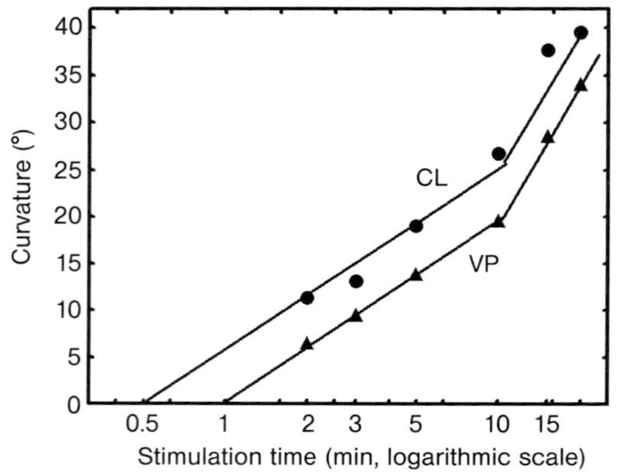

Fig. 4.24 Determination of the presentation time in lentil roots grown in the vertical position on the ground (VP) or grown on a clinostat in simulated weightlessness (CL). The curves are biphasic. The presentation time is estimated by extrapolation of the regression line that fits best the experimental points for stimulation times shorter than 10 min (first phase of the curve). The presentation time is 60 s for roots grown vertically and 25 s for roots grown on the clinostat.

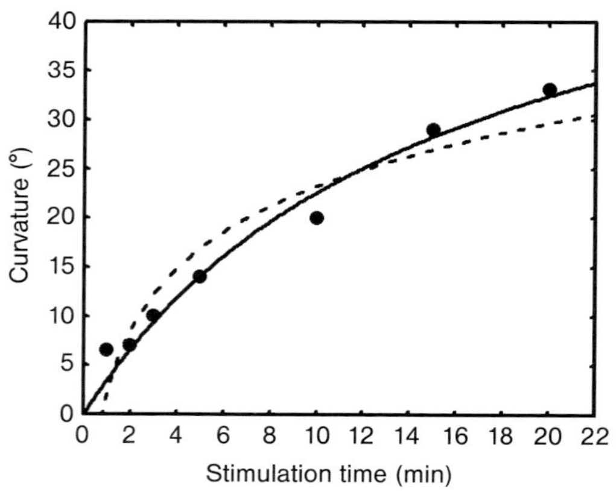

Fig. 4.25 Fitting of the logarithmic (dashed line) and the hyperbolic (solid line) models to the experimental data obtained on lentil roots grown in the vertical position on Earth (see Fig. 4.24, VP).

4.4.2.2 Models for Dose–Response Curves

When the dose–response curve of lentil roots grown in the vertical position was plotted as a function of the stimulation time (Fig. 4.25) and not as a function of the logarithm of the stimulation time as in Fig. 4.6, it was obvious that the logarithmic model (dashed line, Fig. 4.25) did not fit quite so well the experimental points, especially for short (<2 min) and long (>15 min) stimulation times. Since this lack of fitting was also observed in many studies by other authors [32] on different species, it was proposed that the logarithmic model was perhaps not the best one. Among several mathematical models that could have been applied, it seemed that the hyperbolic model is the best since it is related to the binding of a ligand to a receptor. This model is Eq. (3):

$$R = (a \cdot d)/(b + d) \tag{3}$$

where R is the rate of curvature in ° min^{-1}, a and b are constants and d is the stimulation dose (in g×s or g×min).

For almost all published dose–response curves [32], the hyperbolic model was indeed better than the logarithmic model. The main difference between the two models was that the logarithmic model implied a presentation time, i.e. there should have been a threshold of stimulation below which no response should be seen. The hyperbolic model implied, on the contrary, that the presentation time was null or very short. Thus, in principle a stimulation of less than 1 s could have led to a response. This hypothesis was supported by the fact, known for decades, that a stimulation of 1 s at 1×g could induce a response if it was repeated several times, which showed that a slight stimulation of 1 s at 1×g was perceived even if the root did not respond to the stimulus by a curvature. The lack of response to low gravistimuli could be due to the resistance of the organ to binding [68] because of the process of cell elongation and should not be strictly related to perception of gravity.

To estimate gravisensitivity with the hyperbolic model, it was proposed to consider the derivative of the model at the origin, which corresponded to graviresponsiveness to very low doses of gravistimulus. Thus, the results obtained in Space led to the view that the presentation time calculated by extrapolation of the logarithmic model does not represent a good estimate of gravisensitivity.

4.4.2.3 **Difference in Gravisensitivity**

The difference in sensitivity of roots cultivated on a 1×g centrifuge or in microgravity was first observed on *Lepidium sativum* roots [69, 70]. The study of gravisensitivity of lentil roots was carried out during the S/MM-05 mission of the Shuttle (1997). Lentil seedlings were cultivated for 26 h in minicontainers in the Biorack incubator on board of Spacehab. They were then subjected to centrifugal accelerations, ranging from 0.39 to 0.93×g, on a minicentrifuge (Fig. 4.26A) for 9 or 22 min. The advantage of this minicentrifuge was that this device was mounted inside the photobox so that after stimulation the lentil seedlings were photographed without any manual transfer, which had been the disadvantage of the IML1 experiment. Moreover, lower stimulation doses were used (from 3.5 to 20.5 g×min). The curvature was followed in microgravity for 3 h. The rate of curvature was plotted against the stimulation dose and the hyperbolic model, which best fitted the experimental points, was determined for roots grown in microgravity and roots grown on the 1×g centrifuge (Fig. 4.27A). The figure shows that there was indeed a strong difference in gravisensitivity in both kinds of roots. The roots grown in microgravity were more sensitive to the gravistimulus than those grown on the

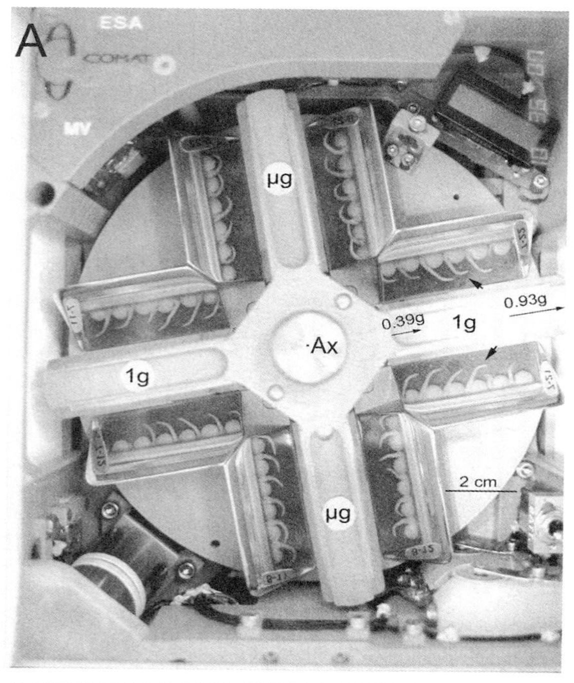

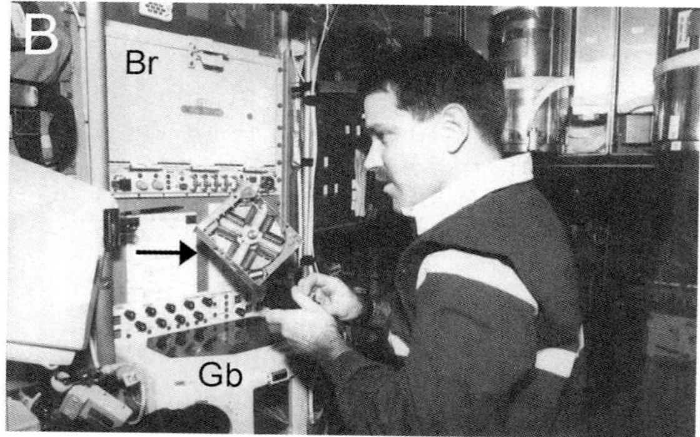

Fig. 4.26 Minicentrifuge (S/MM-05 Spacehab mission). (A) This device was used to stimulate the lentil seedlings that were grown first in microgravity (two minicontainers, µg) or on the Biorack 1×g centrifuge (two minicontainers, 1g). The picture was taken 3h after a stimulation of 22min on the minicentrifuge, followed by a 3h period in microgravity. Both sides of the four minicontainers were photographed through mirrors orientated at 45° with respect to their transparent covers (Fig. 4.8A). The seedlings that were the closest to the axis of rotation (Ax) were subjected to 0.39×g and those that were the furthest away from Ax were subjected to 0.93×g. The curvature of the roots grown first in microgravity is much larger than that of roots grown first in 1×g. (B) The astronaut, John Grunsfeld, performing the experiment with the minicentrifuge (arrow) in Space. Br, Biorack; Gb, Biorack glovebox.

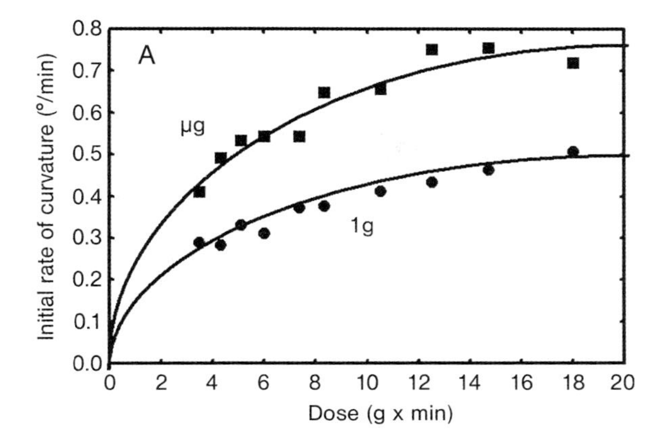

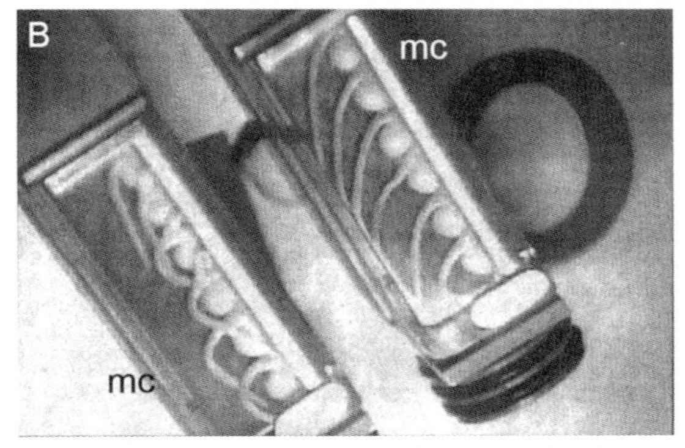

Fig. 4.27 (A) Dose–response curves of the gravitropic reaction of lentil roots grown first on the Biorack 1×g centrifuge (1g) or in microgravity (μg) and then stimulated on the minicentrifuge (S/MM-05 mission). The hyperbolic model is used to fit the experimental data. (B) Difference of curvature (after the 22 min stimulation) between the roots grown in microgravity (left-hand side) and those grown on the 1×g centrifuge was so spectacular that the newspaper *Florida Today* published this photograph taken by the crew. mc, minicontainer.

1×g centrifuge. Thus, the results obtained by means of the clinostat (Fig. 4.24) were confirmed by this Space experiment. The difference was so spectacular after 2 days in microgravity that the picture shown in Fig. 27(B), taken by astronaut John Grunsfeld, was published in *Florida Today* (19/01/1997).

4.4.2.4 **Cause of the Difference in Gravisensitivity**

This difference in gravisensitivity could have several causes. It could be possibly related to any of the phases of the graviresponse (the perception of the stimulus, its transduction, its transmission or curvature). However, notably, growth was identical in microgravity and on the 1×g centrifuge as well as on the clinostat and in the vertical position. The slight differences observed in cell elongation could not explain a greater sensitivity in microgravity. The only difference that has been seen was in the position of the amyloplasts within the statocyte [48, 71]. As shown in Fig. 4.14(A) and (B), these organelles were mainly located in the distal part of these cells on the 1×g centrifuge and mainly located in their proximal part in microgravity. Interestingly, the same difference in the position of the amyloplasts was observed in roots grown in the vertical position and in clinorotated roots, respectively.

Greater sensitivity should have also been due to a difference of volume or density of starch within the amyloplasts of roots grown in microgravity. However, an analysis of sections of root caps stained by periodic acid-Schiff showed that the statolithic apparatus had the same characteristics in microgravity and on the 1×g centrifuge [71].

For the S/MM-05 mission, the location and the movement of the amyloplasts were examined under several g conditions (Fig. 4.28A–C) [71]. When 1×g-grown roots were placed on the minicentrifuge (at 0.93×g for 22 min), the amyloplasts, which were mostly in the distal part of the statocyte, moved towards the lower side of the cell (Fig. 4.28B). When the roots were grown in microgravity and stimulated on the minicentrifuge (at 0.93×g for 22 min), the statoliths that were located in the proximal part of the statocyte also moved towards the lower side (Fig. 4.28C). Thus, after this stimulation, the amyloplasts had almost the same distribution in roots that had been grown in microgravity as in roots grown on the 1×g centrifuge.

Some roots of both samples were fixed after stimulation and a period of 3 h of microgravity, during which gravitropic response occurred. In this case, the statoliths that were sedimented to the lower side of the statocyte (Fig. 4.28B and C) moved to its centre and their position was similar after the period of 3 h of microgravity to that observed in roots grown continuously in microgravity (Fig. 4.28C). This result demonstrated that the mechanism, which maintained the amyloplasts in the proximal part of the statocyte for a root continuously grown in microgravity or which pulled them towards this position in roots grown in 1×g and then placed in microgravity, was also able to move the amyloplasts obliquely from the lower side of the statocyte to the cell centre [71]. As shown above, these movements were due to the presence of a thin actin network, but the movement seen in roots stimulated and placed in microgravity for 3 h yielded new information: the actin filaments should have an overall oblique orientation to be able to pull the statoliths from a lateral position to a central position in the cell.

4.4.2.5 Model of Gravisensing

The results obtained in Space led us to propose a model for gravisensing in roots (Fig. 4.29A) [3] in which the actin network plays a role [65, 72]. In this model, the actin filaments are obliquely orientated and are attached to proteins that are involved in the opening of stretch-activated Ca^{2+} channels (SACs) that can be regulated by tension [73]. The efflux of calcium ions from the protoplasm of the statocytes has not yet been demonstrated [74, 75] and the molecular identity of these channels is still under investigation [76], but the involvement of calcium in the early steps of gravitropism was documented [77]. These proteins should be linked together by bridging filaments. The amyloplasts are connected to the actin

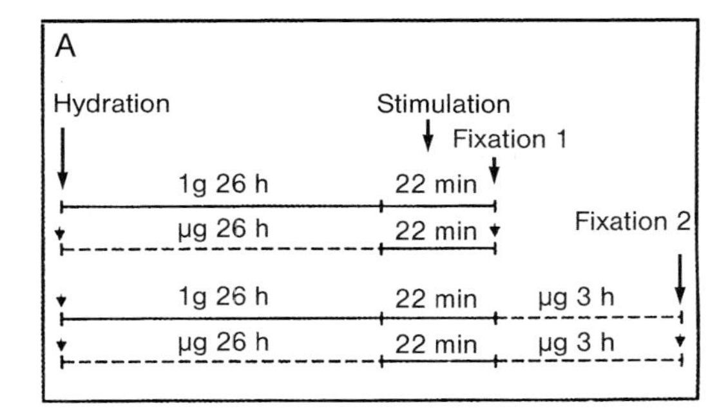

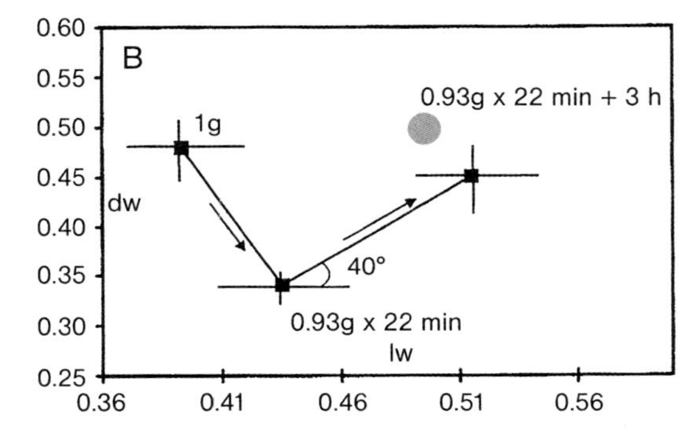

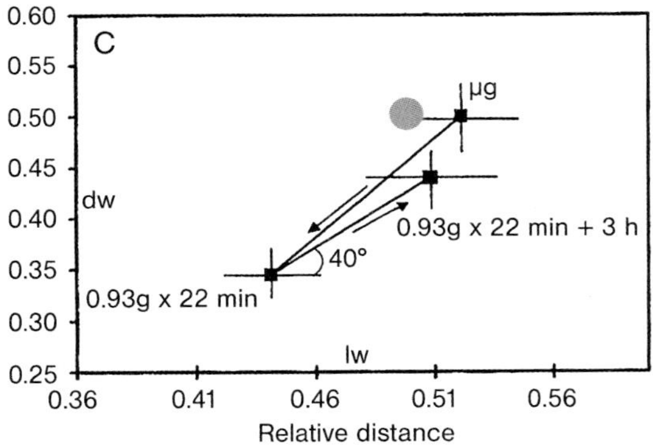

Fig. 4.28 (A) Flow diagram of the S/MM-05 Space experiment. After hydration of the lentil seeds, the seedlings were grown for 26 h on either the Biorack 1×g centrifuge (1g) or in microgravity (μg) and then subjected for 22 min to centrifugal accelerations on the minicentrifuge (Fig. 4.26). The lentil seedlings were chemically fixed (1) just after stimulation or (2) after stimulation and a 3-h period in microgravity. (B and C) Average position of amyloplasts in statocytes for roots grown on the Biorack 1×g centrifuge in Space (B, 1g) or in microgravity (C, μg). *X*-axis: Relative distance with respect to the distal wall of the statocyte (dw). *Y*-axis: Relative distance with respect to the longitudinal wall (lw) or lower wall for the stimulated samples. Two samples grown first in 1×g (B) or in microgravity (C) were subjected to a stimulation of 0.93×g × 22 min and chemically fixed. Two samples grown first in 1×g (B) or in microgravity (C) and stimulated at 0.93×g × 22 min were kept in microgravity for 3 h (0.93×g × 22 min + 3 h) and then chemically fixed. Bars represent confidence intervals of the mean at the 5% level. Arrows indicate the direction of the movements of statoliths. Grey circle: centre of the statocyte.

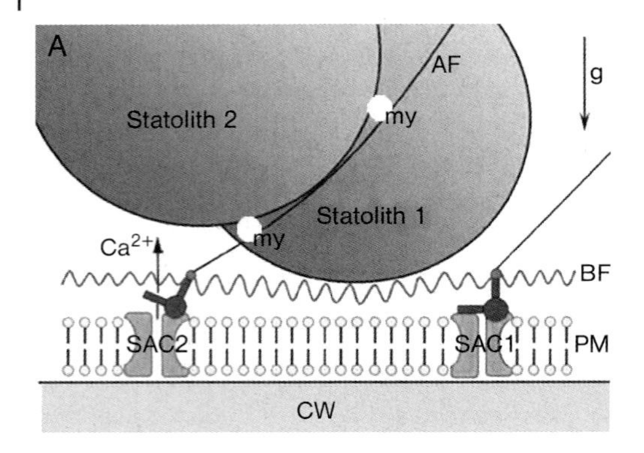

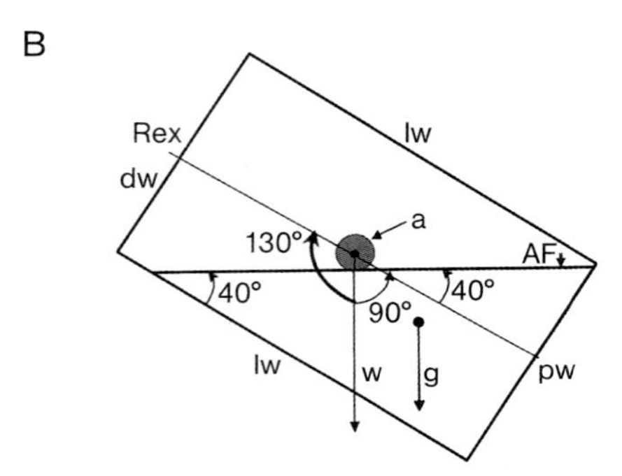

Fig. 4.29 (A) Model of mechanosensing in root statocytes. The statoliths can open the stretch-activated Ca^{2+} channels (SACs) by exerting tension on the actin filament (AF; statolith 2) or by pressing on bridging filaments (BF; statolith 1) that link SACs. Activating SACs by pressing on bridging filaments should be more efficient because statocytes treated with latrunculin B (which perturbs the polymerization of actin filaments) are more sensitive. This treatment could reduce the cytoplasmic viscosity, increase the velocity of statolith sedimentation, and hence activate SACs more rapidly. CW, cell wall; g, direction of gravity; my, myosin; PM, plasma membrane. (B) Orientation of 40° of an actin filament (AF) with respect to the lower longitudinal wall (lw) of a statocyte. The root extremity (Rex) is orientated at 130° with respect to gravity (g). In this particular orientation the tension exerted by the weight (w) of one amyloplast (a) on the actin filament is optimal since w is perpendicular to this filament. dw, distal wall; pw, proximal wall.

filaments by motor proteins (myosin). In 1×g conditions, the motor proteins should not be able to move the statoliths along the actin filaments, because the force they can exert is not sufficient to overcome the effect of gravity. In microgravity, myosin can pull the amyloplasts along the actin filaments towards the proximal wall.

When roots are stimulated (on Earth or in Space) the amyloplasts move to the physically lower side of the statocyte, and these organelles exert forces on the actin network that are transmitted to the SACs, which provokes their opening and could increase the sub-micromolar concentration of calcium ions in the cytosol [78]. As shown on the Fig. 4.29(A), the statoliths can also induce the activation of SACs by exerting a pressure on the bridging filament network. The only difference between the two structures that can activate the SACs is that actin filaments are at the beginning of the stimulation in the vicinity of the amyloplasts, whereas these organelles have to sediment to enable a contact with the bridging filaments. However, the mechanism of activation of SACs should be similar. This model can explain several characteristics of gravisensing in roots. For instance, as the actin filaments have an oblique orientation (on average 40°, Fig. 4.21D; Fig. 4.28B and C), it can be expected that the optimal angle of stimulation for roots should not be 90° (in the horizontal position), since the force exerted by the amyloplasts should be maximal when the gravity vector is perpendicular to the actin filaments (Fig. 4.29B), that is when the root tip is placed in an upward position (at an angle of 130°). This peculiarity has indeed long been noted and is referred to as "the deviation of the sine rule".

This model takes also into account the fact that actin filaments are not absolutely necessary, since bridging filaments can activate SACs. This is supported by the action of cytochalasins or latrunculin, which perturb actin polymerization but do not suppress gravitropic response (Section 4.4.3.2). Importantly, in this model, the amyloplast sedimentation is also not a prerequisite for gravisensing, since the forces exerted by these organelles can be transmitted to the SACs. This can eventually explain why there is a gravitropic reaction in roots of starch-depleted mutants of *Arabidopsis* [17, 18], in which plastids do not sediment within statocytes of roots that have been stimulated in the horizontal position. Finally, this model implies that any structure that can exert a force or a pressure on the bridging filaments can activate SACs. It is thus not impossible that, to a lesser extent, the whole protoplast could have this function.

4.4.3 Gravitropic Response

4.4.3.1 Absence of Counter-reaction

When a lentil root is stimulated on the ground in the horizontal position, its tip bends quickly for approximately 2 h and then the rate of curvature diminishes strongly [66], although the root tip has not yet reached the vertical position (Fig. 4.3). On the S/MM-05 mission the curvature of roots, grown on the Biorack 1×g centrifuge or in microgravity and then stimulated on the minicentrifuge, was followed for 3 h in microgravity by time lapse photography in the photobox (Fig. 4.30). Figure 4.30 shows a typical gravitropic response for roots that were stimulated for 22 min at 0.67×g (which represented a dose of stimulation of 14.7 g×min). In both culture conditions (microgravity or 1×g), the rate of curvature was almost constant for at least 3 h in microgravity and no counter-reaction has been observed. For the roots grown in microgravity the curvature was 110°, which meant that the root tip overshot the direction of the centrifugal acceleration that was at the origin of the gravitropic reaction (Fig. 4.30, dotted line). These results clearly prove that there is a mechanism of regulation of the gravitropic reaction on the ground, which avoids the overshooting of the vertical position by the root tip. This mechanism is gravity dependent, since it does not occur in roots simulated for 22 min and placed

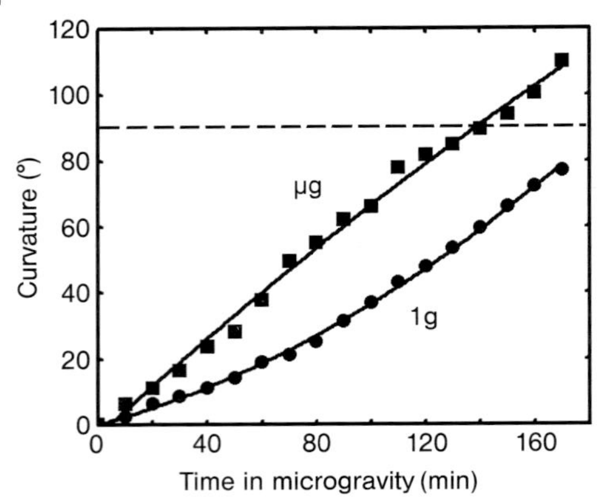

Fig. 4.30 Absence of counter-reaction in microgravity. Roots grown either on the Biorack 1×g centrifuge (1g) or in microgravity (µg), stimulated on the minicentrifuge (22 min at 0.67×g = 14.7×g × min) and then placed in microgravity for 3 h (S/MM-05 mission). The curvature was followed by time lapse photography in the photobox (Fig. 4.22) during the 3 h in microgravity. In both cases there is no counter-reaction. The angle of curvature of roots grown first in microgravity (µg) is larger than that of roots grown in 1×g.

in microgravity. However, the S/MM-05 experiment could not indicate which phase of the gravitropic reaction chain is regulated on the ground.

Regulation of the curvature could have two reasons. It might take place because the stimulus overshoots a certain limit, beyond which it becomes inhibitory, or there could be a second signal related to the position of the root tip, which stops the stimulus or its effects. Some information about the mechanism of regulation was obtained by using cytochalasins.

4.4.3.2 Comparison with the Effect of Cytochalasin Treatments

Cytochalasins perturb actin polymerization. To study the role of the actin network in the perception phase of the gravitropic reaction, a solution containing cytochalasin B was supplied to dry lentil roots grown on the ground for 28 h. Then they were stimulated in the horizontal position. Figure 4.31 shows the results obtained. After 3 h in the horizontal position, the tip of roots grown in water did not reach the vertical position (Fig. 4.31A), but for roots treated by cytochalasin B (Fig. 4.31B) the curvature was stronger and their tip overshot the vertical position, i.e. their normal GSPA.

Thus, cytochalasin B was able to suppress the inhibition of root curvature. The effect of cytochalasin should be due to its action on the reaction zone, since the whole root was in contact with the drug solution. However, it has been shown more recently that a stronger reaction of latrunculin B-treated *Arabidopsis* roots was also observed, when this drug was supplied only to the very root tip [79, 80].

These results showed that the mechanism of regulation should take place in the root cap and, therefore, should be linked to the perception phase or the transduction phase.

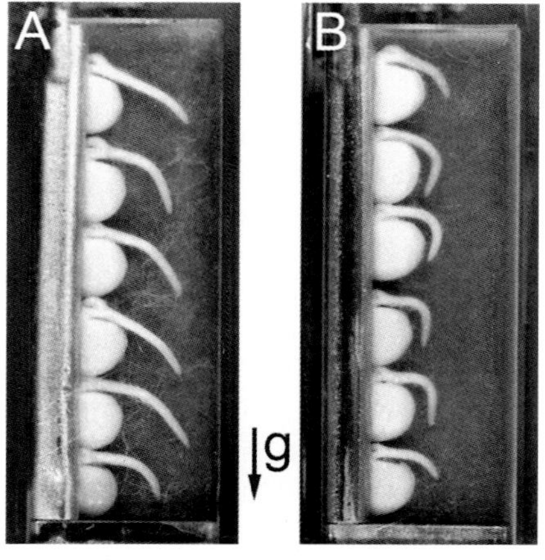

Fig. 4.31 Lentil roots, hydrated with pure water (A) or with a solution of cytochalasin B (CB; 25 µg mL^{-1} in 1% DMSO) (B), grown during 28 h in the vertical position on the ground and then stimulated in the horizontal position. The photograph was taken after 3 h in the horizontal position and shows clearly that for the CB-treated roots the curvature is stronger and that in most cases the root tip overshoots the gravity direction (g).

When lentil seedlings (grown in microgravity) were placed for periods of stimulation longer than 30 min on the 1×g Biorack centrifuge and then placed in microgravity for 3 h (IML1 mission), the overshooting was observed [66] but it was much less than for roots stimulated for 22 min at 0.67×g (Fig. 4.30) [71]. Therefore, it can be proposed that the regulation of the gravitropic response is triggered when the stimulus is too strong and that no second signal is needed. As cytochalasins or latrunculin can suppress this regulation, one can hypothesize that actin filaments are involved in the termination of the gravitropic stimulation. A role of these actin filaments in the opening of the SACs has been mentioned (Fig. 4.29). It is possible that they are also involved in their closing, which could regulate the gravitropic reaction [79, 80].

4.5 Conclusion

4.5.1 Action of Microgravity on Root Growth

Analysis of the root growth of young lentil seedlings showed that, for short periods of cultivation, root elongation was the same in microgravity as on the 1×g flight control (and ground control) [9, 48]. However, when the lentil roots started to grow in microgravity their tip bent away from the cotyledons and then straightened out for several hours (Fig. 4.12). This movement should be considered as nastic, since it had an internal cause and was also seen on the clinostat [50]. On the ground, this bending could be useful for the root to find its way into the soil, but it is strongly reduced because of the presence of gravity. In microgravity the bending was responsible for a difference in cell elongation in the elongation zone of the root after 25 h of growth (Fig. 4.13A), but the difference was much less after 29 h (Fig. 4.13B), i.e. when the root tip has straightened out (Fig. 4.12).

The most surprising results concerned the mitotic index and the cell cycle in the primary lentil root (Tables 4.2 and 4.4) [9, 10]. As the period of growth was short, cells of the root meristem were partly synchronized, since in dry seeds all of them were in the G1 phase [10]. A peak of mitosis occurred on the ground at about 25 h (Fig. 4.2B) and all cells were cycling (Fig. 4.2A) [10]. Experiments carried out in the frame of the IML1 and IML2 missions (with a period of growth of 28 h and 29 h, respectively) have shown that the percentage of cells that had entered the second cell cycle was less in microgravity than in the 1×g ground control (Table 4.4). The results obtained with the 1×g control in Space during the IML1 flight were similar to those obtained for the ground control [9]. On the contrary, the Space control was different to the ground control in the IML2 mission [10]. This was because the 1×g centrifuge was stopped during this experiment after 25 h (to take photographs for the analysis of root growth) and, therefore, the roots were subjected to microgravity for about 30 min. This transfer induced a change in the cell cycle: fewer cells entered the second cell cycle, since more cells remained in the G2 phase of the first cell cycle. These experiments showed that there was a delay in the cell cycle in the meristem of roots growing continuously in microgravity and that the transfer for a short period (30 min) at 25 h from 1×g to microgravity and their return on the 1×g centrifuge provoked a strong delay in the termination of the first cell cycle. Notably, at 25 h the majority of cells were entering the second cell cycle, but a large number of meristematic cells were still in the G2 phase (or in mitosis to a lesser extent). It is well known that at the end of the G2 phase there is a so-called checkpoint just before mitosis [81]. It is possible that a constant growth in microgravity, or the transfer of the seedlings from 1×g to microgravity for a short period, provokes a delay in the entry in mitosis because of an action on the checkpoint. The effects of microgravity on the cell cycle could have many causes. However, it seems that the distribution of auxin should be directly involved in the modification of the cell cycle in microgravity (or under a gravity stress), since the distribution of this hormone is responsible for the gravitropic response [2, 40, 44] and that it is known to play a role in the cell cycle [82–84]. This hypothesis should be tested in a near future in the frame of an ISS experiment (GENARA) by using *Arabidopsis* seedlings harbouring a reporter gene that is expressed in the presence of this hormone.

Finally, the study of growth of the primary root showed that germination occurred normally in microgravity and that the early step of development was not strongly perturbed. However, a delay in the cell cycle in the meristem [9, 10] has been observed, which must have an impact on root development if it should increase for longer periods of growth. Many experiments performed in Space or with a clinostat showed that root growth was modified in microgravity or in simulated microgravity (on a clinostat), but only after several days [52, 53, 85].

4.5.2
Gravisensing Cells and Perception of Gravity by Roots

4.5.2.1 Statocyte Polarity and Movement of Organelles

Before experimentations in Space it was generally accepted that amyloplasts were able to move freely within the statocyte by the phenomenon of buoyancy. These organelles are denser than the surrounding cytoplasm [6] and sediment towards the physically lower cell wall because of gravity. However, the polarity of the statocyte could not be due only to this factor of the environment, since the aggregates of the endoplasmic reticulum, which are located beneath the amyloplasts in the vertically oriented root, have a lower density than these organelles. Moreover the nucleus, which is situated close to the proximal wall, should sediment under the influence of gravity due to its density. This organelle is strongly attached near the proximal wall and very large forces (Fig. 4.17, 19×g) are necessary to make it move from this position [57]. Surprisingly, however, the nucleus did have a slightly different position in lentil roots grown in microgravity. This organelle was located a little further away from the proximal wall (Fig. 4.15) and in a more central position (closer to the cell longitudinal axis of the statocyte) [55, 57]. This finding triggered several experiments in Space and on the ground to understand the mechanism responsible for the positioning of the nucleus. It was expected that actin filaments would be responsible for the location of this organelle and cytochalasins were used to perturb the polymerization of actin. In cytochalasin-treated roots on the ground (Fig. 4.16), the nucleus settled down on the amyloplasts and the endoplasmic reticulum was located close to the proximal wall (Fig. 4.16). This new polarity was only due to gravimetry. Notably, the cytochalasin or latrunculin treatments did not suppress the gravitropic reaction, which was even larger than in the non-treated roots (Fig. 4.31) [79, 80, 86, 87]. This result clearly demonstrated that the statocyte polarity was not necessary for gravisensing.

Thus, the actin network maintains the nucleus in the proximal half of the statocyte, but interestingly, in microgravity the lack of tension within this network leads to a kind of relaxation, which makes the nucleus move away from the distal wall [57]. This behaviour indicates a particular sensitivity of the actin network to tensions.

It has sometimes been claimed that the amyloplasts were distributed randomly in the statocyte of roots grown in microgravity. In lentil roots exposed to microgravity, it was demonstrated that these organelles were mainly located in the proximal part of the statocyte, near the nucleus [48], but any gravitropic stimulation on a 1×g centrifuge in Space made them settle down towards the physically lower wall. This indicated that these organelles were held in position by some mechanism, the force of which was less than the gravity force. This force was measured, in the frame of the S/MM-03 mission, by calculating the rate of displacement of the amyloplasts in roots placed in microgravity after the growth of the lentil seedlings on a 1×g centrifuge (Fig. 4.18). It has been estimated to be seven times lower (0.016 pN) than the gravity force which moves one amyloplast on earth (0.11 pN) [64]. During the same mission it was shown that treatment with cytochalasin almost completely suppressed this movement [64] and that the actin network was responsible for the positioning of the amyloplasts in microgravity. The effect of this force cannot be detected on the ground, since it is overcome by the gravity force. It was also demonstrated that myosins (motor proteins) could be responsible for the movement of the amyloplasts along the actin filaments towards the proximal wall. The force exerted by a molecule of myosin should be much greater (1–10 pN) [88] than the estimated force necessary to pull one amyloplast (0.016 pN) [64]. Moreover, it was observed that the general orientation of the actin network should be oblique, making an angle of about 40° with respect to the longitudinal wall (Fig. 4.28B and C) [71], which demonstrated that if the overall movement of the bulk of the amyloplasts took place from the distal half of the statocyte towards its proximal half, the individual movements of these organelles should be more complex. There should be many contacts between the amyloplasts, which could account for a loss of a large part of the energy dissipated by the myosins to move these organelles. It is also possible that the amyloplasts have to break partially the actin network to move within the statocyte [89], even if the actin network is thin in the centre of the statocyte [64, 90].

4.5.2.2 **Gravisensing**

In the 1980s, there was a popular hypothesis stating that the endoplasmic reticulum was involved in the transduction of gravistimulus [4]. Sensitivity to gravistimulus was greater in roots grown in microgravity than in roots grown on the 1×g centrifuge (Fig. 4.27A) [69–71], although there were much fewer contacts between the amyloplasts and the endoplasmic reticulum in the former than in the latter case due to the difference of structural polarity of the statocytes (Fig. 4.14). These results did not confirm the hypothesis of a role of the endoplasmic reticulum in the transduction of a gravistimulus.

The analysis of gravisensing in Space led to a re-analysis of the dose–response curve of the gravitropic response of roots [32]. For decades a logarithmic model was used to extrapolate the curve down to zero curvature [30, 31] and to estimate the presentation time (minimal time to induce a slight but significant curvature in a plant organ). However, it has been demonstrated that the hyperbolic model best fits all but one of the published dose–response curves [32]. This showed that the presentation time was certainly much shorter than believed and perhaps of the order of 1 s.

Analysis of the movement of the amyloplasts of lentil roots grown in microgravity or on a 1×g centrifuge and stimulated on a minicentrifuge for 9 or 22 min (by centrifugal accelerations from 0.39 to 0.93×g) has indicated that the movement was not the same in both cases [71]. In roots grown in microgravity before stimulation, the bulk of amyloplasts was located in the centre of the statocyte (Fig. 4.28B), whereas in roots grown on the 1×g centrifuge before stimulation these organelles were located mainly in the distal part of the cell. Under the effect of the centrifugal accelerations the statoliths moved towards the physically lower wall of the statocyte, but in two different ways (Fig. 4.28B and C). When some roots of the two samples were placed in microgravity for 3 h (after stimulation on the minicentrifuge), the amyloplasts were pulled towards the proximal wall in both cases and were located in the centre of the statocyte. This indicated that the actin filaments could not be oriented longitudinally, since in this case they should have been pulled and maintained along the longitudinal wall (Fig. 4.28B). The overall direction of the movement of the amyloplasts with respect to the longitudinal wall made an angle of about 40°, and so it has been hypothesized that the actin filaments should be obliquely oriented in the statocytes [71].

As root growth and cell elongation in the cell elongation zone were similar on the clinostat and almost similar in microgravity as in their respective controls, it was concluded that higher sensitivity in microgravity or in simulated microgravity (on a clinostat) should be related to early phases of the gravitropic reaction (perception phase or transduction phase) [71]. In the model of gravisensing shown in Fig. 4.29(A), it is proposed that one amyloplast can exert tension on the actin filaments, which are attached to stretch-activated ion channels, but they can also activate these channels by sedimenting on bridging filaments that link the channels. When the lentil seedlings are grown in microgravity or on a clinostat and then stimulated, the amyloplasts could slide along the actin filaments and, therefore, be more efficient (Fig. 4.28C) than in roots grown in 1×g where the stimulation induces a movement that does not follow the orientation of the actin filaments (Fig. 4.28B). This could be why lentil roots grown in microgravity are more sensitive to a gravistimulus. This model implies that actin filaments increase gravisensitivity, but are not absolutely necessary for gravisensing, which has been discovered by the use of cytochalasins or latrunculin, which perturb polymerization of actin [79, 80, 86, 87], or by studying gravisensitivity in the early development of the embryo of flax [91]. Recently, an experiment was carried out to physically break the actin network by making the bulk of the amyloplasts move from the proximal half of the statocyte to its distal wall for 7 min and all the way back, so that the position of the amyloplasts was almost the same as in roots grown in the vertical position. It was shown [92] that graviresponse was delayed in roots placed in the upside down position and replaced in the normal position, which confirmed a role of the actin network in gravisensing.

In the frame of the S/MM-05 experiment it has been observed that in seedlings grown in microgravity and stimulated on a minicentrifuge for 22 min (at less than 1×g) and then placed in microgravity for 3 h, the curvature was stronger than for roots grown in microgravity but stimulated at 1×g for 3 h [48]. Moreover, the tip of the former roots overshot the direction of the centrifugal acceleration, which was not observed in the latter case (or on roots stimulated on the ground). Thus, long periods of centrifugal accelerations (or gravity on Earth) regulate the gravitropic response to avoid overshooting the direction of stimulus. Interestingly, overshooting was also observed in cytochalasin-treated roots on Earth (Fig. 4.31), which indicates that the actin network should be involved in the perception of gravity and also in the regulation of gravistimulus [79, 80]. In the model shown in Fig. 4.29, it is proposed that the amyloplasts can activate the stretch ion channels as long as they are not completely settled along the longitudinal wall. If all these organelles were exerting a pressure on the bridging filaments, most of the stretch activated ion channels should be closed. This can eventually explain why a continuous stimulation at 1×g is less efficient than a 22 min stimulation with less than 1×g.

The Space experiments on the gravitropic reaction, coupled with the numerous analyses conducted on the ground, have shown that the position and the movement of the amyloplasts before and during gravistimulus play a role in the intensity of the response. They have also shown that the way of measuring gravisensitivity must be completely reconsidered. In this way, the presentation time, which has been estimated for decades, should not be related only to the perception of gravity but also to the minimal asymmetrical distribution of auxin, which can induce a differential growth. The stimulus can, clearly, be sensed within 1 s, even if such stimulation does not provoke a differential growth.

This work has led to a new model for gravisensing that takes into account a dual role of the actin network in the perception and in the regulation of the gravitropic stimulus. It must be continued by the study of calcium influx and proton efflux during gravistimulation [93, 94]. A future experiment in Space will analyse, on the International Space Station, the threshold acceleration for gravisensing and the calcium distribution within the statocyte (experiments GRAVI-1 and GRAVI-2; D. Driss-Ecole and V. Legué are the Principal Investigators of these projects). The distribution of auxin under various stimuli should also be studied with *Arabidopsis* seedlings harbouring a reporter gene, the expression of which is upregulated by an auxin activated promoter.

Acknowledgments

The authors thank the crew members of the D1, IML1, IML2, S/MM-03, S/MM-05 and S/MM-06 missions, who have successfully carried out the Space experiments, as well as the NASA team at KSC for assistance. They have also received considerable help from the ESA team working on the microgravity projects. This work would not have been possible without the scientific, technical and financial support of CNES and the technical help of GSBMS (Université Paul Sabatier, Toulouse).

Dr Giovanna Lorenzi, Dr Valérie Legué, Dr Jawad Aarrouf and Dr Fugen Yu were strongly involved in the analysis of the Space experiments and must be thanked for their work. Mrs Monique Prouteau and Mr Philippe Julianus are acknowledged for technical assistance.

References

1 Rosen, E., Chen, R., Masson, P.H. *Trends Plant Sci.* **1999**, *4*, 407–412.

2 Boonsirichai, K., Guan, C., Chen, R., Masson, P.H. *Annu. Rev. Plant Biol.* **2002**, *53*, 421–447.

3 Perbal, G., Driss-Ecole, D. *Trends Plant Sci.* **2003**, *8*, 498–504.

4 Volkmann, D., Sievers, A., Graviperception in Multicellular Organs, in: *Encyclopedia of Plant Physiology*, Vol. 7, Haupt, W.,

Feinleib, M.E. (Eds.), pp. 573–600, Springer, Berlin, **1979**.
5 Juniper, B.E., Groves, S., Landau-Schachar, B., Audus, L.J. *Nature* **1966**, *209*, 93–94.
6 Sack, F.D. *Int. Rev. Cytol.* **1991**, *127*, 193–252.
7 Blancaflor, E.B., Fasano, J.M., Gilroy, S. *Plant Physiol.* **1998**, *116*, 213–222.
8 Wayne, R., Staves, M.P. *Physiol. Plant.* **1996**, *98*, 917–921.
9 Driss-Ecole, D., Schoëvaërt, D., Noin, M., Perbal, G. *Biol. Cell* **1994**, *81*, 59–64.
10 Yu, F., Driss-Ecole, D., Rembur, J., Legué, V., Perbal, G. *Physiol. Plant.* **1999**, *105*, 171–178.
11 Darbelley, N., Perbal, P., Perbal, G. *Physiol. Plant.* **1986**, *67*, 460–464.
12 Ishikawa, H., Evans, M.L. *Plant Physiol.* **1995**, *109*, 725–727.
13 Firn, R., Wagstaff, C., Digby, J. *Trends Plant Sci.* **1999**, *4*, 252.
14 Haberlandt, G. *Ber. Dtsch. Bot. Ges.* **1900**, *18*, 261–272.
15 Němec, B. *Ber. Dtsch. Bot.Ges.* **1900**, *18*, 241–245.
16 Aarrouf, J., Perbal, G. *Bot. Acta* **1996**, *109*, 278–284.
17 Kiss, J.Z., Hertel, R., Sack, F.D. *Planta* **1989**, *177*, 198–206.
18 Caspar, T., Pickard, B. *Planta* **1989**, *177*, 185–197.
19 MacCleery, S.A., Kiss, J.Z. *Plant Physiol.* **1999**, *120*, 183–192.
20 Kiss, J.Z., Wright, J.B., Caspar, T. *Physiol. Plant.* **1996**, *97*, 237–244.
21 Kuznetzov, O.A., Hasenstein, K.H. *Planta* **1996**, *198*, 87–94.
22 Wolverton, C., Mullen, J.L., Ishikawa, H., Evans, M.L. *Planta* **2002**, *215*, 153–157.
23 Sack, F.D. *Planta* **1997**, *203*, 63–68.
24 Kiss, J.Z. *Crit. Rev. Plant Sci.* **2000**, *19*, 551–573.
25 Perbal, G. *Adv. Space Res.* **1999**, *24*, 723–729.
26 Barlow, P.W. *Plant Cell Environ.* **1995**, *18*, 951–962.
27 Driss-Ecole, D., Lefranc, A., Perbal, G. *Physiol. Plant.* **2003**, *118*, 305–312.
28 Driss-Ecole, D., Vassy, J., Rembur, J., Guivarc'h, A., Prouteau, M., Dewitte, W., Perbal, G. *J. Exp. Bot.* **2000**, *51*, 521–528.
29 Evans, M.L., Moore, R., Hasenstein, K.H. *Sci. Am.* **1986**, *255*, 112–119.
30 Larsen, P. *Physiol. Plant.* **1957**, *10*, 12–163.
31 Shen-Miller, J. *Planta* **1970**, *92*, 152–163.
32 Perbal, G., Jeune, B., Lefranc, A., Carnero-Diaz, E., Driss-Ecole, D. *Physiol. Plant.* **2002**, *114*, 336–342.
33 Larsen, P. Orthogeotropism in Roots, in: *Encyclopedia of Plant Physiology*, Vol 17, W. Ruhland (Ed.), pp. 153–159, Springer, Berlin, **1962**.
34 Hejnowicz, Z., Sondag, C., Alt, W., Sievers, A. *Plant Cell Environ.* **1998**, *21*, 1293–1300.
35 Shen-Miller, J., Hinchman, R., Gordon, S.A. *Plant Physiol.* **1968**, *43*, 338–344.
36 Barlow, P.W., Rathfelder, E.L. *Planta* **1985**, *165*, 134–141.
37 Evans, M.L. *Plant Physiol.* **1991**, *95*, 1–5.
38 Marchant, A., Kargul, J., May, S.T., Mulle, P., Delbarre, A., Perrot-Rechenmann, C., Bennet, M.J. *EMBO J.* **1999**, *18*, 2066–2073.
39 Muday, G.K., DeLong, A. *Trends Plant Sci.* **2001**, *6*, 535–542.
40 Friml, J., Wisniewska, J., Benkova, E., Mendgen, K., Palme, K. *Nature* **2002**, *415*, 806–809.
41 Noh, B., Murphy, A.S., Spalding, E.P. *Plant Cell* **2001**, *13*, 2441–2454.
42 Noh, B., Bandyopadhyay, A., Peer, W.A., Spalding, E.P., Murphy, A.S. *Nature* **2003**, *423*, 999–1002.
43 Ottenschläger, I., Wolff, P., Wolverton, C., Bhalerao, R.P., Sandberg, G., Ishikawa, H., Evans, M., Palme, K. *Proc. Natl. Acad. Sci. U.S.A.* **2003**, *100*, 2987–2991.
44 Rashotte, A.M., Brady, S.R., Reed, R.C., Ante, S.J., Muday, G.K. *Plant Physiol.* **2000**, *122*, 481–490.
45 Swarup, R., Kramer, E.M., Perry, P., Knox, K., Leyser, H.M., Haseloff, J., Beemster, G.T., Bhalerao, R., Bennett, M.J. *Nat. Cell Biol.* **2005**, *7*, 1057–1065.
46 Abas, L., Benjamins, R., Malenica, N., Paciorek, T., Wisniewska, J., Moulinier-Anzola, J.C., Sieberer, T., Friml, J., Luschnig, C. *Nat. Cell Biol.* **2006**, *8*, 249–256.
47 Wolverton, C., Ishikawa, H., Evans, M.L. *J. Plant Growth Regul.* **2002**, *21*, 102–112.
48 Perbal, G., Driss-Ecole, D., Rutin, J., Salle, G. *Physiol. Plant.* **1987**, *70*, 119–126.

49 Legué, V., Yu, F., Driss-Ecole, D., Perbal, G. *J. Biotechnol.* **1996**, *47*, 129–136.

50 Legué, V., Driss-Ecole, D., Perbal, G. *Physiol. Plant.* **1992**, *84*, 386–392.

51 Antonsen, F., Johnsson, A., Perbal, G., Driss-Ecole, D. *Physiol. Plant.* **1995**, *95*, 596–603.

52 Aarrouf, J., Schoevaert, D., Maldiney, R., Perbal, G. *Physiol. Plant.* **1999**, *105*, 708–718.

53 Claasen, D.E., Spooner, B.S. *Int. Rev. Cytol.* **1994**, *156*, 301–373.

54 Darbelley, N., Driss-Ecole, D., Perbal, G. *Plant Physiol. Biochem.* **1989**, *27*, 341–347.

55 Perbal, G., Driss-Ecole, D. *Physiol. Plant.* **1989**, *75*, 518–524.

56 Smith, J.D., Todd, P., Staehelin, L.A. *Plant J.* **1997**, *12*, 1361–1373.

57 Lorenzi, G., Perbal, G. *Biol. Cell* **1990**, *68*, 259–263.

58 Perbal, G., Driss-Ecole, D., Effect of Gravity on the Distribution of Electron Dense Chromatin in the Nucleus of Root Statocytes, in: *Proceedings of the Fourth European Symposium on Life Sciences Research in Space*, David, V., (Ed.), pp. 517–520, ESA SP-307, ESA Publications Division, Noordwijk, **1990**.

59 Volkmann, D., Buchen, B., Hejnowicz, Z., Tewinkel, M., Sievers, A. *Planta* **1991**, *185*, 153–161.

60 Sievers, A., Buchen, B., Volkmann, D., Hejnowicz, Z., Role of the Cytoskeleton in Gravity Perception, in: *The Cytoskeletal Basis of Plant Growth and Form*, Loyd, C.W. (Ed.), pp. 169–182, Academic Press, London, **1991**.

61 Volkmann, D., Baluska, F., Lichtscheidl, I., Driss-Ecole, D., Perbal, G. *FASEB J.* **1999**, *13* (Suppl), S143–S147.

62 Sievers, A., Buchen, B., Hodick, D. *Trends Plant Sci.* **1996**, *1*, 273–279.

63 Braun, M., Buchen, B., Sievers, A. *J. Plant Growth Regul.* **2002**, *21*, 137–145.

64 Driss-Ecole, D., Jeune, B., Prouteau, M., Julianus, P., Perbal, G. *Planta* **2000**, *211*, 396–405.

65 Baluska, F., Hasenstein, K.H. *Planta* **1997**, *203*, S69–S78.

66 Perbal, G., Driss-Ecole, D. *Physiol. Plant.* **1994**, *90*, 313–318.

67 Perbal, G., Driss-Ecole, D., Tewinkel, M., Volkmann, D. *Planta* **1997**, *203*, S57–S62.

68 Pickard, B.G. *Can. J. Bot.* **1973**, *51*, 1003–1021.

69 Volkmann, D., Tewinkel, M. *Plant Cell Environ.* **1996**, *19*, 1195–1202.

70 Volkmann, D., Tewinkel, M. *J. Biotechnol.* **1996**, *47*, 253–259.

71 Perbal, G., Lefranc, A., Jeune, B., Driss-Ecole, D. *Physiol. Plant.* **2004**, *120*, 303–311.

72 Blancaflor, E.B. *J. Plant Growth Regul.* **2002**, *21*, 120–136.

73 Ping Ding, J., Pickard, B.G. *Plant J.* **1993**, *3*, 83–110.

74 Legué, V., Blancaflor, E., Wymer, C., Perbal, G., Fantin, D., Gilroy, S. *Plant Physiol.* **1997**, *114*, 789–800.

75 Fasano, J.M., Massa, G.D., Gilroy, S. *J. Plant Growth Regul.* **2002**, *21*, 71–88.

76 Perrin, R.M., Young, L.S., Murthy, U.M.N., Harrison, B.R., Wang, Y., Will, J. L., Masson, P.H. *Ann. Bot.* **2005**, *96*, 737–743.

77 Plieth, C., Trewavas, A.J. *Plant Physiol.* **2002**, *129*, 786–796.

78 Sanders, D., Brownlee, C., Harper, J.F. *Plant Cell* **1999**, *11*, 691–706.

79 Hou, G., Mohamalawari, D.R., Blancaflor, E.B. *Plant Physiol.* **2003**, *131*, 1360–1373.

80 Hou, G., Kramer, V.L., Wang, Y.S., Chen, R., Perbal, G., Gilroy, S., Blancaflor, E.B. *Plant J.* **2004**, *39*, 113–125.

81 Inzé, D. *EMBO J.* **2005**, *24*, 657–662.

82 Hartig, K., Beck, E. *Plant Biol.* **2006**, *8*, 389–396.

83 Jiang, K., Feldman, L.J. *Annu. Rev. Cell Dev. Biol.* **2005**, *21*, 485–509.

84 Kepinski, S., Leyser, O. *Curr. Biol.* **2005**, *15*, 208–210.

85 Halstead, T.W., Dutcher, F.R. *Annu. Rev. Plant Physiol.* **1987**, *38*, 317–345.

86 Blancaflor, E.B., Hasenstein, K.H. *Plant Physiol.* **1997**, *113*, 1447–1455.

87 Staves, M.P., Wayne, R., Leopold, A.C. *Am. J. Bot.* **1997**, *84*, 1530–1535.

88 Shepherd, G.M.G., Corey, D.P., Bock, S.M. *Proc. Natl. Acad. Sci. U.S.A.* **1990**, *87*, 8627–8631.

89 Yoder, T.L., Zheng, H-Q., Todd, P., Staehelin, A. *Plant Physiol.* **2001**, *125*, 1045–1060.

90 Baluska, F., Kreibraum, A., Vitha, S., Parker, J.S., Barlow, P.V., Sievers, A. *Protoplasma* **1997**, *196*, 212–223.
91 Ma, Z., Hasenstein, K.H. *Plant Physiol.* **2006**, *140*, 159–166.
92 Lefranc, A., Jeune, B., Driss-Ecole, D., Perbal, G. *Adv. Space Res.* **2005**, *36*, 1218–1224.
93 Fasano, J.M., Swanson, S.J., Blancaflor, E.B., Dowd, P.E., Kao, T.H., Gilroy, S. *Plant Cell* **2001**, *13*, 907–921.
94 Johannes, E., Collings, D.A., Rink, J.C., Allen, N.S. *Plant Physiol.* **2001**, *127*, 119–130.

5
Biology of Adherent Cells in Microgravity

Charles A. Lambert, Charles M. Lapière, and Betty V. Nusgens

5.1
Why Cell Biology Research in Microgravity?

It is obvious that a journey in an orbiting spacecraft induces in most astronauts (cosmonauts) a significant reduction of bone mass. This is accompanied by a large redistribution of the body fluids and oedema of the upper part of the body and various disorders of the vascular system by alteration of the balance of the body fluids. Altered renal function and bone resorption were suggested to be responsible for an increased propensity in forming kidney stones. The immune resistance has been considered as another side effect that is potentially responsible for various diseases, infectious and other, that could affect the crew of spacecrafts.

Uni- and multicellular organisms are sensitive to environmental factors, chemical and physical, that modulate their vital functions. In space flights, the absence of the gravity stimulus under which all organisms have developed on Earth has been claimed to alter the precise equilibrium of physiological processes that is called health. Besides the lack of gravity, several additional disturbing elements need also to be considered, among which is increased cosmic radiations as a cause of cell damage by the induction of reactive oxygen species and mutations. Psychological stress, mainly in humans, is potentially responsible for endocrine disturbances that could also modify the operational activity of the cells. In all organisms, an exchange of information between cells modulates their functions. In multicellular organisms this situation is more complex since the different types of cells and their interactions, homo- and heterotypic, are in operation. This situation can be simplified by dissociating the pluricellular organs and organisms: we can consider experimentally the participation of a single cell type and each extracellular message in the complex network of intracellular signalling that precisely controls the multiple pathways required for life. This analytical approach is most suitable for pinpointing the pieces of the puzzle that could be considered as a molecular target for pharmacological intervention. Both the pathology induced by Space exploration and its counterpart on Earth will equally benefit from these investigations.

The present chapter provides an overview of some specific facets of cell biological research in microgravity, their complexity and the need to continuously adapt

Biology in Space and Life on Earth. Effects of Spaceflight on Biological Systems. Edited by Enno Brinckmann

ISBN: 978-3-527-40668-5

the concepts to the progressing knowledge of the physiological functions of the cells. It is aimed at those who have not yet tasted the challenge and enjoyed the reward of biological research in microgravity. This opinion has already been stated by someone who has accompanied her own biological experiments during a space flight, Millie Hughes-Fulford [1]. This chapter also represents a token of gratitude to the scientific and technical teams of the ESA Microgravity Department for their logistic support as well as friendly and helpful advice.

5.2 Medical Disturbances in Astronauts

5.2.1 Similarity to Diseases on Earth

All diseases on Earth are potentially also present in space and possibly modulated by the ambient conditions, microgravity, ionizing radiations, stress, and so forth. Osteoporosis results on Earth from an imbalance between bone formation and degradation. It is a common situation that affects a large proportion of the aging female population and develops to a lesser extent, and later, in males. In most male and female astronauts, a similar osteopenia occurs at a rate one order of magnitude faster than on Earth [2]. Disuse bone reduction induced experimentally on Earth in Man by bed rest and in animals [3] indicates that a mechanical stimulus is operational to maintain the equilibrium between bone formation, a function of the osteoblasts, and degradation, a process resulting from the activity of the osteoclasts. Altered muscle mass and function occur in similar conditions, too [4]. Disuse and space atrophy are, however, not identical [5]. The blood and lymphatic vasculature participates in a large variety of medical conditions on Earth. The endothelial cell is the most important component of this system. It is affected by microgravity during space flights [6]. The known and still to be discovered properties of the endothelial cells are regulated through membrane receptors for diffusible and stromal ligands and largely by mechanical messages, hydrodynamic at the blood exposed site and tissular at the interphase with their support. The endothelium is supported by a basement membrane and surrounded by contractile cells of the smooth muscle cell family in arteries, veins, capillaries and lymphatics. These cells, which are responsible for the vascular tone and the control of permeability of the endothelium, participate in the disturbed distribution of the body fluids during the initial exposure of humans to microgravity [7] and are probably involved in cardiovascular deconditioning affecting astronauts [8]. In venous hypertension and cardiac insufficiency on Earth, the oedema is gravity-dependent and located in the lower limbs. Renal stones are also a common pathology on Earth. Astronauts are exposed to this pathology by a reduced water flow in the urinary tract and the release of calcium from their bone storage [9]. A certain degree of immune depression seems to affect the crew of spacecrafts. This is a common feature in aged people and those who receive drugs against cancer.

Some investigators interpret space-induced bone loss as an accelerated ageing process [10].

The common point between Earth and Space pathologies is clinical. Progress in the understanding of their pathomechanisms is required for the development of logical therapies [11]. Obviously, the involvement of the gravity stimulus present on Earth but not in Space and other conditions of the extraterrestrial environment will add to a better understanding of these mechanisms. Earth helps Space and vice versa. Space designed biomedical materials have already found applications on Earth [12, 13].

5.2.2
Cell Types Potentially Involved

The biology considered in this chapter with respect to the disorders observed in astronauts concerns mainly the types of cells that adhere to a support *in vivo* and *in vitro*. In fact all cells of a multicellular organism at a defined stage of their development are adherent cells, including the circulating blood cells during maturation in the bone marrow and the lymph nodes before migration to their ultimate destination. Some common effects of microgravity on cell biology could exist, whatever might be the mechanism responsible for the cell–cell or cell–matrix relationship and its resulting signalling. All cells are basically similar, since they display a membrane containing adhesion proteins, ligand receptors, signalling mechanisms and a cytoskeleton (CSK) of fibrillar polymers, a membrane associated system for protein synthesis and maturation and a nucleus containing an identical genome. What defines their specificity is the composition and function of each of these structural elements, which depends on a differentiated expression of the genome. The reason why space flights affect more significantly some specific systems of the multicellular organisms might depend on the relative impact of the mechanical signalling in the function of their constitutive organs, roots in plants or bones, muscle and blood vessels in vertebrates. The immunological disorders, potential differences in nutrition, metabolism and pharmacology, increased risk of renal stone, and hormonal disturbances might result from more complex mechanisms involving microgravity, reactive oxygen species and psychological factors. The fact that all organs are not similarly affected also depends on the extent of their homeostatic reserves. Epithelial cells in microgravity display a modified cytoskeleton [14] but the epidermis does not seem clinically affected.

Even though the dermis of astronauts is not visibly modified by space flights, except for transient accumulation of interstitial fluids (oedema) in the upper part of the body, we have selected skin fibroblasts for investigating the effect of microgravity. This choice is based on two considerations. First the effect of microgravity on cells should depend upon (a) basic mechanism(s) operational in all types of adherent cells *in vitro* and, second, we have an extended knowledge of the biology of these cells, including their receptivity–reactivity to mechanical stimuli and the potential to create experimental models suitable for Space investigations (Section 5.4.2).

5.3
Mechano-receptivity and -reactivity of Adherent Cells in Culture

Nature provides clear examples of defined mechano-receptors in eukaryotes as the statoliths in plants and the otoliths of the inner ear in most species of vertebrates. Similar specialized cells of the sense organs detect pressure (touch) and vibrations and communicate these physical stimulations to the nerves of the afferent pathway up to the brain. Mechano-receptors at the muscular–tendinous junction are most significant for the integrity of the myoblast. All cells within all organs also communicate by specialized structures in the cell membrane that are known to convey mechanical signals by the cell–matrix and cell–cell interacting system (Sections 5.3.1. and 5.3.2).

The fibroblast is a largely distributed cell type that produces the extracellular matrix (ECM) of all supporting tissues (skin, tendons, ligaments, fascia, and so forth) and allows the organization of the epithelial and parenchymal cells to form organs. The phenotype of the fibroblasts is basically similar in all locations in terms of generic ECM macromolecules production (the various isotypes of collagens, proteoglycans, glycoproteins, elastin, proteolytic enzymes, and so forth). It is, however, clear that these cells are differentiated, as evidenced by their characteristic transcriptome. The precise nature of these secreted proteins plays a significant and specific role during development and growth and later in life to preserve the organization of the cells in organs. Myofibroblasts occurring during wound healing, smooth muscle cells of the large blood vessels and osteoblasts in bone should be considered as specialized members of the interstitial fibroblasts family.

In connective tissues, resident cells are embedded in and closely interact with a three-dimensional network of polymeric structures. These cell–matrix interactions are mediated by cell adhesion receptors that operate in the sensing of their microenvironment, chemical and mechanical, and in their response to it. Mechanical forces, either "self-generated" by the activity of the cytoskeleton or issued from the extracellular environment, play a significant role in development, morphogenesis and function of the tissues [15]. This is most apparent in specialized structures such as the blood vessel wall and bone, and during physiological processes such as wound closure, maintenance of muscle mass and perception of touch and sound. Pathological conditions such as atherosclerosis, hypertension, and cardiac hypertrophy are often related to cellular reactivity to mechanical signals. Most cell types, even in non-specialized tissues, are strongly influenced by mechanical variables that regulate adhesion, proliferation, survival, contractility, migration, ECM architecture and gene expression.

Two types of cellular structures mediate the receptivity and the reactivity to mechanical forces: membrane associated receptors to ECM and focal adhesions (FA) that connect the cytoskeleton to the surrounding ECM (Fig. 5.1). Adherens junctions (AJ) provide a link between neighbouring cells, as in sheets of epithelial or endothelial cells (for a review, see Ref. [16]). Both cell–matrix and cell–cell adhesions and the cytoskeleton are major players in the mechano-transduction pathway,

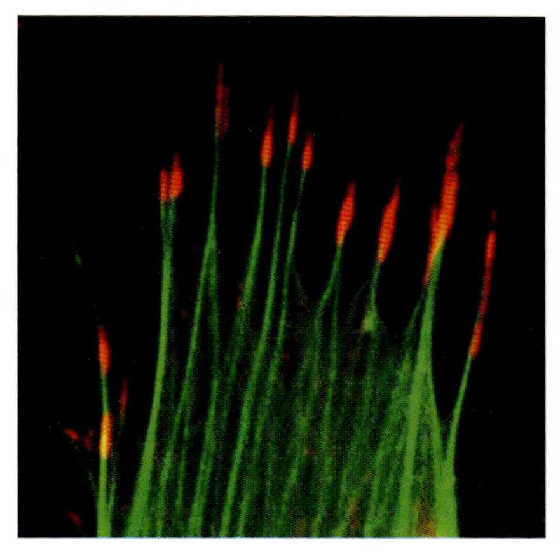

Fig. 5.1 Part of a fibroblast on a rigid adhesive surface, illustrating the structures that allow the cell to sense its chemical and mechanical environment. In red, vinculin, a structural protein of the focal adhesions, participates in the link between the transmembrane integrins that recognize the support and the actin stress fibres (in green) of the cytoskeleton. (Photo: Pierre Mineur, Laboratory of Connective Tissues Biology, Liège, Belgium.)

although other pathways such as stretch-activated ion channels might also participate in specific cell types [17].

5.3.1
Mechano-transduction at the Cell–Matrix Contacts

Cells of the mesenchymal lineage possess *in vitro* specific structures and mechanisms to sense and generate mechanical forces [18–20]. Whether similar mechanisms operate *in vivo* is beginning to be unravelled. Structures containing focal adhesions proteins have indeed been observed in several tissues *in vivo* [21–23]. The mechanical information issued from the cell environment as well as the mechanical tension generated by the cytoskeleton converges to transmembrane receptors, the integrins.

The recognition of specific sequences in the ECM proteins by integrins induces their activation (Fig. 5.2A), clustering in membrane microdomains and leads to the recruitment at their cytoplasmic tail of a battery of structural and signalling proteins that form the FA (Fig. 5.2B). Structural proteins, such as talin, vinculin, α-actinin, paxillin, act as scaffolding adaptor proteins that strengthen the cell adhesion by anchoring FA to the cytoskeleton either by direct binding to actin or by interaction with other actin-binding proteins [24, 25]. Formation, stabilization and maintenance of FA requires either endogenous tensile forces generated by contraction of actin microfilament against the ECM cell contacts [26] or external forces applied to the bound ECM and counteracted by the cytoskeleton [20, 27–29]. Many signalling molecules recruited in FA appear to participate in the transduction of

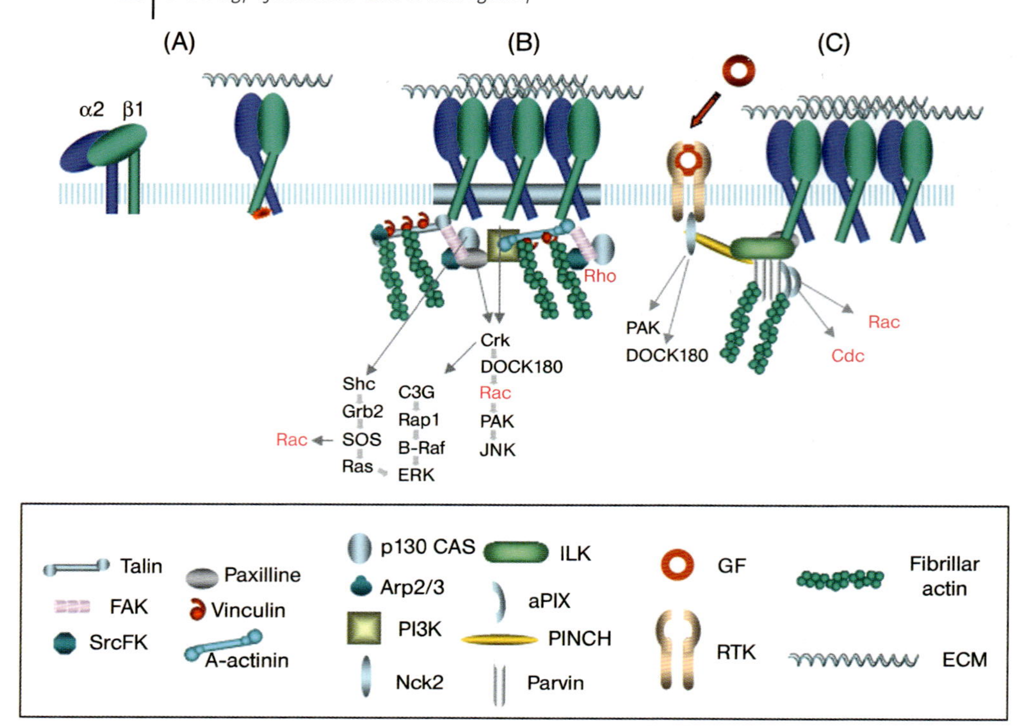

Fig. 5.2 Schematic representation of the signalling produced by the integrins. The free integrin dimmer (α2β1) becomes operational by recognizing its ECM ligand (A). Multiple integrins form clusters upon mechanical stimulation and induce the association of the components of the focal adhesion, polymerization of actin and activation of the signalling. (B) Clusters of activated integrins are also involved in the signalling by diffusible ligands recognized by specific membrane receptors (C). The RhoGTPases (identified in red) largely participate in these signalling pathways. See text for further details.

the mechanical signals into biochemical events. These include tyrosine kinases such as FAK, PYK2, src family kinases (srcFK), Ras (for reviews, see Refs. [30, 31]), ILK [32] and tyrosine phosphatases (for a review, see Ref. [33]). Phosphorylation and dephosphorylation events accompanied by conformational changes and protein–protein interactions trigger a cascade of reactions involving several signalling pathways, ultimately targeting gene transcription through activation of transcription factors. Additional mechanisms, including modulation of protein conformation by stretching as demonstrated for vinculin and src [34], may also operate in mechano-transduction. It is worth mentioning that receptors to growth factors are also localized at FA, where they interact with integrins [35, 36] in a close cross-talk through connecting pathways such as ILK/PINCH/parvin (Fig. 5.2C). Most types of non-transformed cells require integrin-mediated attachment to ECM to respond to growth factor stimulation for proliferation and survival [37]. Mechanical tension appears as a major survival signal, as stated by Ruoslahti [38]: "Stretching is good for a cell".

Several types of accessory molecules directly or indirectly cooperate with integrins to modulate their activity and signalling pathways and ultimately the mechanical functions. Among them, the syndecans, transmembrane or GPI membrane-anchored heparan sulfate proteoglycans, act as receptors for heparin-binding molecules, including extracellular matrix components, and as low-affinity co-receptors to soluble growth factors such as FGF [39]. Discoidin domain receptors 1 (DDR1) and 2 (DDR2) are tyrosine kinase receptors activated by native collagens that can also cross-talk with integrins or signal independently [40]. The tetraspanins are palmitoyl-anchored membrane proteins that tend to self-associate and cluster with numerous protein partners, including integrins, in membrane microdomains. They are involved in the modulation of several integrin-controlled processes, from signalling to motility (for a review, see Ref. [41]).

Numerous molecules are currently described that interact directly with the intracellular domain of integrins or that are indirectly linked and participate in the formation of large multi protein complexes connecting the ECM to the cytoskeleton. Such complexes enable many levels of regulation and highly graduated and specific responses of the cell to external stimuli [25].

5.3.2 Mechano-transduction at the Cell–Cell Contacts

At least three types of cell–cell adhesions are recognized: the adherens junctions (AJ), the tight junctions, and the gap junctions, mediated by, respectively, the cadherins, the occludins and the connexins. The cadherin-based AJ, typically found in epithelium and endothelium, are critical for the development and maintenance of multicellular organisms. The extracellular domains of cadherins participate in homophilic interactions while the cytoplasmic domains binds to $p120^{ctn}$ [42] and to β-catenin that provides a link to α-catenin and the actin cytoskeleton [43]. Upon tyrosine phosphorylation, both $p120^{ctn}$ and β-catenin also play a significant role in signalling when translocated to the nucleus to regulate cell proliferation [44]. If AJ have not been shown to directly modulate mechano-transduction, they act at least indirectly. As a general rule, FA and AJ antagonistically control cell behaviour: increasing cell–cell interaction competes with and reduces cell–matrix adhesion [45].

5.3.3 The Cytoskeleton Network and Its Control by the Small RhoGTPases

The cytoskeleton and the dynamic mechanical balance that exists between the cells and their ECM support appear as major players in several mechano-transduction pathways [46]. The CSK is a network of three interconnected filament systems: the actin microfilaments, the microtubules, and the intermediary filaments. They condition the shape of the cells and the major mechanical functions such as adhesion, polarization, directional migration, as well as proliferation, survival or apoptosis, gene expression and architectural organization of their supporting scaffold.

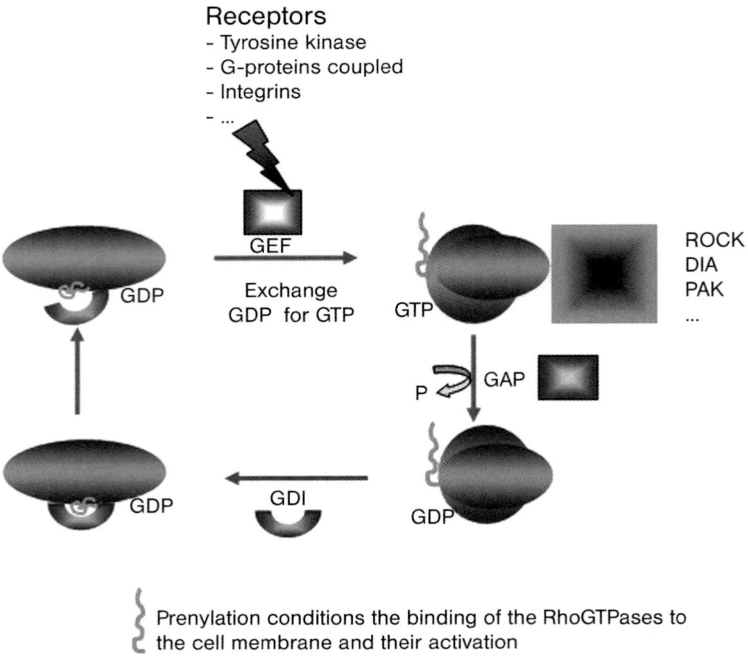

Fig. 5.3 Cycling of a RhoGTPase between its inactive form bound to GDP and its active form when the GDP is exchanged for a GTP upon the activity of a GEF at the cell membrane. The GTP-bound RhoGTPase is then able to interact with effectors that trigger signalling cascades. It becomes inactivated by its intrinsic hydrolytic activity activated by GAP and is sequestered by chelation of its lipid tail in a complex with GDI.

The microtubules play additional roles in eukaryotic cells. They are key structural elements of the mitotic spindle apparatus during mitosis and during interphase and serve as tracks onto which motor proteins transport vesicles and other components throughout the cell [47]. Besides their role in the contractile actomyosin stress fibres, actin filaments are also major structures involved in the formation of membrane protrusions such as lammellipodia, filopodia, vesicles trafficking, endocytosis and exocytosis. The actin and microtubules dynamics are highly regulated by the small GTPases of the Rho family, Rho, Rac1 and Cdc42 (for reviews, see Refs. [48, 49]).

The RhoGTPases function as GDP/GTP-regulated binary switches operating in the relay between extracellular signals elicited by integrins and receptors for soluble ligands and the intracellular effectors pathways (Fig. 5.3). In their inactive GDP-bound forms, the RhoGTPases are sequestered by RhoGDIs (Guanine Dissociating Inhibitors) in the cytoplasm while upon signalling elicited by external stimuli they translocate and anchor to the cell membrane through their prenylated C-terminal sequence. The post-translational prenylation (farnesylation or geranylgeranylation) is critical for their interactions with the membrane. These post-

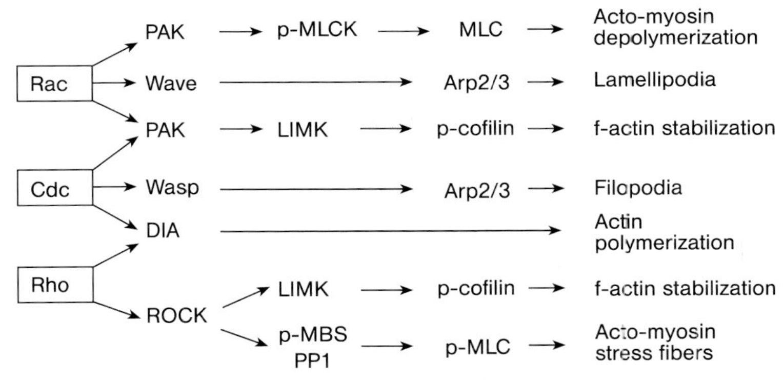

Fig. 5.4 Simplified view of the implication of the RhoGTPases in the organization of the actin-based cytoskeleton through their effectors ROCK, DIA, PAK and WASP/WAVE (for a review, see Ref. [48]).

translational modifications are prevented by inhibitors of the HMG CoA reductase such as statins and of the prenyl-PP synthases by nitrogen-containing bisphosphonates [50]. Upon translocation to the membrane, the GDP is replaced by a GTP under the activity of Guanine nucleotide Exchange Factors, the GEFs. The GTP-bound forms can then interact with and activate downstream effectors that are key-regulators in the organization and dynamics of the CSK and the control of cellular pathways regulating gene expression, cell survival and multiplication. The active Rho-GTP returns to its inactive form upon activation of its own GTPase activity by GAPs (Guanine nucleotide Activating Proteins).

The GTP-bound RhoGTPases relay the activation to a series of effectors that directly or indirectly control actin polymerization (Fig. 5.4). The WASP/WAVE proteins family, targeted by Rac1 and Cdc42, activate the Arp2/3 complex, a nucleator of actin polymerization, to form a branching filament network as seen in lamellipodia at the front of migrating cells. The Diaphanous-related formins, Dia 1, 2 and 3 induce the extension of actin filaments in filopodia. PAK 1-3 (p21-activated kinases) phosphorylate multiple cytoskeletal proteins. ROCK 1-2 (Rho-kinases), together with PAK, activate the actin/myosin filament assembly through an increased phosphorylation of the myosin II light chain, resulting in the formation of stress fibres and enhanced cell contractility (for a review, see Ref. [48]). Further to their activity on the dynamics of the cytoskeleton, the Rho effector kinases (PAK, ROCK, Citron kinase, MLK, PKN, and others) are involved in the regulation of transcription through the MAPKs cascades, death and survival signalling and cell-cycle progression (for a review, see Ref. [51]).

Our laboratory has contributed to the understanding of the regulatory functions of the RhoGTPases in the context of ground and microgravity research programmes. Specific small interfering RNA (siRNA) targeting these molecules has been created [52, 53] and used to transfect primary cells and established cell lines. These tools have already provided significant information, e.g. on the negative regulation operated by Cdc42, through a Rac1/MAPK pathway, on MMP1 expression, the collagenase responsible for the degradation of fibrillar collagens of the connective tissues and bone [53]. The counteracting biological tools, i.e. cells expressing constitutively active forms of the RhoGTPases, have been created and pertinent clones selected [54]. They allowed us to clarify the role of individual RhoGTPase in the mechanical functions of human fibroblasts [55, 56]. Rac1 has been shown to play a pivotal role in fibroblast migration [56] and is similarly involved in osteoblast migration, a function abolished by statins [57]. We recently demonstrated that zoledronic acid, a bisphosphonate used to treat osteoporosis, up-regulates bone sialoprotein expression in osteoblasts through the inactivation of RhoA by interfering with its prenylation and membrane translocation [58].

5.3.4 Cells React to Mechanical Stress and Relaxation

Various *in vitro* experimental models have been created to demonstrate that changes in the geometry or the mechanical compliance of the supporting ECM control key cellular functions in development, growth, physiology and many pathological conditions [15, 59–61]. Cultured in monolayer on a coat of type I fibrillar collagen, human dermal fibroblasts show a robust network of stress fibres (Fig. 5.1) and express high levels of mRNA for structural proteins, while the mRNA level for ECM degrading enzymes (MMPs) is low. Within a free floating 3D-collagen gel, fibroblasts progressively retract the gel, resulting in dissipation of their internal tension ("pre-stress" according to Ingber [62]) until reaching mechanical equilibrium (isotonic tension). In this low-stress condition, the cytoskeleton is profoundly disorganized and the phenotypic expression is drastically reversed [63–67]. Fibroblasts stop dividing and produce large amounts of MMPs and an autocrine loop of the pro-inflammatory cytokines IL-1 and IL-6 [68; personal unpublished data] through a tyrosine-kinase dependent regulation [65]. When the collagen gel is restrained to retract (isometric tension), the phenotype is intermediary between that observed in the monolayer and the retracted gel [64]. Interestingly, fibroblasts in monolayer treated with cytochalasin D, suppressing the actin stress fibres, present a phenotype similar to that in the retracted gel [65]. Colcemide disrupting the microtubules does not induce this catabolic phenotype (personal observations). These features indicate that the signalling related to relaxation passes through the actin cytoskeleton and the focal adhesions, to which they are connected. Similar experimental models, miniaturized to fit in ESA's Type I experiment containers, have been used to test the effect of microgravity on the mechanical reactivity of fibroblasts during the STS-95 flight in the Space Shuttle Discovery (Section 5.4.2).

Alternative experimental models analysing mechano-transduction and -reaction use static or cyclic stretch as stimulus. Fibroblasts, smooth-muscle cells, cardiac myocytes and other cell types have been shown to react to these signals by altered growth, survival and biosynthetic phenotype. Most notably, cyclic stretch up-regulates the production of ECM proteins either directly or indirectly by stimulating the release of paracrine growth factors such as TGF-β, PDGF, CTGF. The involvement of integrins in the sensing and response to cyclic stretch is supported by the activation of c-src, phosphorylation of FAK, paxillin, p130^{cas} followed by the activation of PI3K, ERK1/2, p38 or JNK, in an ECM-specific manner [69–72]. The RhoGTPases are largely implicated in these mechanisms. Although these pathways are triggered by various other external stimuli, mechanical stress induces the transcription of a distinct set of genes specific to the cell type and the nature of the stress. This suggests that specific mechano-responsive cis-acting elements may exist in the promoter of these genes, as proposed for tenascin-C [73].

5.4 Microgravity, the Loss of a Force, Leading to Cellular Disturbances

The weight of a cell on Earth exerts a very small force on its support, of the order of 10^{-12} N. The forces developed by the cytoskeleton are several orders of magnitude stronger. The traction force of a fibroblast within a contracting collagen gel ranges from 10^{-8} to 10^{-7} N [74]. The absence of gravity does not modify (or does so insignificantly) the sedimentation pattern of the organelles or clusters of molecules – the pattern largely depends on the thermal noise and the activity of the cytoskeleton. The absence of convection might, however, affect the movement of the fluids in the pericellular atmosphere of cultured cells, thereby altering the potential for ligands to find their receptors and permitting an accumulation of secretion products, catabolites and heat, in the close vicinity of the cell membrane. This should be considered as an indirect effect of microgravity *in vitro*. It does not apply *in vivo* due to the continuous circulation of the fluid in the extracellular space. The absence of gravity has, however, various effects on many types of adherent cells in culture that await a mechanistic demonstration.

5.4.1 Biological View of the Biophysical Concepts

Defined processes such as the chemical oscillations occurring in the Belousov–Zhabotinsky reactions are reduced in microgravity by the lack of convection [75]. The effect of microgravity on several reactions present in living cells has been investigated similarly. Controversial data [76, 77] indicate that isolated systems such as the catalytic activity of enzymes are or are not modified by microgravity, while the molecular interacting system operating in protein crystallization is improved [78, 79]. More complex reactions involving reaction–diffusion processes are clearly gravity-dependent. To define this concept we quote the words of James Tabony and co-workers from one of their recent publications [80]:

> Weak external fields, such as gravity and vibrations, are
> not normally considered as intervening in chemical and
> biochemical reactions. One possible manner by which they

> can is through certain types of reaction-diffusion processes. Since the late 1930s, theoreticians have predicted that under appropriate conditions, a nonlinear coupling of reactive processes with molecular diffusion can lead to the progressive development of a stationary chemical pattern. Some chemical systems have been established as behaving this way. In some reaction-diffusion systems, it has been predicted that self-organization can depend on weak external fields, such as gravity, which break the symmetry of the process. Their presence, at a critical moment or bifurcation time early in the process, can determine the morphology that subsequently develops. The formation of microtubules, a major element of the cytoskeleton, shows this type of behaviour.

In the signalling cascade, similar effects have been claimed, although at a level that perhaps needs to sum up to become efficient. Recent publications have addressed the kinetics of receptor coupled signalling. After binding of a ligand to its receptor, the multiple steps of activation by a kinase, deactivation by a phosphatase, modulation of the shape of an adaptor protein, each require around 1 s. Taking into account the concentration of each member of the cascade and the rate constant of each reaction, the dynamics of such a cascade have been modelled *in silico* on the basis of cell data supported by observation of fluorescent probes [81]. The importance of non-equilibrium thermodynamics in analysing such a processing of the biological information [82] and of the biological switches consisting of phosphorylation–dephosphorylation have been demonstrated [83]. The concepts proposed by Mesland [84], illustrated by the work of Tabony's group [85], could therefore apply to signalling, suggesting that microgravity can alter its efficiency.

5.4.2
Short Time Microgravity and Space Flights

Various facilities, as described in the Introduction of this book, offer near free-fall conditions (microgravity), varying in duration from seconds (drop towers, parabolic flights) to minutes (sounding rockets), days and weeks (manned and unmanned satellites), months and years (Space Station) (Fig. 5.5).

The duration of actual weightlessness determines the type of experiments that can be accommodated in the various flight facilities and the selected end-points. Table 5.1 summarizes a selection of representative experiments, complementing those compiled by Moore and Cogoli [86].

Altered CSK organization, focal adhesions distribution, number and maturation, cell shape and spreading are common observations made in various, but not all, of cells in a microgravity environment (Table 5.1). In MC3T3-E1 osteoblastic cells the nucleus became oblong and smaller, and displayed an altered architecture

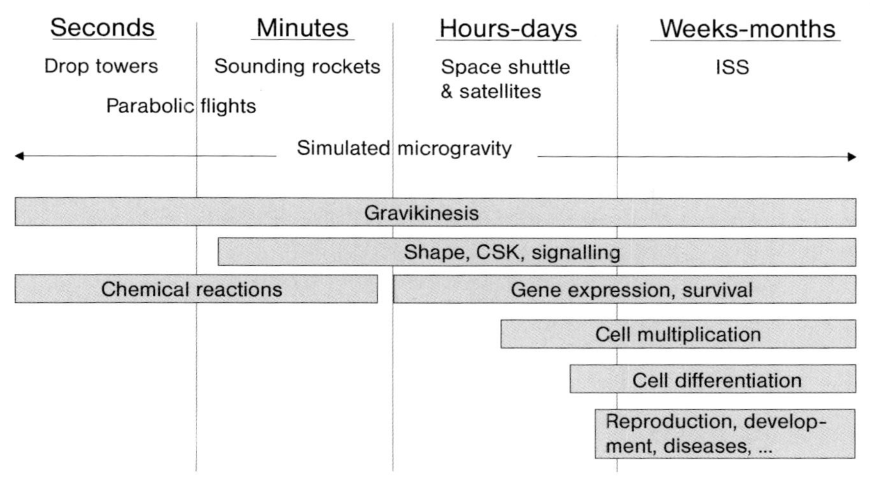

Fig. 5.5 Duration of actual weightlessness condition and type of experiments that can be accommodated in the various flight facilities and for the selected end-points.

as compared with an in-flight 1×*g* control [92]. The chromatin structure of MCF-7 cells was also affected [108]. Rounding and loss of polarity of rat vestibular hair cells in three-dimensional cultures in microgravity were reported, too [120]. Typically, experiments were conducted during space flights lasting hours up to days. However, it was reported that MG-63 osteoblasts displayed a decreased cell area and altered shape and focal contacts organization after experiencing 15 to 30 parabolas during a parabolic flight of 5–10 min of weightlessness interrupted by periods of hypergravity and normogravity [87, 88]. Similarly, the CSK of A431 cells had remodelled after a few minutes of microgravity in sounding rocket [113]. In other experiments the actin CSK of kidney cells was unaffected after prolonged time in microgravity [121]. These data suggest that all cells are not equally sensitive to CSK remodelling by microgravity. Microgravity-induced disorganization of the CSK and disassembly of focal contacts was mainly observed in post-mitotic cells [70], suggesting that loss of gravity would affect more extensively the structures involved in dynamic processes such as cell division. Whether the alterations observed in the CSK of cells living in microgravity are related to the alteration of the microtubules patterning during polymerization of purified tubulin, as shown in simulated and short-term microgravity by Tabony and co-workers [85, 123], remains to be established. The observations of Vassy et al. [14], however, suggest that such a reaction could occur in a cell.

A reduction in cell number in microgravity has often been reported (Table 5.1), but an increased cell number was also reported in a few cases. It is not always possible to discriminate between alteration of proliferation rate or survival. Microgravity reduces the expression of PCNA, a marker of replication and c-myc in

Table 5.1 Effect of microgravity on various cell types and cell-free systems during different flight opportunities. Abbreviations: ALP: alkaline phosphatase; BSP: bone sialoprotein; Col I: collagen type I; Col III: collagen type III; COX: cyclooxygenase; GH: growth hormone; iNOS: inducible nitric oxide synthase; MMP: matrix metalloproteinase; OC: osteocalcin; PKC: protein kinase C; PRL: prolactin; TPA: tetradecanoyl phorbol acetate; vit.: vitamin.

Cell type	*Flight/duration*	*Effect of microgravity*	*Ref.*
ROS	Parabolic flight	Altered shape and size. FA reorganization. Increased PGE2 synthesis	87, 88
ROS	Bion-10 Foton 11 Foton 12 <6 days	CSK remodelling. FA disassembly. Proliferation unaltered OC, Col I and ALP unaltered	70
MC3T3-E1 mouse osteoblasts	STS-76 STS-81 STS-84 <4 days	CSK remodelling. FA disassembly. Altered cell and nuclear shape. Altered nuclear architecture Decreased proliferation and PCNA expression PGE2 unaltered or increased Decreased OC expression Decreased serum-induced Bax and Bcl-2, c-myc, COX-2 TGF-b, FGF-2	89–92
MC3T3-E1 mouse osteoblasts	Sounding rocket TR-1A6	Decreased EGF-induced c-fos expression. MAPK activation unaltered	93
Rat osteoblasts	STS-65 4–5 days	Decreased proliferation Decreased Vit. D-induced expression of OC Increased Vit. D-induced expression of BSP Reduced expression of HSP70, HSP47, HSC73, IGF-I, IRS-1, IGFBP-5, PDGF-receptor, shc and c-fos Reduced secretion of TGF-β1 Induced expression of c-jun, JNK, IL-6, PGHS-2, iNOS, GTPCH, IGFBP-3, PAF receptor, PKC-α, ζ and -θ, Gαq and glucocorticoid receptor Increased PGE2 synthesis	94–102
Chicken calvariae osteoblasts	STS-59 8 days	Decreased Col I and OC expression	103
Chicken calvariae osteoblasts	STS-77 <3 days	Reduced cell number, not related to launch conditions	104
MG-63	Foton 10 <9 days	Decreased Col I, OC and ALP expression	105, 106
Pituitary cells	STS-46 8 days	Altered cell shape Reduced GH and PRL production	107

Table 5.1 *Continued*

Cell type	*Flight/duration*	*Effect of microgravity*	*Ref.*
MCF-7	Foton <48 h	CSK remodelling. Altered cell shape, microtubules and chromatin structure Reduced proliferation. Prolonged mitosis Reduced phosphotyrosine level	108
A431	Sounding rocket Maser-3,-4,-5,-6	Altered cell shape. CSK remodelling. Increased F-actin content Reduced EGF- and TPA-induced c-fos and c-jun expression Unaltered EGF-receptor redistribution	109–113
Mouse fibroblasts	Cosmos-2229 2 days	Altered nucleus size and shape Reduced proliferation	114
Human skin fibroblasts	STS-95	CSK remodelling. Altered cell-substratum adhesion and FA distribution. Reduced FA (phosphorylation)	115, 116
	≤4 days	Increased ECM degradation and MMP1 and IL-6 expression Col I, Col III expression unaltered	
Human fibroblasts	Spacelab D2 <20 hours	Increased collagen and total protein synthesis	117, 118
WI-38	STS-93 5 days	Altered TNF and IL signalling	119
Rat utricular cells	Foton-12 <3 days	CSK remodelling. Altered cell shape and polarity	120
JTC-12 monkey kidney cells	STS-47 8 days	Unaltered morphology and urokinase expression. Delayed attachment on substratum Reduced proliferation	121
–	Sounding rocket Maxus-3	Altered self-ordering of microtubules	85, 123
–	STS-91	T4 DNA polymerase activity unaffected	124
–	Sounding rocket Maser-7	Isocitrate lyase activity unaffected	125
–	Parabolic flight	Reduced K_m of lipoxygenase-1	77
–	Drop tower	Reduced propagation of chemical pattern of Belousov–Zhabotinsky reaction	75

MC3T3-E1 osteoblasts [89], and extends the duration of mitosis in MCF-7 cells [108]. Direct observation of apoptosis has also been reported [122] and anti-apoptotic factors, such as bcl-2 and HSP70, appear to be reduced in osteoblasts in microgravity [90, 98], while the expression of iNOS, involved in the synthesis of the pro-apoptotic molecule NO, was largely increased [99]. These data suggest that the reduced cell number observed in microgravity may result from both a reduced proliferation rate and an increased apoptosis.

The differentiated cell phenotype is determined by the regulated expression of several genes. A reduced expression of markers of osteoblast differentiation has often been reported and is discussed extensively in Chapter 6. Similarly, microgravity did reduce the growth hormone and prolactin production by pituitary cells [107].

Several mechanisms, not mutually exclusive, may explain how microgravity impacts cell shape, proliferation, survival and differentiation. They include modulation of the paracrine/autocrine cellular regulation, the intracellular signalling and the CSK function. The expression of several growth factors, cytokines and their receptors, as well as their downstream signalling events is altered in microgravity. Microgravity reduces the serum-induced expression of TGF-β and FGF-2 in mouse osteoblasts [89], vitamin D-induced secretion of TGF-β as well as the level of IGF-1 and PDGF-B receptor [96, 97, 102] in rat osteoblasts. However, reduced expression of growth factor and their receptors is not the rule as the vitamin D-induced level of IGF-I, PAF and glucocorticoids receptors was reported to be increased [95, 97, 100]. In rat osteoblasts the expression of IGFBP-5, able to increase the circulating half-life and cellular association of IGF-1, is also reduced [95]. IL-6 expression is increased in rat osteoblasts [94] and human primary fibroblasts [116]. Altered TNF-α and IL-1 signalling has been reported in the fibroblastic cell line WI-38, too [119].

Several downstream pathways are also affected by microgravity. The expression of immediate early genes as c-fos and c-jun in EGF-or PMA-treated A431 cells is reduced in microgravity [109, 110, 112, 113], although the binding of EGF to its receptor and its clustering were not affected, suggesting that inhibition occurs downstream of the EGF-binding, possibly at the level of PKC activation. A similar reduction of EGF-induced c-fos was observed in MC3T3-E1 in microgravity, while MAPK phosphorylation was not affected [93], indicating that microgravity may target the signalling downstream of the MAPK pathway. C-fos expression was also reduced in rat osteoblasts exposed to microgravity for 4–5 days [102]. Increased PGE2 synthesis and mRNA level of COX-2 have been reported by several authors [87, 91, 94]. In the study of Guignandon et al. [87], an increase in PGE2 was observed already after 5 min of parabolic flight, suggesting that it may happen independently of a stimulation of COX-2 expression. Indomethacin, an inhibitor of PGE2 synthesis, reduced the cell shape modifications observed in parabolic flights [87].

The CSK is the primary determinant of cell shape and large-scale cell movements; it ensures the translocation of organelles throughout the cell and participates in chromosomes segregation and cytokinesis during mitosis. A continual assembly–disassembly of the microtubules and acto-myosin filaments is required during these processes. Probably, reorganization of the cytoskeleton directs the modifications of size and shape of cells and nucleus as well as the patterning, number and maturation of focal adhesions commonly observed in microgravity. Guignandon et al. [88] reported that nocodazole, an inhibitor of tubulin polymerization, prevented the alterations of focal adhesions observed in ROS cells during parabolic flights. The disorganization of microtubules seen in microgravity may lead to a reduced rate of chromosomes segregation during mitosis, while alterations of actin microfilaments and focal adhesions may also slow down cytokinesis. The increased duration of mitosis, as observed in MCF-7 cells in microgravity, might depend on such alterations [108]. The CSK also modulates signalling pathways, gene expression and survival [126]. This suggests that microgravity-induced modulations of cell differentiation and apoptosis may also depend, at least in part, on CSK remodelling.

One of the experiments that we conducted in microgravity was based on the observations that selected genes were responsive to mechanical stress and relaxation in the *in vitro* models described in Section 5.3.4. Similar models, monolayer on collagen coat (MONO), tethered collagen gel (TCG) and freely retracting gel (RCG), were designed and miniaturized to fit into the culture chambers of the plunger-units in the ESA Type I containers (Fig. 5.6). A detailed description of the experimental conditions and schedule can be found in ESA's Erasmus Experiment Archives (EEA) [http://spaceflight.est.int/eea]. Briefly, fibroblasts cultured in MONO, TCG and RCG configuration were kept at 37 °C for two and four days at 0×g and on the 1×g centrifuge before being fixed by appropriate reagents for morphological [127] and mRNA analysis.

The number of FA (vinculin-positive spots) per cell was reduced; a decreased number of stress-actin fibres and a reduced phosphorylation of the focal contact associated proteins was shown in the microgravity samples and normalized by centrifugation at 1×g in-flight to the level found in the control samples kept on ground [115].

IL-6 is a systemic alarm cytokine that has pleiotropic activity on many cells, including osteoclasts. In the samples cultured at 1×g on ground used as reference and in the 1×g samples in flight, the IL-6 mRNA displayed the expected modulations as a function of the mechanical tension imposed to the cells, i.e. very low at high tension in MONO, intermediate at medium tension in the TCG and higher at low tension in the RCG [116]. The IL-6 mRNA was significantly increased in

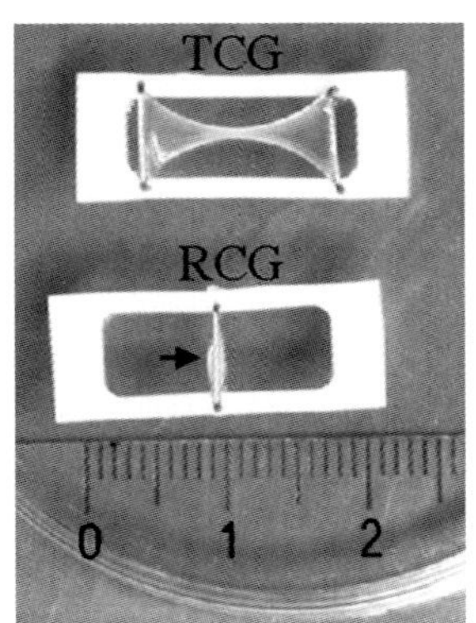

Fig. 5.6 Miniaturized 3D-collagen gels containing human dermal fibroblasts restrained to retract by tethering two end-sides (TCG, isometric tension) or freely retracted (RCG, isotonic tension, arrow) were designed for fitting into the plunger units (CCM, Nuenen, NL). The same cells in monolayer on a 2D-collagen coat (not shown) were used in the experiment during the STS-95 flight. This experiment aimed to evaluate the effect of microgravity on cells maintained under three levels of mechanical stress and cytoskeleton organization: high in MONO (depicted in Fig. 5.1), medium in TCG and low in RCG. Scale in cm.

microgravity in the three types of culture conditions by a similar factor: ×2.4 ($p<0.06$) in MONO, ×2.2 ($p < 0.01$) in TCG and ×3.0 ($p<0.03$) in RCG. This enhanced expression was suppressed in MONO ($p<0.04$) and in RCG ($p<0.03$) on the 1×g centrifuge onboard, but barely reduced in TCG. At the protein level measured by ELISA in the conditioned medium of the MONO samples, a significant stimulation (×7) of the secretion of IL-6 was observed in microgravity, mostly in the medium in contact with the cells prior and during the first 12 h of the flight. This might indicate that the transition from Earth gravity to microgravity is a more powerful stimulus than a stable weightlessness environment. Interestingly, Stein and Schluter [128] have reported that IL-6, measured in daily-collected urines of astronauts, is increased only during the first 24 hours after launch. The transcriptional activation of the IL-6 gene depends on several transacting factors, with NF-κB and AP-1 being the most active. The production and/or activation of these factors as well as their trafficking and translocation to the nucleus are potential targets of the mechanical and gravitational stimuli, possibly by alterations of the microtubules.

In human fibroblasts, stress relaxation or disruption of the CSK stimulates the level of MMP1 and reduces that of fibrillar collagens. The MMP1 mRNA in MONO and TCG was not significantly modified by microgravity, while in RCG it was significantly increased by a factor of two and was reduced in the centrifuge on board below the level observed on ground. The mRNA level of α1 (I) collagen polypeptide was not altered by microgravity.

Altogether, the data obtained from the STS-95 flight suggest that microgravity is interpreted by the cells as stress relaxation, superimposed to the basal level of stress to which they were experimentally submitted. These observations favour a process of potentiation rather than the additive effect of two separate mechanisms. It does not mean, however, that microgravity and mechanical stress relaxation operate through an identical pathway but signify, at least, that they converge to similar end-points.

5.4.3 Modelled Altered Gravity

The limited number of opportunities to conduct cell biology experiments onboard spacecrafts combined with the space constraints and difficulties in carrying out sophisticated experiments called for the development of devices able to simulate, at least partly, the Space environment on Earth. Microgravity and weightlessness, respectively, may be simulated by two means: nullifying the g-vector and simulating free fall. A few selected examples are discussed, with particular emphasis on adherent mammalian cells.

5.4.3.1 Averaging the g-Vector

Clinorotation refers to the continuous rotation of an object around an axis perpendicular to the g-vector. The physical considerations on clinorotation are summarized in dedicated publications [129–131].

A clinostat had been used already at the end of the nineteenth century by von Sachs [132] and later by Morgan to investigate the gravitropism of plant roots. In the mid-1960s, the same concept was applied to cultured cells, adherent on a support or in suspension (for a review see Ref. [133]). For cell cultures the clinostat usually consists in a small diameter tubular container, fully filled with culture medium and rotated at constant speed. The random positioning machine (RPM) [134] is similar to the clinostat but allows continuous rotation about orthogonal axes together, and is, therefore, called a three-dimensional clinostat. The rotating wall vessel (RWV) is a specialized version of the clinostat. It consists of a horizontally rotated, fluid-filled culture vessel equipped with a gas permeable membrane that allows optimum gas exchange and oxygen supply.

In agreement with several observations in true microgravity, disorganization of the CSK by *g*-vector averaging is documented for several cell types [135–146] (Table 5.2). In some experiments, effects were recordable as soon as after 30 min of clinorotation and less dramatic after 20 h or more [141, 143, 144], suggesting that an adaptive process reverses the effects of simulated microgravity, possibly through the expression of heat shock proteins [144]. However, CSK remodelling was also observed after prolonged periods of clinorotation. *G*-vector averaging also impacts on the distribution of integrins [135] and focal contacts [147, 148] in osteoblastic cell lines, accelerates spreading and migration of endothelial cells [140] and decreases neurite extension in neuron-like cells [142]. Moreover, it prevents the self-patterning of purified tubulin [80].

Recently it was demonstrated that a complete disruption of actin stress fibres was induced in human mesenchymal stem cells cultured for seven days in RWV, due to a largely reduced activity of the small GTPase RhoA and phosphorylation of cofilin [137]. Forced expression of constitutively active RhoA restored actin stress fibres and phosphorylation of cofilin. Similar effects were demonstrated in bovine brain microvascular endothelial cells after three days in simulated microgravity, possibly due to inhibition of the expression of leukaemia-associated Rho guanine nucleotide exchange factor (LARG), a GEF for RhoA [149]. These studies suggest that the small GTPases participate in the cell response to (micro-) gravity.

As observed for several cells in true microgravity, *g*-vector averaging reduces the proliferation rate of several cell lines (Table 5.2). In MC3T3-E1 cells this reduction correlates with reduced expression of EGF-induced c-fos, in agreement with a similar effect observed in true microgravity using the same cells [93]. A similar reduction of c-fos was observed in A431 cells in true and simulated microgravity [110]. Signs of apoptosis have been recorded in several cells under *g*-vector averaging. They include DNA fragmentation, annexin V staining, loss of mitochondrial membrane potential, low level of anti-apoptotic signals as Akt and Bcl-2 expression, and increased levels of apoptotic signals as Fas-L, p53, Bax and caspase-3 activity. However, the reported kinetics of induction of apoptosis is largely variable. *G*-vector averaging induced morphological and biochemical evidences of apoptosis in C6 glioma cells and Leydig cells are seen after 30 min, but these signs were much reduced after 20–32 h [143, 144]. Again this recovery could depend on the

Table 5.2 Effects of modelled microgravity in ground laboratories on various cell types and cell-free systems. Abbreviations: ALP: alkaline phosphatase; BMP: bone morphogenetic protein; BSP: bone sialoprotein; Col I: collagen type I; Col III: collagen type III; eNOS: endothelial nitric oxide synthase; FN: fibronectin; iNOS: inducible nitric oxide synthase; LM: laminin; nNOS: neuronal nitric oxide synthase; OC: osteocalcin; OPN: osteopontin; PLA2: phospholipase A2; PLC: phospholipase C; PLD: phospholipase D; rec: receptor; SM: sphingomyelinase; TPA: tetradecanoyl phorbol acetate.

Cell type	*Effects of modelled microgravity*	*Ref.*
HU09 osteoblast-like	Reduced ALP and osteocalcin expression Increased expression of ALP by 1,25-dihydroxyvitamin D	151
MC3T3-E1	Reduced proliferation and EGF-induced c-Fos expression	93
ROS	CSK remodelling and altered distribution of β1 integrin Increased apoptosis (DNA fragmentation, annexin V staining)	135
ROS	Reduced proliferation. Increased ALP expression and activity, and OPN, OC, BMP-4, IL-6 expression. Osteonectin, BSPII and BMP-2 unaltered. Increased apoptosis, but p53, Bax/Bcl-2 and caspase-8 unaltered	156
ROS	Proliferation unaltered. Reduced focal contacts number and clustering	147, 148
ROS	Increased collagen expression, at the transcriptional level	157
MG-63	Reduced ALP, OC, Col 1A1 expression. Reduced responsiveness to 1,25-dihydroxyvitamin D. The analogue EB1089 partly compensates these effects	152
MC3T3-E1	Reduced ALP, OC and runx2 expression and AP-1 transactivation. Reduction in runx2 is likely due to hypoxia	150, 155
MC3T3-E1	Increased helical lysyl hydroxylase expression and activity	158
MC3T3-E1	Loss of mitochondrial membrane potential, low level of Akt and Bcl-2, and increased sensitivity to apoptogens Reduced osteoblast phenotype	159
2T3 preosteoblasts	Unaltered proliferation and shape. Decreased ALP expression and activity, and expression of runt-related TF2, osteomodulin, PTH-Receptor 1 Increased cathepsin K	160
Human osteoblasts cell line	Increased cell size. Delayed ALP detection, decreased CBFA1 and OC expression, p38 activity and bone nodule formation and calcification	136
Primary rat osteoblasts	Proliferation unaltered	
Normal human osteoblasts	Cell proliferation unaffected. Apoptosis unaffected. Increased Bax/Bcl-2 (proapoptotic) and XIAP (anti-apoptotic). No DNA fragmentation. No alteration in caspase-3 level	161
Human mesenchymal stem cells	Reduced ALP, Collagen type I, osteonectin and runt-related TF-2 expression Reduced ERK activation and induced p38 activation Increased markers of adipocyte differentiation	153

Table 5.2 *Continued*

Cell type	*Effects of modelled microgravity*	*Ref.*
Human mesenchymal stem cells	Increased expression of α2β1 integrin. Reduced phosphorylation of FAK and PYK2 Reduced expression of collagen type I Reduced activation of Ras and ERK	162
Human mesenchymal stem cells	Complete disruption of stress fibres and reduced phosphorylation of cofilin. Reduced expression of ALP, runt-related TF-2, collagen type I and RhoA activity. Increased adipocytic differentiation. Forced expression of active RhoA reverses the process	137
Rat marrow mesenchymal cells	Proliferation unaltered. Reduced ALP and ECM formation	163
Human bone marrow mesenchymal stem cells	Decreased proliferation. Increased size	138
Immortalized human fibroblasts	Altered filopodia	164
Human fibroblasts	Decreased Erb-B2 and p21 expression. Increased XRCC1 expression	165
Cardiac fibroblasts	Increased bFGF, angiotensin II and ERK1/2 expression, but decreased ERK1/2 activity. TGF-β1 unaltered	166
A431	Altered cell shape. Reduced EGF- and TPA-induced c-fos expression	110
Porcine aortic endothelial cells	Decreased cell number and PCNA expression. Reduced migration induced by VEGF. Increased expression of pro-apoptotic signals (Fas-L, p53, Bax). Reduced expression of Bcl-2 (antiapoptotic). Mitochondrial disassembly. Increased expression of Rank and Rank-L	167
Endothelial cells	Induced proliferation. CSK remodelling. Decreased expression of actin. No apoptosis. Increased expression of HSP70 and TIMP, and decreased expression of IL-1α. Unaltered MMP expression. Increased NO and prostacyclins production	139
Bovine aortic endothelial cells	Increased tight junction proteins expression and NO production	168
Human vascular endothelial cells	CSK remodelling. Accelerated spreading and migration Decreased proliferation	140
Human EA.hy296 (endothelial cells)	CSK remodelling. Morphological and biochemical signs of apoptosis. Increased Col I, FN, OPN, LM and Flk-1 expression	169
Bovine brain microvascular endothelial cells	CSK remodelling. Increased RhoA expression but reduced activity. Reduced LARG expression	149
1G11 (microvascular)	Reduced proliferation. Increased expression of eNOS and p21, and production of NO. Reduced expression of IL-6	170

Table 5.2 *Continued*

Cell type	*Effects of modelled microgravity*	*Ref.*
Human umbilical chord blood stem cells	Increased proliferation. Increased endothelial differentiation	171
ML-1 (follicular thyroid carcinoma)	Signs of apoptosis. Increased expression of Fas, p53, Bax and caspase-3 and decreased expression of Bcl-2 Increased expression of vimentin, vinculin, Col I, Col III, LM, FN, chondroitin sulfate and thyroid stimulating hormone receptor, and reduced secretion of fT3 and fT4	172, 173
ONCO-DG1 (papillary thyroid cancer cells)	Early CSK remodelling, reversed by 48 h. Signs of apoptosis. Increased expression of Col I, Col III, chondroitin sulfate, OPN, CD44, TGF-β1 and TGF-β receptor type II	141
PC12 and SH-SYSY (neuron-like cells)	Increased neurite extension. Cell aggregation. Induced survival. Induced NO production	142
PC12	Increased NO production, protein nitration and nNOS expression	174
C6 glial cells	CSK remodelling. Altered shape Signs of apoptosis. Effects recordable after 30 min. and less obvious after 20 h	143
Schneider S-1 (*Drosophila*)	Increased apoptosis. Altered mitochondria function. Mitochondria clustering, likely due to failure of transport along microtubules	175
Chicken embryo chondrocytes	CSK remodelling. Decreased ALP and $(Ca^{2+})i$. ATP synthase activity unaltered. Decreased ECM density and disordered ECM matrix fibres	145
DU145 (human prostate carcinoma)	Decreased proliferation. Reduced PLD activity. Increased PI-PLC, PLA2 and SMase activity. Reduced EGF, EGF-rec and TGF-β-rec expression. Induced collagen and laminin expression	176
L6 (rat myoblastic cells)	Unaltered proliferation. Reduced differentiation into myotubes, expression of myogenin and myosin heavy chain, and global protein ubiquitinylation. Reduced NF-κB binding activity and increased IκB level	154
SP-2/0 and 1D6 hybridoma cells	Reduced proliferation. Altered glycosaminoglycans accumulation. Increased $(Ca^{2+})i$. Nicotinic receptor desensitization but no down-regulation	177
Rat luteal cells	Signs of apoptosis. Altered mitochondria distribution activity. Decreased total cellular proteins expression and progesterone secretion	178
Mouse splenocytes	Reduced cell proliferation. Effect suppressed by nucleoside/nucleotide supplementation	179
Leydig cells	CSK remodelling. Reduced proliferation. Mitochondria disruption and signs of apoptosis. Reduced expression of 3β- and 17β-hydroxysteroid dehydrogenase. Increased HSP expression	144
STe (swine testis)	CSK remodelling. Reduced ALP expression	146
–	Reduced self-organization of microtubules	80

ability of the cells to synthesize protective heat shock proteins, and might indicate adaptation to altered gravity.

Numerous studies of the effect of g-vector averaging on cell phenotype (Table 5.2) affirm an altered expression of phenotypic markers and/or impaired differentiation from progenitors, in agreement with data from spaceflight experiments (Table 5.1). Reduced alkaline phosphatase activity and expression, osteocalcin and type I collagen by osteoblasts progenitors and/or osteoblastic cells have been described in several reports. These effects may be triggered in part by the reduced expression of transcription factors such as CBFA1/Runx2 [136, 150]. Alternatively, or additionally, they may be due to a reduced responsiveness to dihydroxyvitamin D3 [151, 152]. G-vector averaging commits mesenchymal stem cells to adipocytic differentiation rather than osteoblastic differentiation [137, 153] in relation with a reduced RhoA activity [137]. Reduced differentiation of rat L6 myoblastic cells into myotubes and expression of myogenin and myosin heavy chain have also been reported and correlate with an altered NF-kB pathway [154].

In a few reports [93, 110], the effects induced by g-vector averaging and those observed in real microgravity were similar, supporting the validity of the procedure, with subtle differences in the effects of microgravity and g-vector averaging being often interpreted as "approximated simulation". Moreover, negative results that may be unpublished could constitute a bias. Our personal experiments of clinorotation indeed failed to reproduce the induction of IL-6 expression observed in human dermal fibroblasts during the STS-95 flight [116]. Finally, the similarities between the effects obtained from true and simulated microgravity might be coincidental or artificial. The inhibition of CBFA1/runx2 expression, a master transcriptional regulator of osteoblast differentiation, in differentiating mouse osteoblasts cultured in RWV was reported to depend on hypoxia rather than on g-vector averaging [155]. These data open the question of adequate controls. For clinorotation of adherent cells, some authors and our self have used rotation along the vertical axis as a control providing similar motion, vibrations and shear-stress. Although other forces like hydrostatic pressure still differ in samples and controls, such a simple strategy is worth including whenever possible.

5.4.3.2 Free-fall Simulation

The free-fall machine (FFM) designed by Mesland and collaborators [180] is manufactured by CCM (Nuenen, The Netherlands). It produces periods of free-fall lasting approximately 800 ms followed by bounces to the top of the device at ~20×g on average and lasting 20–80 ms. The principle of this procedure is based on the assumption that the perception time (that is the minimal time required to sense the acceleration) of the cells is shorter than the duration of the free fall but longer than the duration of the bounces. The perception time for the *Avena* coleoptile and the *Lepidium* root is below 1 s [181]. It might explain how FFM induces in protoplasts an impaired distribution of the microtubules [182]. However, the perception time in animal cells, if existing, is unknown. Investigations on lymphocytes showed that FFM did not reproduce the lack of mitogenic activation observed in true microgravity [183].

5.4.3.3 Diamagnetic Levitation

Magnetism can impact any atom or molecule, provided the magnetic field is strong enough. Impressive records of levitating insects, strawberries or frogs are available on the web at http://www.hfml.ru.nl/levitation-movies.html. As they are levitating the objects are experiencing free fall, the gravitational force being compensated by that of the magnetic field acting on all atoms. Diamagnetic levitation altered self-organization of tubulin *in vitro* and produced patterns similar to that observed in weightlessness [80]. In addition to the instability of the levitating samples that could induce unwanted accelerations, there is concern on a possible impact of the electromagnetic field on the cells.

5.4.3.4 Hypergravity

Experimental strategies have been designed to identify and elucidate cell gravi-sensing mechanisms on Earth. Although it is unclear whether graviperception and mechanoperception use similar mechanisms and signalling pathways, several studies show similarities in the response of some cell types to hypergravity and mechanical stimulation. Besides the fact that a short period of hypergravity (1–5×g) is experienced by living organisms and cells during lift-off and re-entry of space-crafts, many studies indicate that hypergravity induces effects opposite to those generated by microgravity, supporting the "principle of continuity" [184]. Tissues and cells that react to hypergravity would be more likely to react to microgravity. Ground-based investigations by using low speed centrifugation are, therefore, recommended to select suitable biological materials and end-points for microgravity experiments.

Proliferation rates of mammalian cells are often, but not systematically [185, 186], increased in hypergravity [187–189] – sometimes at the expense of differentiation [190] and with a bi-directional response according to the magnitude of the applied hypergravity [191]. Several signalling pathways are stimulated under hypergravity, including early genes, transcription factors and gene expression [186, 188, 192–194]. Notably, hypergravity induces, as opposed to mechanical stimulation, a decrease in collagen and fibronectin synthesis [117, 118] and, surprisingly, an increased expression and activity of enzymes involved in collagen post-translational modifications [158]. Simulated microgravity induces the inverse effects [157]. In addition to these alterations of the biosynthetic phenotype, hypergravity produces changes in cell shape related to an increased density and thickness of actin fibres, increased focal adhesions and forced alignment of fibroblasts on grooves [195–197].

5.5 From Ground Research to Investigations in Microgravity

Much relevant biological information is required before having a chance of testing a potential effect of space flight on cell functions. This expertise should allow the definition of precise end-points suited for post-flight analyses.

5.5.1
Testable Hypotheses

In principle, all types of cells composing multicellular and unicellular organisms that exist on Earth should react to some extent to the loss of gravity. If this statement is correct, it means that (a) common cellular mechanism(s) is (are) involved in the process. Since each type of cells is differentiated, the end-point should be linked to its organ-specific trait. Analysis of cell signalling and pharmacology teaches us that parallel pathways often operate to create a specific phenotype. Similarly the activation of signalling pathways resulting in similar end-points might depend on multiple signals, with each recognized by specific membrane receptors for the ECM, for cytokines, for growth factors, and so forth. These different pathways cooperate after integration within the cells to control most of their functions [198]. The testable hypothesis can therefore use an end-point that means the variation of one function such as migration, multiplication, biosynthetic phenotype, to detect gravisensitive genes or measurable parameters of the signalling involved in these functions. A recent publication provides an excellent example of such an approach [149]. Simulated microgravity alters the organization of the actin cytoskeleton of microvascular endothelial cells that is visible (end-point 1) and modifies the activation of Rho (end-point 2) by down-regulating the expression of one GEF (end-point 3) identified as LARG. The proof of concept of the involvement of this GEF in the organization of the actin fibres could be obtained by its suppression by a small interfering RNA. Better knowledge of the mechanism operating in LARG or additional GEF down-regulation would create more end-points, allowing a closer approach to the mechanism(s) of graviperception.

Additional strategies can be used to test hypotheses, such as, for example, the use of cells that have been engineered to analyze the implication of one specific protein in the signalling pathway potentially targeted by microgravity. Cells expressing the different members of the RhoGTPases family under their constitutively activated form or deleted by siRNA have been created for this purpose in our research team [53, 54, 56]. RhoGTPases-dependent end-points [55] known to be modulated at 1×g (migration for example) can be measured after exposure to microgravity to determine their implication in the receptivity–reactivity of the cells. Pharmacological modulation of cells [58] can also create additional models, provided a defined end-point susceptible to microgravity is available.

The ultimate goal of these investigations is a clear definition of the pathway(s) by which gravity or its absence modulate the activity of defined cells so as to devise progressively more specific means to counteract their deleterious effects in Space and their implication for similar diseases on Earth.

5.5.2
Experimental Strategy and Constraints

Many aspects of the experimental strategy need to be considered when investigating microgravity on adherent cells in culture. To date, the cultures prepared on ground (at 1×g) need to consider the goal of the experiment in terms of the mechanism that will be submitted to the effect of microgravity. There are several types of culture units to be used in various types of incubators, allowing the control of temperature and atmosphere (see Introduction). Most of these set-ups are designed to allow one or more changes of medium including fluid for preserving the biological material at the end of the experimental period in microgravity. Centrifuges producing 1×g are available in some of these incubators and are most useful as an internal control. This should allow the identification of confounding variables produced, for example, by the high level of vibrations and acceleration during take off, the low level of vibration in space and the effect of the cosmic radiations on the cell behaviour [199]. The sealed experimental culture units are transported to the launch site at the lowest temperature (between 22 and 25 °C), allowing survival of the biological sample for often an extended period of time (several days). The maintenance of cells in a "sleeping mode" at low temperature might protect them from disturbing events such as hypergravity and vibrations occurring during launch. Recovery of the full biological functions of the quiescent cells, when microgravity has been reached, is obtained by raising the temperature to 37 °C. This recovery period will take some time and is, therefore, a confounding factor in estimating the effect of microgravity. The delay required for this recovery and the impact on the quality of the samples needs to be ascertained on ground. Medium refreshment, addition of biological compounds, washing and fixation needs to be precisely scheduled. Finally, depending on the experimental goal the preserved samples will require a suitable conservation temperature. It will depend on the facilities on board. Most often the time between landing and collection of the samples will be hours to days. The effect of this delay on the quality of the recovered samples also needs much attention.

5.5.3
The Future

The use of the BIOLAB facility in the European Columbus Module on board the International Space Station will allow experiments starting from frozen cells. This will partly suppress the stress related to the preparation on Earth and the vibrations and acceleration in the transportation vessel. It will, however, introduce new constraints.

The procedure for cryopreservation of the cells needs to be adapted to the modalities of their use in space. No standard procedure will suit all strains of cells [200, 201]. It should be investigated and adapted to the experimental protocol.

A second control experiment will have to be performed for investigations using adherent cells. On Earth the gravitational force (1×g) allows a rapid sedimentation of the cells, their attachment to the support and spreading. In microgravity, centrifugation will be required to achieve plating at a suitable efficiency. To prevent compromising the expected effect of microgravity, centrifugation should be restricted to a period of time that has to be experimentally determined beforehand. Morphological investigations on Earth represent examples of end-points that can be used to test this parameter. Observation of similar changes in microgravity upon centrifugation requires the use of a suitable microscope and online transfer of the image to the experimenter. This will, hopefully, soon be available. Morphological examination in microgravity will open the possibility of viewing the movements of the cells and the mobility of their organelles by labelling constitutive proteins with visible tags. The endless imagination of scientists promises a bright future for microgravity research.

References

1 Hughes-Fulford, M. *J. Gravit. Physiol.* **2004**, *11*, 105–109.
2 Vico, L., Collet, P., Guignandon, A., Lafage-Proust, M.H., Thomas, T., Rehaillia, M., Alexandre, C. *Lancet* **2000**, *355*, 1607–1611.
3 David, V., Lafage-Proust, M.H., Laroche, N., Christian, A., Ruegsegger, P., Vico, L. *Am. J. Physiol. Endocrinol. Metab.* **2006**, *290*, E440–E447.
4 Stein, T.P., Wade, C.E. *J. Nutr.* **2005**, *135*, 1824S–1828S.
5 Nikawa, T., Ishidoh, K., Hirasaka, K., Ishihara, I., Ikemoto, M., Kano, M., Kominami, E., Nonaka, I., Ogawa, T., Adams, G.R., Baldwin, K.M., Yasui, N., Kishi, K., Takeda, S. *FASEB J.* **2004**, *18*, 522–524.
6 Carlsson, S.I., Bertilaccio, M.T., Ballabio, E., Maier, J.A. *Biochim. Biophys. Acta* **2003**, *1642*, 173–179.
7 Drummer, C., Gerzer, R., Baisch, F., Heer, M. *Pflugers Arch.* **2000**, *441*, R66–R72.
8 Aubert, A.E., Beckers, F., Verheyden, B. *Acta Cardiol.* **2005**, *60*, 129–151.
9 Zerwekh, J.E. *Nutrition* **2002**, *18*, 857–863.
10 Wang, E. *FASEB J.* **1999**, *13*, S167–S174.
11 Williams D.R. *Annu. Rev. Med.* **2003**, *54*, 245–256.
12 Mortimer, A.J., DeBakey, M.E., Gerzer, R., Hansen, R., Sutton, J., Neiman, S.N. *Acta Astronaut.* **2004**, *54*, 805–812.
13 Gramse, V., De Groote, A., Paiva, M. *Ann. Biomed. Eng.* **2003**, *31*, 152–158.
14 Vassy, J., Portet, S., Beil, M., Millot, G., Fauvel-Lafeve, F., Gasset, G., Schoevaert, D. *Adv. Space Res.* **2003**, *32*, 1595–1603.
15 Ingber, D.E. *Int. J. Dev. Biol.* **2006**, *50*, 255–266.
16 Chen, C.S., Tan, J., Tien, J. *Annu. Rev. Biomed. Eng.* **2004**, *6*, 275–302.
17 Lehoux, S., Tedgui, A. *J. Biomech.* **2003**, *36*, 631–643.
18 Burridge, K., Fath, K., Kelly, T., Nuckolls, G., Turner, C. *Annu. Rev. Cell Biol.* **1988**, *4*, 487–525.
19 Wang, N., Butler, J.P., Ingber, D.E. *Science* **1993**, *260*, 1124–1127.
20 Choquet, D., Felsenfeld, D.P., Sheetz, M.P. *Cell* **1997**, *88*, 39–48.
21 Kano, Y., Katoh, K., Fujiwara, K. *Circ. Res.* **2000**, *86*, 425–433.
22 Cukierman, E., Pankov, R., Stevens, D.R., Yamada, K.M. *Science* **2001**, *294*, 1708–1712.
23 Robles, E., Gomez, T.M. *Nat. Neurosci.* **2006**, *9*, 1274–1283.
24 Miyamoto, S., Teramoto, H., Coso, O.A., Gutkind, J.S., Burbelo, P.D., Akiyama, S.K., Yamada, K.M. *J. Cell. Biol.* **1995**, *131*, 791–805.
25 Brakebusch, C., Fassler, R. *EMBO J.* **2003**, *22*, 2324–2333.
26 Chrzanowska-Wodnicka, M., Burridge, K. *J. Cell Biol.* **1996**, *133*, 1403–1415.
27 Balaban, N.Q., Schwarz, U.S., Riveline, D., Goichberg, P., Tzur, G., Sabanay, I., Mahalu, D., Safran, S., Bershadsky, A., Addadi, L., Geiger, B. *Nat. Cell Biol.* **2001**, *3*, 466–472.
28 Riveline, D., Zamir, E., Balaban, N.Q., Schwarz, U.S., Ishizaki, T., Narumiya, S., Kam, Z., Geiger, B., Bershadsky, A.D. *J. Cell. Biol.* **2001**, *153*, 1175–1186.

29 Galbraith, C.G., Yamada, K.M., Sheetz, M.P. *J. Cell Biol.* **2002**, *159*, 695–705.

30 Mitra, S.K., Hanson, D.A., Schlaepfer, D.D. *Nat. Rev. Mol. Cell Biol.* **2005**, *6*, 56–68.

31 Panetti, T.S. *Front. Biosci.* **2002**, 7, d143–d150.

32 Legate, K.R., Montanez, E., Kudlacek, O., Fassler, R. *Nat. Rev. Mol. Cell Biol.* **2006**, *7*, 20–31.

33 Burridge, K., Sastry, S.K., Sallee, J.L. *J. Biol. Chem.* **2006**, *281*, 15593–15596.

34 Johnson, R.P., Craig, S.W. *Nature* **1995**, *373*, 261–264.

35 Plopper, G.E., McNamee, H.P., Dike, L.E., Bojanowski, K., Ingber, D.E. *Mol. Biol. Cell* **1995**, *6*, 1349–1365.

36 Schneller, M., Vuori, K., Ruoslahti, E. *EMBO J.* **1997**, *16*, 5600–5607.

37 Chan, P.C., Chen, S.Y., Chen, C.H., Chen, H.C. *J. Biomed. Sci.* **2006**, *13*, 215–223.

38 Ruoslahti, E. *Science* **1997**, *276*, 1345–1346.

39 Iozzo, R.V. *J. Clin. Invest.* **2001**, *108*, 165–167.

40 Vogel, W.F., Abdulhussein, R., Ford, C.E. *Cell Signal.* **2006**, *18*, 1108–1116.

41 Hemler, M.E. *Nat. Rev. Mol. Cell Biol.* **2005**, *6*, 801–811.

42 Anastasiadis, P.Z. *Biochim. Biophys. Acta* **2007**, *1773*, 34–46.

43 Brembeck, F.H., Rosario, M., Birchmeier, W. *Curr. Opin. Genet. Dev.* **2006**, *16*, 51–59.

44 Van Roy, F.M., McCrea, P.D. *Nat. Rev. Cancer* **2005**, *5*, 956–964.

45 Ryan, P.L., Foty, R.A., Kohn, J., Steinberg, M.S. *Proc. Natl. Acad. Sci. U.S.A.* **2001**, *98*, 4323–4327.

46 Alenghat, F.J., Ingber, D.E. *Sci. STKE* **2002**, *119*, PE6.

47 Walczak, C.E. *Opin. Cell. Biol.* **2000**, *12*, 52–56.

48 Ridley, A.J. *Trends Cell Biol.* **2006**, *16*, 522–529.

49 Narumiya, S., Yasuda, S. *Curr. Opin. Cell Biol.* **2006**, *18*, 199–205.

50 Walker, K., Olson, M. *Curr. Opin. Genet. & Develop.* **2005**, *15*, 62–68.

51 Zhao, Z.S., Manser, E. *Biochem. J.* **2005**, *386*, 201–214.

52 Deroanne, C., Vouret-Craviari, V., Wang, B., Pouyssegur, J. *J. Cell. Sci.* **2003**, *116*, 1367–1376.

53 Deroanne, C., Hamelryckx, D., Ho, G.T.T., Lambert, C., Catroux, P., Lapière, Ch.M., Nusgens, B. *J. Cell Sci.* **2005**, *118*, 1173–1183.

54 Servotte, S., Zhang, Z., Lambert, Ch.A., Ho, T.T.G., Chometon, G., Eckes, B., Krieg, Th., Lapière Ch.M., Nusgens, B.V., Aumailley, M. *Protoplasma* **2006**, *229*, 215–220.

55 Nusgens, B.V., Chometon, G., Guignandon, A., Ho, G., Lambert, Ch., Mineur, P., Servotte, S., Zhang, Z., Deroanne, C., Eckes, B., Vico, L., Krieg, Th., Aumailley, M., Lapière Ch.M. *J. Gravit. Physiol.* **2005**, *12*, 269–270.

56 Zhang, Z., Lambert, Ch.L., Servotte, S., Chometon, G., Eckes, B., Krieg, Th., Lapière, Ch.M., Nusgens, B.V., Aumailley, M. *Cell. Mol. Life Sci.* **2006**, *63*, 82–91.

57 Fukuyama, R., Fujita, T., Azuma, Y., Hirano, A., Nakamuta, H., Koida, M., Komori, T. *Biochem. Biophys. Res. Commun.* **2004**, *315*, 636–642.

58 Chaplet, M., Detry, C., Deroanne, C., Fisher, L.W., Castronovo, V., Bellahcene, A. *Biochem. J.* **2004**, *384*, 591–598.

59 Pelham, R.J., Jr., Wang, Y. *Proc. Natl. Acad. Sci. U.S.A.* **1997**, *94*, 13661–13665.

60 Katz, B.Z., Zamir, E., Bershadsky, A., Kam, Z., Yamada, K.M., Geiger, B. *Mol. Biol. Cell* **2000**, *11*, 1047–1060.

61 Deroanne, C., Lapière, Ch.M., Nusgens, B. *Cardiovasc. Res.* **2001**, *49*, 647–658.

62 Ingber, D.E. *FASEB J.* **2006**, *20*, 811–827.

63 Mauch, C., Adelmann-Grill, B., Hatamochi, A., Krieg, T. *FEBS Lett.* **1989**, *250*, 301–305.

64 Lambert, Ch.A., Soudant, E.P., Nusgens, B.V, Lapiere, Ch.M. *Lab Invest.* **1992**, *66*, 444–451.

65 Lambert, Ch.A., Lapière, Ch.M., Nusgens, B.V. *J. Biol. Chem.* **1998**, *273*, 23143–23149.

66 Lambert, Ch.A., Colige, A., Munaut, C., Lapière, Ch.M., Nusgens, B.V. *Matrix Biol.* **2001**, *20*, 397–408.

67 Lambert, Ch.A., Colige, A., Lapière, Ch. M., Nusgens, B.V. *Eur. J. Cell Biol.* **2001**, *80*, 479–485.

68 Eckes, B., Hunzelmann, N., Ziegler-Heitbrock, H.W., Urbanski, A., Luger, T., Krieg, T., Mauch, C. *FEBS Lett.* **1992**, *298*, 229–232.

69 MacKenna, D.A., Dolfi, F., Vuori, K., Ruoslahti, E. *J. Clin. Invest.* **1998**, *101*, 301–310.

70 Guignandon, A., Lafage-Proust, M.H., Usson, Y., Laroche, N., Caillot-Augusseau, A., Alexandre, C., Vico, L. *FASEB J.* **2001**, *15*, 2036–2038.

71 Li, W., Duzgun, A., Sumpio, B.E., Basson, M.D. *Am. J. Physiol. Gastrointest. Liver Physiol.* **2001**, *280*, G75–G87.

72 Katsumi, A., Naoe, T., Matsushita, T., Kaibuchi, K., Schwartz, M.A. *J. Biol. Chem.* **2005**, *280*, 16546–16549.

73 Chiquet, M., Renedo, A.S., Huber, F., Fluck, M. *Matrix Biol.* **2003**, *22*, 73–80.

74 Delvoye, P., Wiliquet, P., Leveque, J.L., Nusgens, B.V., Lapière, Ch.M. *J. Invest. Dermatol.* **1991**, *97*, 898–902.

75 Fujieda, S., Mogami, Y., Moriyasu, K., Mori, Y. *Adv. Space Res.* **1999**, *23*, 2057–2063.

76 Miele, A.E., Federici, L., Sciara, G., Draghi, F., Brunori, M., Vallone, B. *Acta Crystallogr. D. Biol. Crystallogr.* **2003**, *59*, 982–988.

77 Maccarrone, M., Bari, M., Battista, N., Finazzi-Agro, A. *Biophys. Chem.* **2001**, *90*, 303–306.

78 Terzyan, S.S., Bourne, C.R., Ramsland, P.A., Bourne, P.C., Edmundson, A.B. *J. Mol. Recognit.* **2003**, *16*, 83–90.

79 Vahedi-Faridi, A., Porta, J., Borgstahl, G.E. *Acta Crystallogr. D. Biol. Crystallogr.* **2003**, *59*, 385–388.

80 Glade, N., Beaugnon, E., Tabony, J. *Biophys. Chem.* **2006**, *121*, 1–6.

81 Fujioka, A., Terai, K., Itoh, R., Aoki, K., Nakamura, T., Kuroda, S., Nishida, E., Matsuda, M. *J. Biol. Chem.* **2006**, *281*, 8917–8926.

82 Mayawala, K., Gelmi, C.A., Edwards, J.S. *Biophys J.* **2004**, *87*, L01–L02.

83 Qian, H., Reluga, T.C. *Phys. Rev. Lett.* **2005**, *94*, 28101.

84 Mesland, D. *Adv. Space Res.* **1992**, *12*, 15–25.

85 Papaseit, C., Pochon, N., Tabony, J. Proc. *Natl. Acad. Sci. U.S.A.* **2000**, *97*, 8364–8368.

86 Moore D., Bie, P., Oser, A. (Eds.), *Biological and Medical Research in Space – An overview of Life Sciences Research in Microgravity, ESA*, pp. 1–106, Springer, Berlin-Heidelberg, **1996**.

87 Guignandon, A., Vico, L., Alexandre, C., Lafage-Proust, M.H. *Cell. Struct. Funct.* **1995**, *20*, 369–375.

88 Guignandon, A., Usson, Y., Laroche, N., Lafage-Proust, M.H., Sabido, O., Alexandre, C., Vico, L. *Exp. Cell. Res.* **1997**, *236*, 66–75.

89 Hughes-Fulford, M., Rodenacker, K., Jutting, U. *J. Cell. Biochem.* **2006**, *99*, 435–449.

90 Hughes-Fulford, M. *J. Gravit. Physiol.* **2001**, *8*, P1–P4.

91 Hughes-Fulford, M. *Adv. Space Res.* **2003**, *3*, 585–593.

92 Hughes-Fulford, M., Lewis, M.L. *Exp. Cell. Res.* **1996**, *224*, 103–109.

93 Sato, A., Hamazaki, T., Oomura, T., Osada, H., Kakeya, M., Watanabe, M., Nakamura, T., Nakamura, Y., Koshikawa, N., Yoshizaki, I., Aizawa, S., Yoda, S., Ogiso, A., Takaoki, M., Kohno, Y., Tanaka, H. *Adv. Space Res.* **1999**, *24*, 807–813.

94 Kumei, Y., Shimokawa, H., Katano, H., Hara, E., Akiyama, H., Hirano, M., Mukai, C., Nagaoka, S., Whitson, P.A., Sams, C.F. *J. Biotechnol.* **1996**, *47*, 313–324.

95 Kumei, Y., Shimokawa, H., Katano, H., Akiyama, H., Hirano, M., Mukai, C., Nagaoka, S., Whitson, P.A., Sams, C.F. *J. Appl. Physiol.* **1998**, *85*, 139–147.

96 Kumei, Y., Akiyama, H., Hirano, M., Shimokawa, H., Morita, S., Mukai, C., Nagaoka, S. *Biol. Sci. Space* **1999**, *13*, 142–143.

97 Kumei, Y., Nakamura, H., Morita, S., Akiyama, H., Hirano, M., Ohya, K., Shinomiya, K., Shimokawa, H. *Ann. New York Acad. Sci.* **2002**, *973*, 75–78.

98 Kumei, Y., Morita, S., Shimokawa, H., Ohya, K., Akiyama, H., Hirano, M., Sams, C.F., Whitson, P.A. *Ann. New York Acad. Sci.* **2003**, *1010*, 476–480.

99 Kumei, Y., Morita, S., Nakamura, H., Akiyama, H., Hirano, M., Shimokawa, H., Ohya, K. *Ann. New York Acad. Sci.* **2003**, *1010*, 481–485.

100 Kumei, Y., Morita, S., Nakamura, H., Akiyama, H., Katano, H., Shimokawa, H., Ohya, K. *Ann. New York Acad. Sci.* **2004**, *1030*, 116–120.

101 Kumei, Y., Morita, S., Nakamura, H., Katano, H., Ohya, K., Shimokawa, H., Sams, C.F., Whitson, P.A. *Ann. New York Acad. Sci.* **2004**, *1030*, 121–124.

102 Akiyama, H., Kanai, S., Hirano, M., Shimokawa, H., Katano, H., Mukai, C., Nagaoka, S., Morita, S., Kumei, Y. *Mol. Cell. Biochem.* **1999**, *202*, 63–71.

103 Landis, W.J., Hodgens, K.J., Block, D., Toma, C.D., Gerstenfeld, L.C. *J. Bone Miner. Res.* **2000**, *15*, 1099–1112.

104 Kacena, M.A., Todd, P., Landis, W.J. *In Vitro Cell. Dev. Biol. Anim.* **2003**, *39*, 454–459.

105 Carmeliet, G., Nys G., Bouillon, R. *J. Bone Miner. Res.* **1997**, *12*, 786–794.

106 Carmeliet, G. Nys, G., Stockmans, I., Bouillon, R. *Bone* **1998**, *22*, 139S–143S.

107 Hymer, W.C., Shellenberger, K., Grindeland, R. *Adv. Space Res.* **1994**, *14*, 61–70.

108 Vassy, J., Portet, S., Beil, M., Millot, G., Fauvel-Lafeve, F., Karniguian, A., Gasset, G., Irinopoulou, T., Calvo, F., Rigaut, J.P., Schoevaert, D. *FASEB J.* **2001**, *15*, 1104–1106.

109 de Groot, R.P., Rijken, P.J., den Hertog, J., Boonstra, J., Verkleij, A.J., De Laat, S.W., Kruijer, W. *Exp. Cell. Res.* **1991**, *197*, 87–90.

110 Rijken, P.J., de Groot, R.P., Kruijer, W., De Laat, S.W., Verkleij, A.J., Boonstra, J. *Adv. Space Res.* **1992**, *12*, 145–152.

111 Rijken, P.J., de Groot, R.P., van Belzen, N., De Laat, S.W., Boonstra, J., Verkleij, AJ. *Exp. Cell Res.* **1993**, *204*, 373–377.

112 Rijken, P.J., Boonstra, J., Verkleij, A.J., de Laat, S.W. *Adv. Space Biol. Med.* **1994**, *4*, 159–188.

113 Boonstra, J. *FASEB J.* **1999**, *3(Suppl.)*, S35–S42.

114 Tairbekov, M.G., Margolis, L.B., Baibakov, B.A., Gabova, A.V., Dergacheva, G.B. *Izv. Akad. Nauk, Ser. Biol.* **1994**, *5*, 745–750.

115 Guignandon, A., Lambert, Ch., Réga, G., Laroche, N., Heyeres, A., Lapière, Ch.M., Vico, L., Nusgens, B.V. *J. Gravit. Physiol.* **2005**, *12*, 239–240.

116 Lambert, Ch.A., Guignandon, A., Colige, A., Rega, G., Laroche, N., Heyeres A., Munaut, C., Lapière, Ch.M., Nusgens, B.V. in *Space Scientific Research in Belgium, Microgravity Ed. OSTC*, Brussels, **2001**, Vol. *1*, pp. 167–177.

117 Seitzer, U., Bodo, M., Muller, P.K., Acil, Y., Batge, B. *Cell Tissue Res.* **1995**, *282*, 513–517.

118 Seitzer, U., Bodo, M., Muller, P.K. *Adv. Space Res.* **1995**, *16*, 235–238.

119 Semov, A., Semova, N., Lacelle, C., Marcotte, R., Petroulakis, E., Proestou, G., Wang, E. *FASEB J.* **2002**, *16*, 899–901.

120 Gaboyard, S., Blanchard, M.P., Travo, C., Viso, M., Sans, A., Lehouelleur, J. *Neuroreport.* **2002**, *13*, 2139–2142.

121 Sato, A., Kumei, Y., Sato, K., Hongo, T., Hamazaki, T., Masuda, I., Nakajima, T., Ohmura, T., Kaiho, M., Sato, T., Wake, K. *Biol. Sci. Space* **2001**, *15*, S61–S63.

122 Nomura, J., Himeda, J., Chen, Z., Sugaya, S., Takahashi, S., Kita, K., Ichinose, M., Suzuki, N. *J. Radiat. Res. (Tokyo)* **2002**, *43*, S251–S255.

123 Tabony, J., Pochon, N., Papaseit, C. *Adv. Space Res.* **2001**, *28*, 529–535.

124 Takahashi, A., Ohnishi, K., Takahashi, S., Masukawa, M., Sekikawa, K., Amano, T., Nakano, T., Nagaoka, S., Ohnishi, T. *Int. J. Radiat. Biol.* **2000**, *76*, 783–788.

125 Ranaldi, F., Vanni, P., Giachetti, E. *Biophys. Chem.* **2003**, *103*, 169–177.

126 Lambert, Ch.A., Nusgens, B.V., Lapière, Ch.M. *Adv. Space Res.* **1998**, *21*, 1081–1091.

127 Usson, Y., Guignandon, A., Laroche, N., Lafage-Proust, M.H., Vico, L. *Cytometry* **1997**, *28*, 298–304.

128 Stein, T.P., Schluter, M.D. *Am. J. Physiol.* **1994**, *266*, E448–E452.

129 Albrecht-Buehler, G. *ASGSB Bull.* **1992**, *52*, 3–10.

130 Kessler, J.O. *ASGSB Bull.* **1992**, *5*, 11–21.

131 Klaus, D.M. *Gravit. Space Biol. Bull.* **2001**, *14*, 55–64.

132 Sachs, J. *Arb. Bot. Inst. Würzburg* **1873**, *1*, 385–474.

133 Briegleb, W. *ASGSB Bull.* **1992**, *5*, 23–30.

134 Hoson, T., Kamisaka, S., Masuda, Y., Yamashita, M. *Micrograv. Sci. Technol.* **1992**, *6*, 278–281.

135 Sarkar, D., Nagaya, T., Koga, K., Kambe, F., Nomura, Y., Seo, H. *J. Gravit. Physiol.* **2000**, *7*, P71–P72.

136 Yuge, L., Hide, I., Kumagai, T., Kumei, Y., Takeda, S., Kanno, M., Sugiyama, M., Kataoka, K. *In Vitro Cell Dev. Biol. Anim.* **2003**, *39*, 89–97.

137 Meyers, V.E., Zayzafoon, M., Douglas, J.T., McDonald, J.M. *J. Bone Miner. Res.* **2005**, *20*, 1858–1866.

138 Merzlikina, N.V., Buravkova, L.B., Romanov, Y.A. *J. Gravit. Physiol.* **2004**, *11*, P193–P194.

139 Carlsson, S.I., Bertilaccio, M.T., Ascari, I., Bradamante, S., Maier, J.A. *J. Gravit. Physiol.* **2002**, *9*, P273–P274.

140 Buravkova, L.B., Romanov, Y.A. *Acta Astronaut.* **2001**, *48*, 647–650.

141 Infanger, M., Kossmehl, P., Shakibaei, M., Bauer, J., Kossmehl-Zorn, S., Cogoli, A., Curcio, F., Oksche, A., Wehland, M., Kreutz, R., Paul, M., Grimm, D. *Cell Tissue Res.* **2006**, *324*, 267–277.

142 Wang, S.S., Good, T.A. *J. Cell. Biochem.* **2001**, *83*, 574–584.

143 Uva, B.M., Masini, M.A., Sturla, M., Prato, P., Passalacqua, M., Giuliani, M., Tagliafierro, G., Strollo, F. *Brain Res.* **2002**, *934*, 132–139.

144 Uva, B.M., Strollo, F., Ricci, F., Pastorino, M., Mason, J.I., Masini, M.A. *J. Endocrinol. Invest.* **2005**, *28*, 84–91.

145 Zhang, X., Li, X.B., Yang, S.Z., Li, S.G., Jiang, P.D., Lin, Z.H. *Adv. Space Res.* **2003**, *32*, 1577–1583.

146 Strollo, F., Masini, M.A., Pastorino, M., Ricci, F., Vadrucci, S., Cogoli-Greuter, M., Uva, B.M. *J. Gravit. Physiol.* **2004**, *11*, P187–P188.

147 Guignandon, A., Akhouayri, O., Laroche, N., Lafage-Proust, M.H., Alexandre, C., Vico, L. *Adv. Space Res.* **2003**, *32*, 1561–1567.

148 Guignandon, A., Akhouayri, O., Usson, Y., Rattner, A., Laroche, N., Lafage-Proust, M.H., Alexandre, C., Vico, L. *Cell. Commun. Adhes.* **2003**, *10*, 69–83.

149 Higashibata, A., Imamizo-Sato, M., Seki, M., Yamazaki, T., Ishioka, N. *BMC Biochem.* **2006**, *7*, 19.

150 Ontiveros, C., McCabe, L.R. *J. Cell. Biochem.* **2003**, *88*, 427–437.

151 Kunisada, T., Kawai, A., Inoue, H., Namba, M. *Acta Med. Okayama* **1997**, *51*, 135–140.

152 Narayanan, R., Smith, C.L., Weigel, N.L. *Bone* **2002**, *31*, 381–388.

153 Zayzafoon, M., Gathings, W.E., McDonald, J.M. *Endocrinology* **2004**, *145*, 2421–2432.

154 Hirasaka, K., Nikawa, T., Yuge, L., Ishihara, I., Higashibata, A., Ishioka, N., Okubo, A., Miyashita, T., Suzue, N., Ogawa, T., Oarada, M., Kishi, K. *Biochim. Biophys. Acta* **2005**, *1743*, 130–140.

155 Ontiveros, C., Irwin, R., Wiseman, R.W., McCabe, L.R. *J. Cell. Physiol.* **2004**, *200*, 169–176.

156 Rucci, N., Migliaccio, S., Zani, B.M., Taranta, A., Teti, A. *J. Cell. Biochem.* **2002**, *5*, 167–179.

157 Dai, Z., Li, Y., Ding, B., Zhang, Y., Liu, W., Liu, P. *Sci. China C. Life. Sci.* **2004**, *47*, 203–210.

158 Saito, M., Soshi, S., Fujii, K. *J. Bone. Miner. Res.* **2003**, *18*, 1695–1705.

159 Bucaro, M.A., Fertala, J., Adams, C.S., Steinbeck, M., Ayyaswamy, P., Mukundakrishnan, K., Shapiro, I.M., Risbud, M.V. *Ann. New York Acad. Sci.* **2004**, *1027*, 64–73.

160 Pardo, S.J., Patel, M.J., Sykes, M.C., Platt, M.O., Boyd, N.L., Sorescu, G.P., Xu, M., van Loon, J.J., Wang, M.D., Jo, H. *Am. J. Physiol. Cell. Physiol.* **2005**, *288*, C1211–C1221.

161 Nakamura, H., Kumei, Y., Morita, S., Shimokawa, H., Ohya, K., Shinomiya, K. *Ann. New York Acad. Sci.* **2003**, *1010*, 143–147.

162 Meyers, V.E., Zayzafoon, M., Gonda, S.R., Gathings, W.E., McDonald, J.M. *J. Cell. Biochem.* **2004**, *93*, 697–707.

163 Nishikawa, M., Ohgushi, H., Tamai, N., Osuga, K., Uemura, M., Yoshikawa, H., Myoui, A. *Cell Transplant.* **2005**, *4*, 829–835.

164 Larina, O.N., Sidorenko, L.A., Moshkov, D.A., Pogorelov, A.G., Pavlik, L.L., Arutyunyan, A.V., Grivennikov, I.A., Manuilova, E.S., Inozemtseva, L.S., Umarkhodzhaev, R.M., Pivkin, A.N. *J. Gravit. Physiol.* **2002**, *9*, P287–P288.

165 Arase, Y., Nomura, J., Sugaya, S., Sugita, K., Kita, K., Suzuki, N. *Cell. Biol. Int.* **2002**, *26*, 225–233.

166 Yang, F., Li, Y.H., Ma, Y.J., Zhong, P., Song, J.P., Dai, Z.Q. *Space Med. Med. Eng. (Beijing)* **2003**, *16*, 532–537.

167 Morbidelli, L., Monici, M., Marziliano, N., Cogoli, A., Fusi, F., Waltenberger, J., Ziche, M. *Biochem. Biophys. Res. Commun.* **2005**, *334*, 491–499.

168 Sanford, G.L., Ellerson, D., Melhado-Gardner, C., Sroufe, A.E., Harris-Hooker, S. *In Vitro Cell Dev. Biol. Anim.* **2002**, *38*, 493–504.

169 Infanger, M., Kossmehl, P., Shakibaei, M., Baatout, S., Witzing, A., Grosse, J., Bauer, J., Cogoli, A., Faramarzi, S., Derradji, H., Neefs, M., Paul, M., Grimm, D. *Apoptosis* **2006**, *11*, 749–764.

170 Cotrupi, S., Ranzani, D., Maier, J.A. *Biochim. Biophys. Acta* **2005**, *1746*, 163–168.

171 Chiu, B., Wan, J.Z., Abley, D., Akabutu, J. *Acta Astronaut.* **2005**, *56*, 918–922.

172 Grimm, D., Bauer, J., Kossmehl, P., Shakibaei, M., Schoberger, J., Pickenhahn, H., Schulze-Tanzil, G., Vetter, R., Eilles, C., Paul, M., Cogoli, A. *FASEB J.* **2002**, *16*, 604–606.

173 Kossmehl, P., Shakibaei, M., Cogoli, A., Pickenhahn, H., Paul, M., Grimm, D. *J. Gravit. Physiol.* **2003**, *9*, P295–P296.

174 Qu, L., Yang, T., Yuan, Y., Zhong, P., Li, Y. *Nitric Oxide* **2006**, *15*, 58–63.

175 Schatten, H., Lewis, M.L., Chakrabarti, A. *Acta Astronaut* **2001**, *49*, 399–418.

176 Clejan, S., O'Connor, K., Rosensweig, N. *J. Cell. Mol. Med.* **2001**, *5*, 60–73.

177 Skok, M.V., Koval, L.M., Petrova, Y.I., Lykhmus, O.Y., Kolibo, D.V., Romanyuk, S.I., Yevdokimova, N.Y., Komisarenko, S.V. *Acta Astronaut* **2005**, *56*, 721–728.

178 Yang, H., Bhat, G.K., Sridaran, R. *Biol. Reprod.* **2002**, *66*, 770–777.

179 Hales, N.W., Yamauchi, K., Alicea, A., Sundaresan, A., Pellis, N.R., Kulkarni, A.D. *In Vitro Cell Dev. Biol. Anim.* **2002**, *38*, 213–217.

180 Mesland, D.A., Anton, A.H., Willemsen, H., van den Ende, H. *Microgravity Sci. Technol.* **1996**, *9*, 10–14.

181 Sievers, A., Hejnowicz, Z. *ASGSB Bull.* **1992**, *5*, 69–75.

182 Skagen, E.B., Iversen, T.H. *In Vitro Cell. Dev. Biol. Plant.* **2000**, *36*, 312–318.

183 Schwarzenberg, M., Pippia, P., Meloni, M.A., Cossu, G., Cogoli-Greuter, M., Cogoli, A. *J. Gravit. Physiol.* **1998**, *5*, P23–P26.

184 Wade, C.E. *Adv. Space Biol. Med.* **2005**, *10*, 225–245.

185 Hirasaka, K., Nikawa, T., Asanoma, Y., Furochi, H., Onishi, Y., Ogawa, T., Suzue, N., Oarada, M., Shimazu, T., Kishi, K. *Biol. Sci. Space* **2005**, *19*, 3–7.

186 Spisni, E., Bianco, M.C., Blasi, F., Santi, S., Riccio, M., Toni, M., Griffoni, C., Tomasi, V. *J. Gravit. Physiol.* **2002**, *9*, P285–P286.

187 Cogoli, A., Valluchi-Morf, M., Bohringer, H.R., Vanni, M.R., Muller, M. *Life Sci. Space Res.* **1979**, *17*, 219–224.

188 Kumei, Y., Nakajima, T., Sato, A., Kamata, N., Enomoto, S. *J. Cell Sci.* **1989**, *93*, 221–226.

189 Tschopp, A., Cogoli, A. *Experientia* **1983**, *39*, 1323–1329.

190 Miwa, M., Kozawa, O., Tokuda, H., Kawakubo, A., Yoneda, M., Oiso, Y., Takatsuki, K. *Bone Miner.* **1991**, *14*, 15–25.

191 Furutsu, M., Kawashima, K., Negishi, Y., Endo, H. *Biol. Pharm. Bull.* **2000**, *23*, 1258–1261.

192 Nose, K., Shibanuma, M. *Exp. Cell Res.* **1994**, *211*, 168–170.

193 Fitzgerald, J., Hughes-Fulford, M. *Exp. Cell Res.* **1996**, *228*, 168–171.

194 Sumanasekera, W.K., Zhao, L., Ivanova, M., Morgan, D.D., Noisin, E.L., Keynton, R.S., Klinge, C.M. *Cell Tissue Res.* **2006**, *324*, 243–253.

195 Croute, F., Gaubin, Y., Pianezzi, B., Soleilhavoup, J.P. *Microgravity Sci. Technol.* **1995**, *8*, 118–124.

196 Kacena, M.A., Todd, P., Gerstenfeld, L.C., Landis, W.J. *Micrograv. Sci. Technol.* **2004**, *5*, 28–34.

197 Loesberg, W.A., Walboomers, X.F., van Loon, J.J., Jansen, J.A. *Cell. Motil. Cytoskeleton* **2006**, *63*, 384–394.

198 Hughes-Fulford, M. *J. Gravit. Physiol.* **2002**, *9*, P257–P260.

199 Klaus, D.M., Benoit, M.R., Nelson, E.S., Hammond, T.G. *J. Gravit. Physiol.* **2004**, *11*, 17–27.

200 Elliott, T.F., Das, G.C., Hammond, D.K., Schwarzkopf, R.J., Jones, L.B, Baker, T.L., Love, J.E. *Gravit. Space Biol. Bull.* **2005**, *18*, 83–84.

201 Lambert, Ch.A, Deroanne, C., Servotte, S., Mineur, P., Lapière, Ch.M., Nusgens, B.V. *Gravit. Space Biol.* **2005**, *18*, 103–104.

6
Microgravity and Bone Cell Mechanosensitivity

Rommel G. Bacabac, Jack J. W. A. Van Loon, and Jenneke Klein-Nulend

6.1
Overview

Stress derived from bone loading that affects bone cells is likely the strain-induced flow of interstitial fluid along the surface of osteocytes and lining cells. The response of bone cell cultures to fluid flow includes prostaglandin (PG) synthesis and expression of prostaglandin G/H synthase inducible cyclooxygenase (COX-2). Cultured bone cells also rapidly produce nitric oxide (NO) in response to fluid flow as a result of activation of endothelial nitric oxide synthase (ecNOS), the enzyme that also mediates the adaptive response of bone tissue to mechanical loading. Several studies have shown that disruption of the actin-cytoskeleton abolishes the response to stress, suggesting that the cytoskeleton is involved in cellular mechanotransduction.

Microgravity, or near weightlessness, is associated with the loss of bone in astronauts, and has catabolic effects on mineral metabolism in bone organ cultures. This might be explained as resulting from an exceptional form of disuse under near weightlessness conditions. We found earlier that the transduction of mechanical signals in bone cells also involves the cytoskeleton and is related to prostaglandin E_2 (PGE_2) production. Therefore, it is possible that the mechanosensitivity of bone cells is altered under near weightlessness conditions, and that this abnormal mechanosensation contributes to disturbed bone metabolism observed in astronauts.

Detailed investigations with Space experiments also help to understand bone loss diseases known on Earth by evaluating the processes involved in mechanosensitivity on the cellular and tissue level.

6.2
Introduction

The capacity of bone tissue to alter its mass and structure in response to mechanical loading has long been recognized [1], but the cellular mechanisms involved

Biology in Space and Life on Earth. Effects of Spaceflight on Biological Systems. Edited by Enno Brinckmann

ISBN: 978-3-527-40668-5

remained poorly understood. Bone not only develops as a structure designed specifically for mechanical tasks, but it can adapt during life toward more efficient mechanical performance. The mechanical adaptation of bone is a cellular process and needs a biological system that senses the mechanical loading. The loading information must then be communicated to the effector cells that form new bone or destroy old bone [2, 3].

It has been well documented that bone tissue is sensitive to its mechanical environment. Subnormal mechanical stress as a result of bed rest or immobilization results in decreased bone mass and disuse osteoporosis [4]. Space flight produces a unique condition of skeletal unloading as a result of the near absence of gravitational acceleration. Studies of animals and humans subjected to space flight agree that near weightlessness negatively affects the mass and mechanical properties of bone (for a review, see Ref. [5]). Although the exact mechanism whereby bone loss as a result of space flight occurs is still unknown, recent *in vivo* studies suggest that bone cells are directly sensitive to near weightlessness. Using organ cultures of living bone rudiments from embryonic mice, Van Loon et al. [6] showed that 4 days of spaceflight inhibited matrix mineralization (Fig. 6.1), while stimulating osteoclastic resorption of mineralized matrix. Mineral resorption was studied in 17.5-day-old embryonic ^{45}Ca pre-labelled metatarsal bones. In contrast to the decreased mineral formation there was a pronounced increase of 43% in mineral resorption under microgravity conditions versus the in-flight 1×g control. The difference between microgravity and ground controls, however, could not be proven statistically [6]. From one of the first space flight studies with humans [7, 8] and

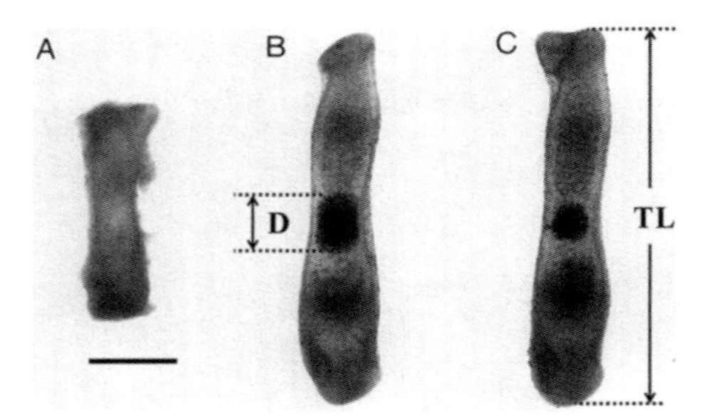

Fig. 6.1 Sixteen-day-old embryonic mouse metatarsal long bones (ED16). (A) Immediately after dissection. (B) Cultured in the in-flight 1×g centrifuge. (C) Cultured under microgravity conditions in Biorack on IML-1. Bar = 0.5 mm. During the 4-day culture period of an ED16 foetal mouse long bone, mineralization starts in the ossifying centre of the rudiment (diaphysis). The mineralized diaphysis is clearly visible as an opaque part in the centre of the rudiment (B and C). D is the length of the diaphysis, TL is total length of the metatarsal. Mineralization was studied in the ED16 bones by measuring the length of the calcified diaphysis, which at this stage of development (ED16 plus 4-days culture) consists of a core or mineralized cartilage surrounded by a primitive bone collar. A 31.3% reduction in length of the diaphysis in the microgravity group was found versus the in-flight 1×g control [6]. (Picture modified after Ref. [104].)

rats [9, 10] it has become clear that osteoblastic bone formation is retarded as compared with Earth controls. Most space flight studies show no effect with respect to bone resorption by osteoclasts [9, 10], but some have indicated an increased resorption activity under near weightlessness conditions *in vivo*, although the various recovery times, as in most space flight studies, must be kept in mind [11–14]. Other experiments have shown that isolated foetal mouse metatarsal long bones respond to changes in increased mechanical loading such as intermittent hydrostatic pressure [15] or in a hypergravity field with increased mineralization. In addition, mineral resorption was decreased in metatarsals cultured under intermittent compression for four days [16]. In the Biorack "BONES" experiment we used similar organ cultures to study growth, glucose utilization, collagen synthesis and mineral metabolism under space flight conditions during a 4-day culture period in microgravity. Using these parameters the hypothesis was tested that weightlessness influences the metabolic activity of skeletal tissues *in vitro*. This microgravity experiment was performed during the first International Microgravity Laboratory mission (IML-1) of the Space Shuttle program in the Biorack facility of Spacelab. Similar results were found during later studies in Bion-10 and the IML-2 mission.

Monolayer cultures of the human osteoblastic cell line MG-63 responded to 9 days of near weightlessness with reduced expression of osteocalcin, alkaline phosphatase, and collagen Iα1 mRNA [17]. Reduced prostaglandin production was found in cultures of MC3T3-E1 osteoblastic cells exposed to 4 days of near weightlessness, probably due to inhibition of serum-induced growth activation [18]. In addition, near weightlessness induced prostaglandin E_2 (PGE_2) and interleukin-6 production in rat bone marrow stroma cultures, an observation that may be related to alterations in bone resorption [19]. These results suggest that mineral metabolism and bone cell differentiation are modulated by near weightlessness, and that bone cells are directly responsive to microgravity conditions.

The *in vivo* operating cell stress derived from bone loading is likely the flow of interstitial fluid along the surface of osteocytes and lining cells (Fig. 6.2) [2, 20]. An *in vitro* model using a parallel-plate fluid flow chamber is supposed to simulate *in vivo* fluid shear stresses on various cell types exposed to dynamic fluid flow in their physiological environment (Fig. 6.3). The metabolic response of cells *in vitro* is associated with the wall shear stress. The response of bone cells in culture to fluid flow includes prostaglandin (PG) synthesis and expression of prostaglandin G/H synthase inducible cyclooxygenase (COX-2) [21, 22]. Cultured bone cells also rapidly produce nitric oxide (NO) in response to fluid flow as a result of activation of endothelial nitric oxide synthase (ecNOS), an enzyme that also mediates the adaptive response of bone tissue to mechanical loading [23, 24]. Earlier studies have shown that disruption of the actin-cytoskeleton abolishes the response to stress, suggesting that the cytoskeleton is involved in cellular mechanotransduction [25]. The cytoskeleton of the cell also determines its mechanical structure and as such, becomes crucial for the mechanical response of cells to environmental

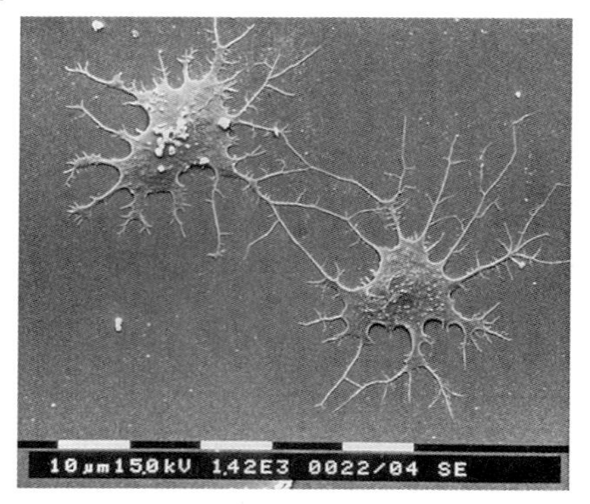

(A)

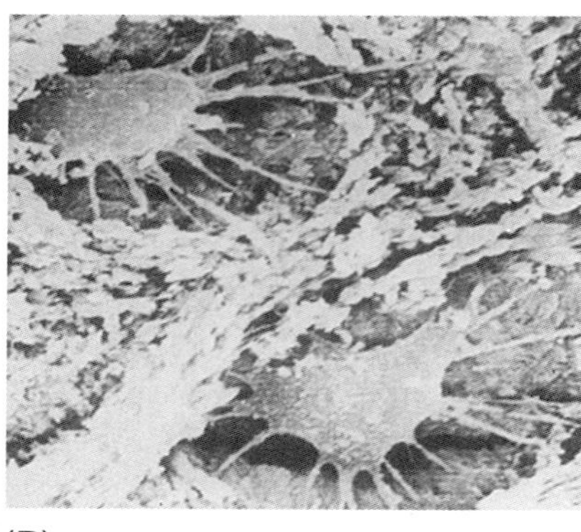

(B)

Fig. 6.2 Osteocyte morphology. (A) Isolated osteocytes in culture. Osteocytes were isolated from chicken by an immunodissection method using MAb OB7.3-coated magnetic beads. After isolation the cells were seeded on a glass support, cultured for 24 h and studied with a scanning electron microscope. After attachment, osteocytes form cytoplasmic extrusions in all directions. (B) Osteocytes embedded in calcified bone matrix. Note the many cell processes, radiating from the osteocyte cell bodies, as visualized using scanning electron microscopy. Magnification: 1000×.

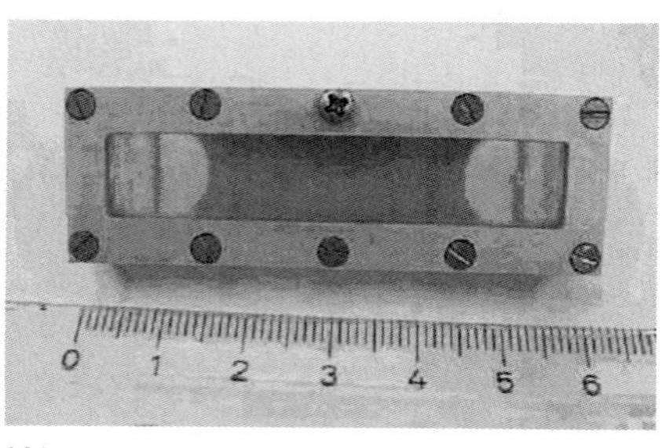

(A)

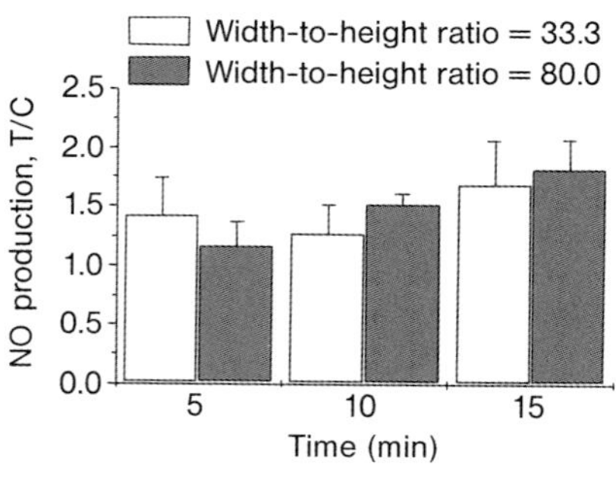

(B)

Fig. 6.3 Parallel-plate flow chamber *in vitro* system. This system simulates *in vivo* fluid shear stresses on various cell types exposed to dynamic fluid flow in their physiological environment. The metabolic response of cells *in vitro* is associated with the wall shear stress. This parallel-plate flow chamber was characterized for dynamic fluid flow experiments. A dimensionless ratio h/λ_v was used to determine the exact magnitude of the dynamic wall shear stress. It was shown that, to expose cells to predictable levels of dynamic fluid shear stress, two conditions have to be met: (1) $h/\lambda_v < 2$, where h is the distance between the plates and λ_v is the viscous penetration depth, and (2) $f_o < f_c/m$, where the critical frequency f_c is the upper threshold for this flow regime, m is the highest harmonic mode of the flow, and f_o is the fundamental frequency of fluid flow. (A) The parallel-plate flow chamber system was scaled to have a width-to-height ratio of 33.3 or 80.0. (B) Bone cells respond similarly to the same flow regime [pulsating fluid shear stress (PFSS), average shear stress = 0.7 Pa, 5 Hz] for different flow chamber width-to-height ratios (33.3 and 80.0) with up to nearly two-fold NO production. T/C, PFF treatment over control ratio; PFF, pulsating fluid flow.

forces. Extracellular matrix receptors such as integrins and CD44 receptors, located in the cellular membrane, are attached to the extracellular matrix as well as to the cytoskeleton. They are prime candidates as mechanotransducers [26]. Thus, the ability of cells to respond to mechanical loading is possibly closely related to its mechanical properties and the transfer of forces via intervening proteins linking the internal structure of the cell to its environment.

The influence of forces on cells has long been recognized and the molecular processes involved are now being uncovered. Living bone is an evident biological system, where the interplay of force and metabolic response is exemplified both at the tissue and the cellular level. Forces are imparted on osteocytes and bone lining cells by the flow of fluid through the lacuno-canalicular system in mechanically loaded bone. Recently, it has been proposed, theoretically, that fluid drag through the extracellular matrix concentric to the osteocytic processes is able to amplify strain by up to two orders of magnitude [27]. Theoretical modelling along with *in vitro* studies revealed that the range of stress capable of soliciting meaningful physiological response from bone cells is within the range 0.1–20 Pa [28]. The response of bone cells to mechanical stress has been studied using varying techniques for imparting mechanical loads *in vitro* [28, 29].

Several studies emphasize the role of osteocytes as the professional mechanosensory cells of bone, and the lacuno-canalicular porosity as the structure that mediates mechanosensing [2, 20, 29]. Strain-derived flow of interstitial fluid through this porosity seems to mechanically activate the osteocytes, as well as to ensure transport of cell signalling molecules and nutrients and waste products. The rapid production of NO in human bone cells in response to fluid flow results from activation of endothelial cells nitric oxide synthase (ecNOS) [24]. These results suggest that the response of bone cells to mechanical stress resembles that of endothelial cells to blood flow [30–32]. In the vascular system, changes in arterial diameter occur in response to changes in blood flow rate, to ensure a constant vessel tone, and endothelial cells are widely recognized as the mechanosensory cells of this response. The early response of endothelial cells to fluid flow *in vivo* includes the release of NO and prostaglandins [32]. Surprisingly, therefore, bone tissue seems to use a similar sensory mechanism to detect and amplify mechanical information as does the vascular system. This concept allows an explanation of the local bone gain and loss, as well as remodelling in response to fatigue damage, as processes supervised by mechanosensitive osteocytes.

Microgravity, as occurs under space flight, has negative effects on the skeleton, leading to bone loss. Several studies suggest that bone tissue is directly sensitive to space flight conditions. Microgravity provides a unique mechanical environment that might directly affect the ability of cells to sense forces. Hence, the question remains as to how the lack of gravity is detected by bone cells. Could microgravity act directly on the bone cells? Or more precisely, could osteocytes and osteoblasts read the gravitational field change directly? To answer these questions, the fundamental properties of the way bone cells respond to stress in general have to be addressed.

6.3
Mechanotransduction in Bone

Bone is an obvious biological system that exemplifies the interplay of mechanical stress and adaptive response both at the tissue and cellular levels [1, 2, 33, 34]. It is currently thought that osteocytes do not directly sense the loading of the bone matrix, but rather respond to the strain-induced flow of interstitial fluid along the network of osteocytes [35, 36].

Mechanotransduction is the process by which the mechanosensor cells convert the mechanical stimulus into intracellular signals. How the mechanical signal is detected and converted into a chemical intracellular response has yet to be established. The composition and structure of the matrix in the periosteocytic sheath and the adherence of osteocytes to their surrounding matrix appear very important. The matrix composition and structure determines the bone's porosity for fluid flow and, therefore, the magnitude of the fluid shear stress [37]. The osteocyte processes and their surrounding matrix possess a range of structural elements, which according to a theoretical model [38] should allow for a dramatic amplification of cellular-level strains [39]. As osteocytes are still capable of producing matrix proteins and proteoglycans, they might even modify their responsiveness to mechanical loading by adapting the matrix around them, and thereby the porosity of the lacuno-canalicular system.

Osteocytes adhere to their surrounding matrix by cytoplasmic membrane receptors such as integrins and CD44 receptors, coupled to the cytoskeleton. These receptors are likely the first step of intracellular signal transduction after mechanical stimulation [26]. Regulation of the number of adhesion sites and/or their coupling to intracellular signal transduction pathways might provide a mechanism by which endocrine modulation of the mechanoregulation of bone occurs. Extracellular Na^+ and Ca^{2+} influxes are necessary for downstream anabolic effects of mechanotransduction in osteocytes [40].

6.4
Signal Transduction in Mechanosensing

To respond to mechanical stimuli with the production of signal molecules which modulate the activities of osteoblasts and osteoclasts, the mechanosensitive osteocytes have to convert the mechanical stimuli into intracellular signals (mechanotransduction). Extracellular matrix receptors, attached to the extracellular matrix as well as to the cytoskeleton, are prime candidates as mechanotransducers [26]. The known intracellular signal transduction pathways, such as the intracellular Ca^{2+}, IP3, or cAMP dependent pathways, shown to play a role in other mechanosensitive cells, are probably involved [41, 42]. Thereafter signal molecules are produced and secreted, such as prostaglandins and nitric oxide (NO) [43]. Osteopontin is rapidly upregulated by osteocytes after acute disuse, which may serve to mediate bone resorption, given that osteopontin acts as an osteoclast chemotaxant, and a modulator of osteoclastic attachment to bone [44].

Prostaglandins are involved in the response of bone tissue and cells to stress [22, 25, 45–49]. Both bone resorption caused by immobilization and bone formation caused by mechanical loading are inhibited by indomethacin *in vivo* [50]. The early upregulation of prostaglandin release in response to mechanical stress was associated with induction of cyclooxygenase-2 (COX-2) [24], and an increase in the number of functional gap junctions [51]. Induction of COX-2 by stress might explain why prostaglandin production is continued for several hours after stress was stopped [52, 53] and could be related to the memory phenomenon described *in vivo* [54].

NO produced by nitric oxide synthase (NOS) is an important mediator of the response of bone to stress. Several studies [24, 29, 53, 55–57] have shown that NO production is rapidly increased in response to mechanical stress in bone cells, including isolated osteocytes (Fig. 6.2). Both endothelial and neuronal NOS isoforms are present in osteocytes [58]. In fracture patients, the proportions of osteocytes expressing ecNOS and nNOS were reduced, suggesting that the capacity to generate NO might be reduced as a result of fracture, leading to impairment of the ability of NO to minimize resorption [59]. The rapid release by mechanically stressed bone cells makes NO an interesting candidate for intercellular communication within the three-dimensional network of bone cells.

6.5 Single Cell Response to Mechanical Loading

Osteocytes' response to different kinds of mechanical loading has predominantly been studied in cell cultures or entire bone. However, single-cell level mechanosensing and chemical signalling in osteocytes is essential for bone adaptation [60]. The mechanosensitive part of the osteocytes, the cell body or the cell processes, has not been conclusively determined yet, but it has been shown that both the cell body and the cell processes might be involved in mechanosensing [61]. Hence, information is needed on the single osteocyte's response to a localized mechanical stimulation. NO is an important marker for studying mechanosensitivity of osteocytes [52], and it mediates the induction of bone formation by mechanical loading *in vivo* [62, 63]. NO is produced as a by-product when L-arginine is converted into L-citrulline in the presence of nitrogen oxide synthase enzyme, molecular oxygen, NADPH and other cofactors [64, 65]. It metabolizes very rapidly and has a physiological half-life of 5–15 s, making its direct online detection difficult. DAR-4M AM has been developed as an NO indicator [66], which has been used to show upregulation of intracellular NO production in single osteocytes after localized mechanical stimulation (Fig. 6.3) [61]. This unique technique, which allows real-time monitoring of chemical signalling at the single osteocyte level, might lead to a better understanding of the dynamic processes involved in mechanosensing.

6.6
Rate-dependent Response by Bone Cells

Several studies have suggested that the rate of the applied mechanical strain is related to bone formation rather than the magnitude of strain (e.g. see Refs. [67, 68]). Low magnitude (<10 $\mu\varepsilon$), high frequency (10–100 Hz) loading has been shown to be capable of stimulating bone growth and inhibiting disuse osteoporosis [69]. Also, high-amplitude, low-frequency stimuli are quite rare in the activities of daily life, whereas high-frequency, low-amplitude stimuli are quite common [70]. High rates of loading also increased bone mass and strength after jumping exercises in middle-aged osteopenic ovariectomized rats [71]. The rate of loading seems to be a decisive factor in bone formation and maintenance. However, the underlying physical picture for understanding how the rate of stress might relate to a meaningful physiological response remains poorly understood.

We found that bone cells release increased amounts of NO, which linearly correlated with the rate of fluid shear stress (Fig. 6.4) [55]. The precise flow regimes to stimulate bone cells were based on the calculation of the frequency spectra after loading of bone (Fig. 6.5). For this calculation the results were used from the studies of Bergmann et al. [72], in which forces on a human hip were measured using strain sensors. This supports the notion that bone formation *in vivo* is stimulated by dynamic rather than static loads [73], and that low-magnitude, high-frequency mechanical stimuli may be as stimulative as high-amplitude, low-frequency stimuli. This rate-dependent response was found to occur, provided that the cells are "kicked" in a pre-conditioned state [74].

Studies on bone cell responses to vibration stress at a wide frequency range (5–100 Hz) showed correlations of NO and PGE_2 with the maximum acceleration rate (Figs. 6.6 and 6.7) [75]. This correlation was related to nucleus oscillations,

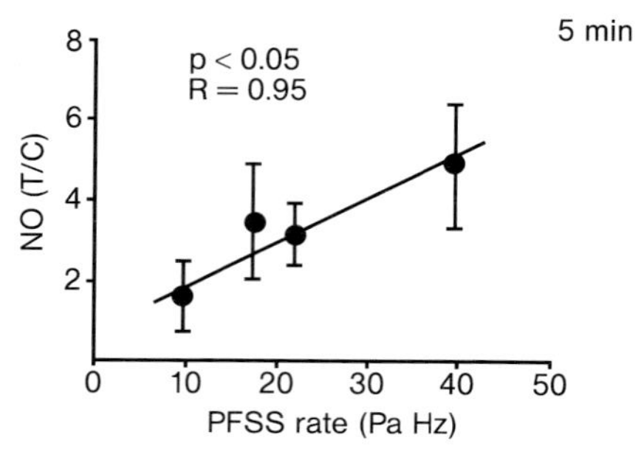

Fig. 6.4 Nitric oxide production by bone cells is linearly proportional to the rate of fluid shear stress. The slope was 0.11 Pa Hz at 5 min of MC3T3-E1 osteoblastic cell treatment with pulsating fluid shear stress (PFSS), but decreased to 0.05 Pa Hz at 10 min, and to 0.03 Pa Hz at 15 min (data not shown). Cells were treated with PFSS, and the cell culture medium was assayed for NO. Precise flow regimes were implemented by controlling the pump of the parallel-plate flow chamber using computer-mediated instrumentation by LabView.

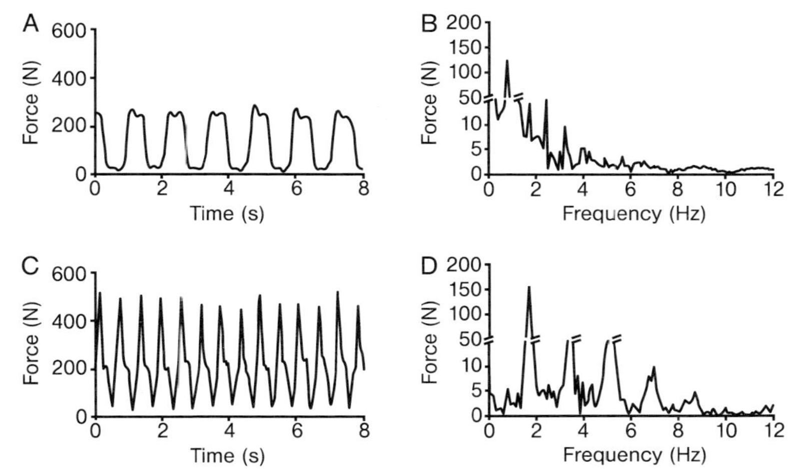

Fig. 6.5 Hip joint forces: Frequency spectra of the resultant forces on the human hip showed a rich harmonic content, ranging between 1–3 Hz for walking cycles, and reaching 8–9 Hz for running cycles. (A) Resultant force on the left hip joint of a male human subject (62 kg) as a function of time, measured by implanted sensors inside a femur at normal walking on a treadmill ($2\,\mathrm{km\,h^{-1}}$). (B) Frequency spectrum of the resultant hip joint force due to normal walking. (C) Resultant force on the hip as a function of time at jogging pace on a treadmill ($8\,\mathrm{km\,h^{-1}}$). (D) Frequency spectrum of the resultant force due to jogging pace. (Data taken with permission from Bergmann and coworkers [72].)

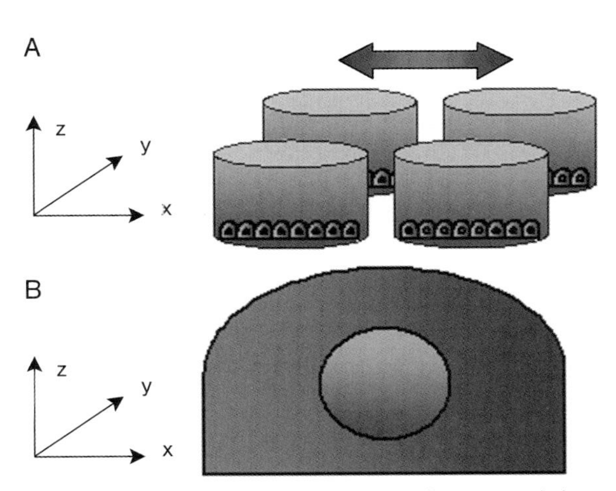

Fig. 6.6 Vibration stress application. (A) Cells were seeded onto the bottom surface of the 24-well plate. Sinusoidal motion is along the *x*-axis (arrow). (B) Simplified model of a cell with an approximately rigid spherical nucleus compared with its viscoelastic cell body.

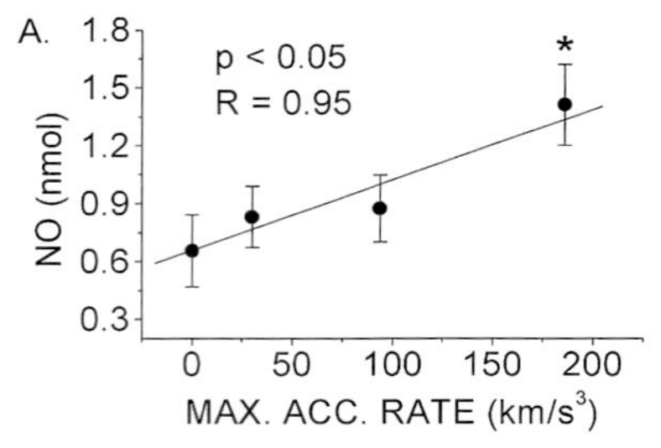

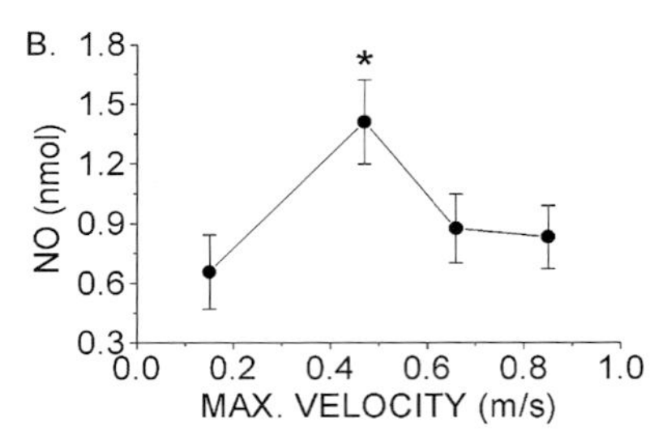

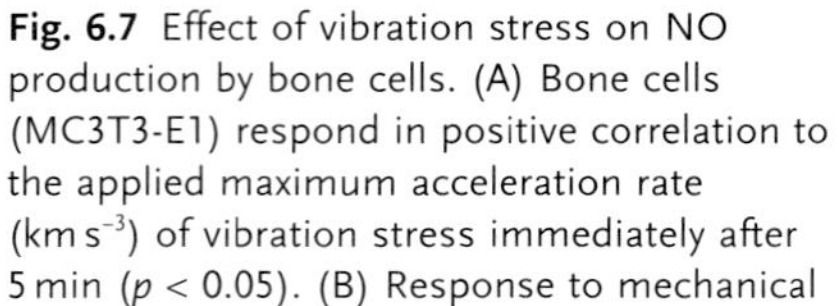

Fig. 6.7 Effect of vibration stress on NO production by bone cells. (A) Bone cells (MC3T3-E1) respond in positive correlation to the applied maximum acceleration rate ($km\,s^{-3}$) of vibration stress immediately after 5 min ($p < 0.05$). (B) Response to mechanical vibration does not correlate linearly to the applied maximum velocity ($m\,s^{-1}$). The response to vibration stress of 100 Hz was significantly larger than the response to 5 Hz and 60 Hz ($^{*}p < 0.03$). Values are mean total amount ± SEM.

providing a physical basis for cellular mechanosensing of high-frequency loading.

The finding that the bone cell's response to fluid shear stress is rate dependent provides an explanation as to why adaptive bone formation can occur despite the sporadic occurrence of high-amplitude strains in daily life [55]. Based on theoretical analysis, it was shown that strain-induced flow in the canalicular system, which in turn induces fluid drag across the extracellular matrix on osteocyte processes that are amplified to two-orders of magnitude [27, 37]. This also provides an explanation for sustained bone formation despite the sporadic occurrence of high amplitude strains in normal physiological loading conditions. The theoretical approach leads to an *extracellular* mechanism for amplifying stress, whereas experimental investigations leading to a rate-dependent response provided a cellular basis for understanding the osteogenic adaptation of bone to mechanical loading. Further understanding of how bone cells respond to stress at the cellular level might provides deeper insight into how maintained bone health copes with low amounts of high amplitude loading. A phenomenological interpretation underlines the importance of both the rate-dependence and the requirement of an initial stress-kick for the stress response of bone cells. It is a phenomenon that bone is able to sustain itself despite the sporadic occurrence of meagre strains, whereas bone cells are known *in vitro* to require stresses imparting higher strains. However, it might be possible that sporadic bouts of responses, in terms of signalling molecule release, could account for sustained bone health. A normal person is in a condition of unloading during sleep. However, this does not necessarily support a predicament of bone loss. Thus, the theoretical attempt to explain strain amplification via extracellular matrix drag, while being important for understanding fluid shear stress stimulation, might not fully support an understanding of sustained bone metabolism despite normal conditions of low strains.

6.7
Implications of Threshold Activation: Enhanced Response to Stochastic Stress

As bone exhibits the property of an adaptive response to mechanical stress, various ways of imparting stress to bone have been shown to achieve enhanced bone formation. For instance, muscular activity might be related for stimulating strain on bone despite low magnitudes. Strong evidence for the relation between muscle and bone was shown for 778 healthy Argentineans by correlating "whole body bone mineral content", as indicator for bone strength, and "lean body mass", as indicator for muscle strength [76]. This undermines a possible excessive requirement for exercise to induce high enough strain magnitudes for stimulating bone cells. Another example is the use of low-magnitude high frequency loading, which has been shown to stimulate osteogenic response from various species [69]. One other example is the use of low intensity pulsed ultrasound, which has osteogenic benefits [77–79]. Techniques of imparting stress by vibrating plates or ultrasound seem to exhibit effective stimulation of cells; however, the underlying physical picture of the transfer of forces to the cellular effectors, despite the soft tissue barrier, is not straightforward. The possible medical benefits of these techniques might overshadow the importance of understanding the underlying mechanisms of how they might work. However, further benefits can only achieved through a deeper understanding. The processes by which bone cells are stimulated by vibrating plates, ultrasound, or even muscular vibration might not be directly the result of a force transfer to the cells themselves. As mentioned, intervening soft-tissue might attenuate the assault of vibratory stress from an already meagre source.

We have shown that the bone cell response to stress is enhanced by noise [80] (Fig. 6.8). The nitric oxide released by bone cells reached a maximum at the application of an optimum noise-level by fluid shear stress. No distinct peak response was conclusive for the prostaglandin E_2 response. Our study used noise by fluid shear stress stimulation to find differences in the response of MLO-Y4 and MC3T3-E1 cells as models for osteocytes and osteoblasts, respectively (Fig. 6.8) [80]. The results indicated differences in threshold-activation for NO and PGE_2 production for both cell types. A peak response is indicative of a small difference between the input signal and the threshold. Hence these results suggest that, at low stress conditions with noise, MLO-Y4 cells could have a peak PGE_2 response, while MC3T3-E1 cells have a peak NO response. At conditions of high stress with noise, MLO-Y4 cells could have a peak NO response, while MC3T3-E1 cells, a PGE_2 response. *In vivo*, osteoclasts are, possibly, driven to be active close to osteocytic regions at low stress conditions with noise. In contrast, osteoclasts are driven to be active near osteoblastic regions at high stress conditions with noise. Clearly, low stress might promote bone loss; however, high stress seems to promote the activity of osteoclasts near osteoblasts. This supports the notion that high stresses, ultimately leading to bone microdamage and osteocyte apoptosis, initiate bone remodelling [81]. Whereas microdamage promotes low fluid flow and osteocyte apoptosis, explaining the recruitment of osteoclasts [60, 81], very high stress by

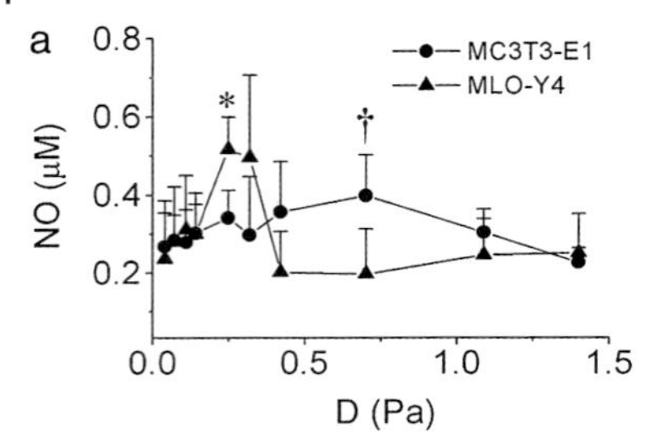

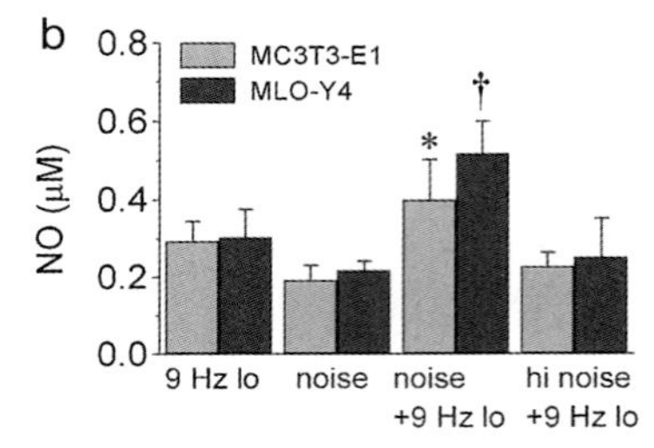

Fig. 6.8 NO production by MLO-Y4 and MC3T3-E1 cells in response to noisy stress. (a) NO release by MLO-Y4 cells in response to noisy stress was optimum at a noise intensity of 0.25 Pa ($^*p < 0.047$, greater than the response at noise intensity 0.04 Pa; $p <$ 0.038, greater than the response at noise intensity 1.40 Pa; $p < 0.023$, greater than the response at noise intensity 0.42 Pa). NO release by MC3T3-E1 cells in response to noisy stress does not result in a statistically significant optimum (†, probable peak at 0.70 Pa). (b) NO release by MLO-Y4 cells at its optimum noise intensity (0.25 Pa noise intensity superposed with 9 Hz lo regime) is significantly larger than the response to noise alone ($^*p < 0.030$, for 0.25 Pa noise alone) and to the highest noise intensity ($^*p < 0.038$, for 1.4 Pa noise intensity superposed to 9 Hz lo). NO release by MC3T3-E1 cells at its possible optimum noise intensity (0.70 Pa noise intensity superposed with 9 Hz lo regime) is significantly higher than the response to noise alone ($^{\dagger}p < 0.030$, for 0.70 Pa noise alone), but not to the highest noise intensity (1.4 Pa noise intensity superposed to 9 Hz lo). Results are mean ± SEM.

itself stimulates osteoclasts via PGE_2 upregulation. Furthermore, the possible role of noisy or stochastic stress in the mechanical adaptation of bone provides a partial explanation for the role of obscure effects of indirect stress applications to stimulate bone formation. The threshold-activation property of bone cells indicates a capacity for enhanced response at stress conditions obscured even by minute stress sources, such as from muscular vibration, ultrasound, or vibratory motion, typified by low-magnitude high frequency loading.

6.8
Stress Response and Cellular Deformation

Studies on the osteogenic activity of bone cells have investigated the effects of stress using varying techniques (fluid flow, substrate strain, hydrostatic pressure, vibration stress). In these studies, the magnitude (and rate) of stress was shown to correlate with the cellular response, in terms of signalling molecules. This correlation was suggested to be deformation-dependent. Relating the effects of fluid flow and substrate straining has shown that the former induces higher release of signalling molecules [82, 83]. A recent numerical study confirmed that the cellular deformation caused by stress induced by fluid flow is fundamentally different from that induced by substrate straining [84]. Fluid shear stress has a larger overturning effect on the bone cells, while the effect of substrate strain is focused on cell–substrate attachments. These recent results confirm the importance of investigating how cells deform in response to stress for understanding the corresponding physiological response of cells.

The excitation mechanism of osteocytes might be due to a unique strain amplification that results from the interaction of the pericellular matrix and the cell process cytoskeleton [27]. This also provides an explanation for sustained bone formation despite the sporadic occurrence of high amplitude strains in normal physiological loading conditions. The theoretical approach leads to an extracellular mechanism for amplifying stress, whereas experimental investigations, leading to a rate-dependent response, provided a cellular basis for understanding the osteogenic adaptation of bone to mechanical loading. Further understanding of how bone cells respond to stress at the cellular level might provide deeper insight into how maintained bone health copes with meagre amounts of high amplitude loading.

The collaboration between experimental investigation and theoretical analysis of the response of bone cells to stress has proven effective for an advanced understanding of underlying processes in bone mechanotransduction. The role of osteocytes, as the mechanosensors in bone *par excellence*, has been elucidated in more detail. Computational models for cells have previously treated the cytoplasm and cytoskeleton as a continuum [85]. Recognizing the importance of the cytoskeleton has led to more recent approaches that treat the cytoskeleton more closely to its physical reality, as consistent of interconnected fibres.

Using a two-particle *in vitro* assay to measure the viscoelasticity of cells, with the recently derived two-particle microrheology, we have probed the mechanoactivity and mechanosensitivity of various bone cell types and fibroblasts. Mechanoactivity is characterized by the induction of force traction on attachment sites by cells, and mechanosensitivity is the ability of cells to sense forces. We found that osteocytic cell types (primary osteocytes from chicken and MLO-Y4 cells) induce a relatively higher traction force on attached particles than osteoblastic cells (primary osteoblasts from chicken and MC3T3-E1 cells) (Fig. 6.9). Fibroblastic cells (CCL-224) are even more mechano-active than MLO-Y4 cells, which explains the propensity of fibroblasts for motility *in vitro* (Fig. 6.10). In our two-particle *in vitro* assay, MLO-Y4 cells release nitric oxide simultaneously with increasing force application. Another typical response to increasing force application is the induction of force traction on the attached beads by cells, simultaneously with morphological adaptation from a spherical to a polar shape defining ends at the attachment points. Clearly, force traction, morphology change, and possibly the release of signalling molecules are all related in similar pathways, in response to environmental stress conditions.

6.9
Towards a Quantitative Description of Bone Cell Mechanosensitivity

We have outlined above two discernible properties of bone cell response to fluid shear stress, i.e. rate-dependence and threshold-activation. The rate-dependent

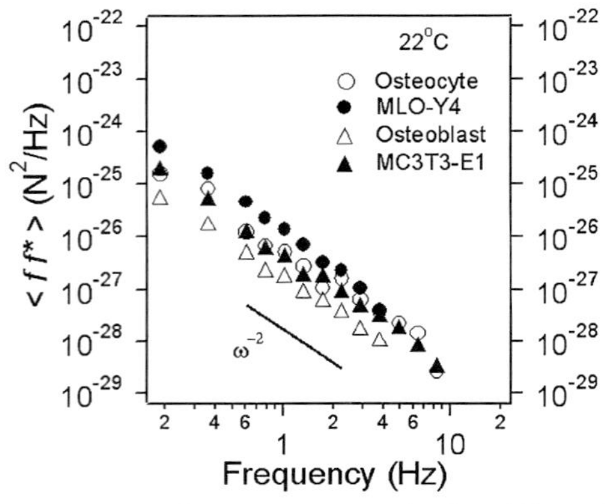

Fig. 6.9 Force fluctuation at 22 °C by four different bone cells. Isolated chicken osteocytes (open circles), MLO-Y4 osteocytic cell line (filled circles), isolated chicken osteoblasts (open triangles) and MC3T3-E1 osteoblastic cell line (filled triangles).

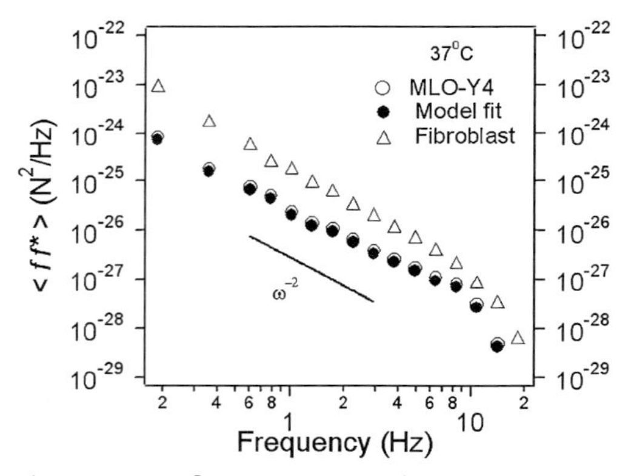

Fig. 6.10 Force fluctuation at 37 °C by MLO-Y4 osteocytes (open circles) and CCL-224 fibroblasts (open triangles).

response was, interestingly, demonstrated by bone cells in a broad range of frequencies induced by vibration stress, showing a distinctive correlation to the acceleration rate [75]. This rate-dependent response showed a negatively correlated release of NO and PGE_2. These signalling molecules modulate the activity of osteoclasts [60]. Hence, loading frequency can fine-tune the localized recruitment and inhibition of osteoclasts. The loading rate is the important parameter for stimulating bone formation [86, 87]. However, the underlying cellular mechanisms are now slowly being revealed. By recognizing the mathematical precision of bone mechanical adaptation, as Wolff did [1], the science of understanding the mechanobiology of bone has also been subject to numerical simulations [88, 89]. The final local effects on bone, under conditions of loading, are further understood by having a quantifiable interplay between cell types, due to the resulting release of signalling molecules. Complex systems generally exhibit power-law phenomena that might prove useful in characterizing the behaviour of biological systems (for a review on complex biological systems and scale invariance phenomena see Ref. [90]). A power-law description is typical for complex systems, which generally exhibit scale invariance. Power-laws and scale invariance have been used in studying phase transitions in materials and in describing systems that spontaneously self-arrange [90].

We can derive an empirical power-law for bone cells. Based on our results [55, 75], we propose an empirical quantifiable property of bone cells, i.e. the correlation of the accumulated amount of signalling molecule to the cell's experienced frequency of stress. Thus, a relation can be formulated:

$$[M] \propto s\left(\frac{\sigma - \theta}{\sigma}\right)\omega^{\beta} \tag{1}$$

where s is the normalized sigmoid function to enforce threshold-activation, with a threshold θ, and the experienced stress σ. [M] is the accumulated amount of the signalling molecule M and the frequency $f = \omega/(2\pi)$, and β determines the power-law relation between stress and response. Equation (1) incorporates, in a power-law, the property of bone cells that the released amount of signalling molecules is related to the frequency of stress, above a threshold. We showed that $\beta = 1$ for fluid shear stress stimulation [55], and we showed that $\beta = \pm 3$ for translational vibration stress acting on attached bone cells [75]. Thus, β is a property associated with the type of stress. Whereas $\beta = 1$ for contact stress (e.g. fluid shear stress, for low frequencies, <10 Hz), $\beta = \pm 3$ for stress induced by body forces (that means motion of the nucleus through cytoplasm, for high frequencies, >10 Hz). The absolute value of β is then useful for finding possible new mechanisms by which bone cells sense loading at different frequency ranges. We have also shown that the release of NO and PGE_2 correlated negatively [75]. We showed that for NO $\beta > 0$ and for PGE_2 $\beta < 0$, since NO release is positively correlated with ω^3, while PGE_2 release is negatively correlated with ω^3. Thus, β is also a property of the specific signalling molecule. The sign of β for the associated signalling molecule indicates upregulation ($\beta > 0$) or downregulation ($\beta < 0$) in relation to the loading frequency.

The threshold activation, enforced by the sigmoid function we introduced in Eq. (1), was partially demonstrated in our study, where we showed that without an initial stress-kick no rate-dependent response occurred [74]. Threshold-activation was also demonstrated by the possibility of stochastic resonance in the way bone cells respond to fluid shear stress [80]. Note, however, that the accumulated signalling molecule released [M] is the sum of all molecules released by a population of cells. In addition, the thresholds for NO and PGE_2 formation were demonstrated to be unequal [80]. Thus, the threshold θ is also a property of the cell population under question in relation to a specific signalling molecule. The threshold is reached when a critical number of cells have produced sufficient amounts of signalling molecules [M]. Further proof of this assumption can be demonstrated by showing that single cells do not have the same stress thresholds for releasing a measurable amount of signalling molecules. Possibly, the release of signalling molecules might be related to a change in the mechanical properties of the cell. Physiological conditions are probably influenced by the activity of collaborating aggregates of cells. Hence, experimental results suggesting power-laws on cell cultures *in vitro* might not necessarily translate to single cell properties, but, considering that a biological system is complex, the property of scale invariance might suggest that single cells possess similar power-laws.

The validity of Eq. (1) is anticipated to be limited to a range of frequencies, possibly below 100 Hz, for an actual mechanical loading regime on the cell. Using frequencies of 5, 30, 60 and 100 Hz, we found a linear correlation between ω^3 and the total NO and PGE_2 released by bone cells [75]. Bone cells are basically elastic until ~10 Hz, having a viscoelastic transition between 10 and 100 Hz, beyond which a viscoelastic stiffening occurs [91]. Transmembrane proteins and, ultimately, the cell cytoskeleton might be related to mechanosensing by cells [26]. Since the mechanical properties of cells depend on the cytoskeleton, changes in cell compliance might be a direct indication of cell mechanosensitive properties. Thus, a linear relation between the physiological responses of bone cells and ω^β is possibly related to changes in the viscoelasticity of cells.

6.10
Implications for the Extreme Condition of Unloading Microgravity

We have shown that bone cells are responsive to dynamic stress [29, 55, 75]. This supports findings that the rate rather than the magnitude of loading is the important parameter for osteogenic properties of mechanical loading to bone [56]. This insight has powerful implications on the local activity of osteocytes for directing the mechanical adaptation of bone. Here we proposed that the released signalling molecules of bone cells are related to the frequency of stress in an empirical relation. Since the activation of bone cells is highly dependent on the frequency of experienced stress, bone cells might be capable of responding to very meagre amounts of strain, after overcoming a stress threshold. Thus, under extreme conditions of unloading, it might be possible to counteract the onslaught of bone loss by sporadic bouts of high impact loading. There is much data on the catabolic effects to bone of prolonged unloading [92–94], and suggested counteractive remedies or preventive measures for bone loss, which are both based on exercise or pharmacological applications [95–97]. Regardless of the counteractive procedure for bone loss, the approach has to eventually target the bone cells. It has been suggested that imbalances in the activity of bone cells for directing bone resorption or formation contributes to bone loss [92]. We have shown here relations between what we consider fundamental parameters of mechanotransduction in bone, which are the applied stress, the amount of signalling molecules released by osteocytes, and their possible roles for directing the local activity of osteoclasts and osteoblasts. Bone loss can be understood as resulting from a disturbance of the homeostasis of these parameters. However, since these parameters are closely related, as we have shown, it might be possible to restore their homeostasis despite the extreme conditions of unloading.

Studies by Tabony et al. [98–100] have shown that microtubule self-organization *in vitro* is gravity-dependent and that this self-organization affects the transport of intracellular particles. Since a reorganization of microtubules might affect the cell viscoelastic properties, gravity or the loss of gravity might affect the way cells sense forces. To address the underlying cellular mechanisms for studies on bone loss in relation to microgravity, Cowin [101] has addressed the question of whether bone cells can read the changes in gravitational field or detect this indirectly via contact stresses. It would seem that the light weight of cells undermines the role of gravity on cellular behaviour. Thus, the question remains as to how cells might detect microgravity directly. We probed the activity of bone cells by measuring the forces induced by cells on attached fibronectin-coated beads [91]. We showed that the force fluctuation $\langle ff^* \rangle$ was proportional to ω^{-2} [91]. This is a signature for continuums with slowly evolving internal processes [102]. Although we probed the traction forces induced by cells on particles outside the cell, this force fluctuation is indicative of intracellular processes. The power-law for cellular force fluctuation is an indication for the diffusive properties of intracellular particles or organelles. Reaction–diffusion processes govern the relation between microtubule re-organization, and possibly intracellular transport, which might be affected by gravity [99, 103]. Thus, intracellular transport might be crucial for mechanosensing. Hence, cells might be able to detect changes in the gravitational field directly by the gain or loss of gravitational forces influencing the re-structuring of self-organizing polymers inside cells, thereby influencing intracellular transport. Thus, it is possible that the signature for force fluctuations inside cells might change under microgravity, influencing bone cell mechanosensitivity, as indicated by a changed release of signalling molecules.

"FLOW" was our entry to the biological experiments carried by the Dutch Soyuz Mission "DELTA" (Dutch Expedition for Life Science, Technology and Atmospheric Research). During DELTA, FLOW was carried by the Soyuz craft, which was launched on April 19, 2004, on its way to the International Space Station (ISS). The main scientific objective of the FLOW experiment was to test whether the production of early signalling molecules, which are involved in the mechanical loading-induced osteogenic response (NO and PGE_2) by osteocytes, is changed under microgravity conditions as compared with 1×g conditions. Since we argue that especially the osteocyte, and not the osteoblast, is the mechanosensitive cell type within bone involved in mechanotransduction, the production of signalling molecules by osteocytes was compared with osteoblasts. Periosteal fibroblasts were used as negative controls. Chicken osteocytes, osteoblasts, and periosteal fibroblasts were incubated in plunger boxes developed by the Centre for Concepts in Mechatronics (CCM), using plunger activation events for single pulse fluid shear stress stimulations. Cultures in-flight were subjected to microgravity (μ-g) and simulated 1×g level by centrifugation. Ground controls were subjected to identical culture environment and fluid shear stress stimulations. Owing to unforeseen hardware complications resulting in a loss of electric power, no results from in-flight cultures could be obtained. Despite the setback from the first flight experiment complications, the preparations for another FLOW experiment and preliminary ground results indicate that the FLOW setup is viable for a future flight opportunity. FLOW has been considered as a potential candidate by the European Space Agency (ESA) for one of the next Soyuz missions to the ISS.

Acknowledgments

The Space Research Organization of the Netherlands supported the work of J.J.W.A. Van Loon (DESC, MG-057 and SRON grant MG-055) and R.G. Bacabac (SRON grant MG-055), who also received financial assistance from the European Space Agency (ESA).

References

1 Wolff, J. *Gesetz der Transformation der Knochen*, Hirschwald, Berlin, **1892**.
2 Burger, E.H., Klein-Nulend, J. *FASEB J.* **1999**, *13*(Suppl.), S101–S112.
3 Smit, T.H., Burger, E.H., Huyghe, J.M. *J. Bone Min. Res.* **2002**, *17*(11), 2021–2029.
4 Houde, J.P., Schulz, L.A., Morgan, W.J., Breen, T., Warhold, L., Crane, G.K., Baran, D.T. *Clin. Orthop.* **1995**, *57*(317), 199–205.
5 Van Loon, J.J.W.A., Veldhuijzen, J.P., Burger, E.H., Bone and Space Flight: An Overview, in: *Biological and Medical Research in Space*, Moore, D., Bie, P., Oser, H. (Eds.), pp. 259–299, Springer, Berlin, **1996**.
6 Van Loon, J.J.W.A., Bervoets, D.J., Burger, E.H., Dieudonne, S.C., Hagen, J.W., Semeins, C.M., Zandieh Doulabi, B., Veldhuijzen, J.P. *J. Bone Miner. Res.* **1995**, *10*(4), 550–557.
7 Vose, G.P. *Am. J. Roentgenol. Radium Ther. Nucl. Med.* **1974**, *121*(1), 1–4.
8 Oganov, V.S., Rakhmanov, A.S., Novikov, V.E., Zatsepin, S.T., Rodionova, S.S., Cann, C. *Acta Astronaut.* **1991**, *23*, 129–133.
9 Morey, E.R., Baylink, D.J. *Science* **1978**, *201*(4361), 1138–1141.
10 Vico, L., Novikov, V.E., Very, J.M., Alexandre, C. *Aviat. Space Environ. Med.* **1991**, *62*(1), 26–31.
11 Kaplansky, A.S., Durnova, G.N., Burkovskaya, T.E., Vorotnikova, E.V. *Physiologist* **1991**, *34*(1 Suppl), S196–S199.
12 Vico, L., Chappard, D., Alexandre, C., Palle, S., Minaire, P., Riffat, G., Novikov, V.E., Bakulin, A.V. *Bone* **1987**, *8*(2), 95–103.
13 Foldes, I., Rapcsak, M., Szilagyi, T., Oganov, V.S. *Acta Physiol. Hung.* **1990**, *75*(4), 271–285.
14 Zerath, E., Holy, X., Malouvier, A., Caissard, J.C., Nogues, C. *Physiologist* **1991**, *34*(1 Suppl), S194–S195.
15 Klein-Nulend, J., Veldhuijzen, J.P., Burger, E.H. *Arthritis Rheum.* **1986**, *29*(8), 1002–1009.
16 Klein-Nulend, J., Veldhuijzen, J.P., van Strien, M.E., de Jong, M., Burger, E.H. *Arthritis Rheum.* **1990**, *33*(1), 66–72.
17 Carmeliet, G., Nys, G., Bouillon, R. Differentiation of Human Osteoblastic Cells (MG-63) In Vivo is decreased under Microgravity Conditions, in: *Proceedings of the 6th European Symposium on Life Science Research in Space, SP-390*, pp. 279–282, ESA Publications Division, ESTEC, Noordwijk, **1996**.

18 Hughes-Fulford, M., Lewis, M.L. *Exp. Cell Res.* **1996**, *224*(1), 103–109.

19 Kumei, Y., Shimokawa, H., Katano, H., Hara, E., Akiyama, H., Hirano, M., Mukai, C., Nagaoka, S., Whitson, P.A., Sams, C.F. *J. Biotechnol.* **1996**, *47*(2–3), 313–324.

20 Cowin, S.C., Moss-Salentijn, L., Moss, M.L. *J. Biomech. Eng.* **1991**, *113*(2), 191–197.

21 Ajubi, N.E., Klein-Nulend, J., Alblas, M.J., Burger, E.H., Nijweide, P.J. *Endocrinol. Metab.* **1999**, *276*(1), E171–E178.

22 Klein-Nulend, J., Burger, E.H., Semeins, C.M., Raisz, L.G., Pilbeam, C.C. *J. Bone Miner. Res.* **1997**, *12*(1), 45–51.

23 Johnson, D.L., McAllister, T.N., Frangos, J.A. *Am. J. Physiol.* **1996**, *271*(1), E205–E208.

24 Klein-Nulend, J., Helfrich, M.H., Sterck, J.G., MacPherson, H., Joldersma, M., Ralston, S.H., Semeins, C.M., Burger, E.H. *Biochem. Biophys. Res. Commun.* **1998**, *250*(1), 108–114.

25 Ajubi, N.E., Klein-Nulend, J., Nijweide, P.J., Vrijheid-Lammers, T., Alblas, M.J., Burger, E.H. *Biochem. Biophys. Res. Commun.* **1996**, *225*(1), 62–68.

26 Wang, N., Butler, J.P., Ingber, D.E. *Science* **1993**, *260*(5111), 1124–1127.

27 Han, Y.F., Cowin, S.C., Schaffler, M.B., Weinbaum, S. *Proc. Natl. Acad. Sci. U.S.A.* **2004**, *101*, 16689–16694.

28 Brown, T.D. *J. Biomech.* **2000**, *33*(1), 3–14.

29 Klein-Nulend, J., van der Plas, A., Semeins, C.M., Ajubi, N.E., Frangos, J.A., Nijweide, P.J., Burger, E.H. *FASEB J.* **1955**, 9(5), 441–445.

30 Frangos, J.A., Eskin, S.G., McIntire, L.V., Ives, C.L. *Science* **1985**, *227*(4693), 1477–1479.

31 Furchgott, R.F., Vanhoutte, P.M. *FASEB J.* **1989**, *3*(9), 2007–2018.

32 Hecker, M., Mulsch, A., Bassenge, E., Busse, R. *Am. J. Physiol.* **1993**, *265*(3), H828–H833.

33 Frost, H.M. *Anat. Rec. A Discov. Mol. Cell. Evol. Biol.* **2003**, *275*(2), 1081–1101.

34 Cowin, S.C., Moss-Salentijn, L., Moss, M.L. *J. Biomech. Eng.* **1991**, *113*(2), 191–197.

35 Piekarski, K., Munro, M. *Nature* **1977**, *269*(5623), 80–82.

36 Cowin, S.C., Weinbaum, S. *Am. J. Med. Sci.* **1998**, *316*(3), 184–188.

37 Weinbaum, S., Cowin, S.C., Zeng, Y. *J. Biomech.* **1994**, *27*(3), 339–360.

38 You, L.D., Cowin, S.C., Schaffler, M.B., Weinbaum, S. *J. Biomech.* **2001**, *34*(11), 1375–1386.

39 You, L.D., Weinbaum, S., Cowin, S.C., Schaffler, M.B. *Anat. Rec. A Discov. Mol. Cell Evol. Biol.* **2004**, *278*(2), 505–513.

40 Mikuni-Takagaki, Y. *J. Bone Min. Metab.* **1999**, *17*(1), 57–60.

41 Iqbal, J., Zaidi, M. *Biochem. Biophys. Res. Commun.* **2005**, *328*(3), 751–755.

42 Watson, P.A. *FASEB J.* **1991**, *5*(7), 2013–2019.

43 Hughes-Fulford, M. *Sci. STKE* **2004**, *2004*(249), RE12.

44 Gross, T.S., King, K.A., Rabaia, N.A., Pathare, P., Srinivasan, S. *J. Bone Miner. Res.* **2005**, *20*(2), 250–256.

45 Bakker, A.D., Klein-Nulend, J., Tanck, E., Albers, G.H., Lips, P., Burger, E.H. *Osteoporos Int.* **2005**, *16*(8), 983–989.

46 Binderman, I., Shimshoni, Z., Somjen, D. *Calcif. Tissue Int.* **1984**, *36*(Suppl 1), S82–S85.

47 Klein-Nulend, J., Sterck, J.G., Semeins, C.M., Lips, P., Joldersma, M., Baart, J.A., Burger, E.H. *Osteoporos Int.* **2002**, *13*(2), 137–146.

48 Mullender, M., El Haj, A.J., Yang, Y., van Duin, M.A., Burger, E.H., Klein-Nulend, J. *Med. Biol. Eng. Comput.* **2004**, *42*(1), 14–21.

49 Rawlinson, S.C., El Haj, A.J., Minter, S.L., Tavares, I.A., Bennett, A., Lanyon, L.E. *J. Bone Miner. Res.* **1991**, 6(12), 1345–1351.

50 Thompson, D.D., Rodan, G.A. *J. Bone Miner. Res.* **1988**, *3*(4), 409–414.

51 Cheng, B., Kato, Y., Zhao, S., Luo, J., Sprague, E., Bonewald, L.F., Jiang, J.X. *Endocrinology* **2001**, *142*(8), 3464–3473.

52 Klein-Nulend, J., Semeins, C.M., Ajubi, N.E., Nijweide, P.J., Burger, E.H. *Biochem. Biophys. Res. Commun.* **1995**, *217*(2), 640–648.

53 Pitsillides, A.A., Rawlinson, S.C., Suswillo, R.F., Bourrin, S., Zaman, G., Lanyon, L.E. *FASEB J.* **1995**, *9*(15), 1614–1622.

54 Sterck, J.G.H., Klein-Nulend, J., Lips, P., Burger, E.H. *Am. J. Physiol. – Endocrinol. Metab.* **1998**, *274*(37–6), E1113–E1120.

55 Bacabac, R.G., Smit, T.H., Mullender, M.G., Dijcks, S.J., Van Loon, J.J.W.A., Klein-Nulend, J. *Biochem. Biophys. Res. Commun.* **2004**, *315*, 823–829.

56 Turner, C.H., Owan, I., Takano, Y. *Am. J. Physiol.* **1995**, *269*(1), E438–E442.

57 Rubin, C.T., Lanyon, L.E. *J. Bone Joint. Surg. Am.* **1984**, *66*(3), 397–402.

58 Caballero-Alias, A.M., Loveridge, N., Lyon, A., Das-Gupta, V., Pitsillides, A., Reeve, J. *Calcif. Tissue Int.* **2004**, *75*(1), 78–84.

59 Caballero-Alias, A.M., Loveridge, N., Pitsillides, A., Parker, M., Kaptoge, S., Lyon, A., Reeve, J. *J. Bone Miner. Res.* **2005**, *20*(2), 268–273.

60 Burger, E.H., Klein-Nulend, J., Smit, T.H. *J. Biomech.* **2003**, *36*(10), 1453–1459.

61 Vatsa, A., Mizuno, D., Smit, T.H., Schmidt, C.F., MacKintosh, F.C., Klein-Nulend, J. J. Bone Miner. Res. **2006**, *21*(11), 1722–1728.

62 Turner, C.H., Owan, I., Jacob, D.S., McClintock, R., Peacock, M. *Bone* **1997**, *21*(6), 487–490.

63 Chow, J.W., Fox, S.W., Lean, J.M., Chambers, T.J. *J. Bone Miner. Res.* **1998**, *13*(6), 1039–1044.

64 Marletta, M.A. *Cell* **1994**, *78*(6), 927–930.

65 Zaman, G., Pitsillides, A.A., Rawlinson, S.C., Suswillo, R.F., Mosley, J.R., Cheng, M.Z., Platts, L.A., Hukkanen, M., Polak, J.M., Lanyon, L.E. *J. Bone Miner. Res.* **1999**, *14*(7), 1123–1131.

66 Kojima, H., Hirotani, M., Nakatsubo, N., Kikuchi, K., Urano, Y., Higuchi, T., Hirata, Y., Nagano, T. *Anal. Chem.* **2001**, *73*(9), 1967–1973.

67 Turner, C.H., Owan, I., Takano, Y. *Am. J. Physiol. Endocrinol. Metab.* **1995**, *269*(3), E438–E442.

68 Turner, C.H., Pavalko, F.M. *J. Orthop. Sci.* **1998**, *3*(6), 346–355.

69 Rubin, C.T., Sommerfeldt, D.W., Judex, S., Qin, Y.X. *Drug Discov. Today* **2001**, *6*(16), 848–858.

70 Fritton, S.P., McLeod, J., Rubin, C.T. *J. Biomech.* **2000**, *33*(3), 317–325.

71 Tanaka, M., Ejiri, S., Nakajima, M., Kohno, S., Ozawa, H. *Bone* **1999**, *25*(3), 339–347.

72 Bergmann, G., Graichen, F., Rohlmann, A., Hip Joint Force Measurements, http://www.medizin.fu-berlin.de/biomechanik/Homefrme.htm. 3-6-**2003**.

73 Lanyon, L.E., Rubin, C.T. *J. Biomech.* **1984**, *17*(12), 897–905.

74 Bacabac, R.G., Smit, T.H., Mullender, M.G., Dijcks, S.J., Van Loon, J.J.W.A., Klein-Nulend, J. *Ann. Biomed. Eng.* **2005**, *33*(1), 104–110.

75 Bacabac, R.G., Smit, T.H., Van Loon, J.J.W.A., Zandieh Doulabi, B., Helder, M.N., Klein-Nulend, J. *FASEB J.* **2006**, *20*(7), 858–864.

76 Zanchetta, J.R., Plotkin, H., Filgueira, M.L.A. *Bone* **1995**, *16*(4), S393–S399.

77 Cook, S.D., Salkeld, S.L., Patron, L.P., Ryaby, J.P., Whitecloud, T.S. *Spine J.* **2001**, *1*(4), 246–254.

78 Sakurakichi, K., Tsuchiya, H., Uehara, K., Yamashiro, T., Tomita, K., Azuma, Y. *J. Orthop. Res.* **2004**, *22*(2), 395–403.

79 Schortinghuis, J., Ruben, J.L., Raghoebar, G.M., Stegenga, B., de Bont, L.G. *Int. J. Oral Maxillofac. Impl.* **2005**, *20*(2), 181–186.

80 Bacabac, R.G., Smit, T.H., Van Loon, J.J.W.A., Klein-Nulend, J. *Trans. 52nd Annu. Meeting, Orthopaedic Res. Soc.* **2006**, Vol 31, abstract #0356.

81 Bentolila, V., Boyce, T.M., Fyhrie, D.P., Drumb, R., Skerry, T.M., Schaffler, M.B. *Bone* **1998**, *23*(3), 275–281.

82 Owan, I., Burr, D.B., Turner, C.H., Qiu, J., Tu, Y., Onyia, J.E., Duncan, R.L. *Am. J. Physiol.* **1997**, *273*(1), C810–C815.

83 You, J., Yellowley, C.E., Donahue, H.J., Zhang, Y., Chen, Q., Jacobs, C.R. *J. Biomech. Eng.* **2000**, *122*(4), 387–393.

84 McGarry, J.G., Klein-Nulend, J., Mullender, M.G., Prendergast, P.J. *FASEB J.* **2005**, *19*(3), 482–484.

85 Kamm, R.D., McVittie, A.K., Bathe, M. *ASME Int. Congr. – Mech. Biol.* **2000**, *242*, 1–9.

86 Mosley, J.R., Lanyon, L.E. *Bone* **1998**, *23*(4), 313–318.

87 Turner, C.H., Owan, I., Takano, Y. *Am. J. Physiol. Endocrinol. Metab.* **1995**, *269*(3), E438–E442.

88 Weinans, H., Homminga, J., van Rietbergen, B., Ruegsegger, P., Huiskes, R. *J. Biomech.* **1998**, *31*(1), 152.

89 Frost, H.M. *Anat. Rec. A Discov. Mol. Cell Evol. Biol.* **2003**, *275*(2), 1081–1101.

90 Gisiger, T. *Biol. Rev. Cambridge Philos. Soc.* **2001**, *76*(2), 161–209.

91 Bacabac, R.G., Mizuno, D., Smit, T.H., Van Loon, J.J.W.A., MacKintosh, F.C., Schmidt, C.F., Klein-Nulend, J. *Trans 52nd Annu. Meeting, Orthopaedic Res. Soc.* **2006**, Vol. 31, abstract #1016.

92 Bikle, D.D., Halloran, B.P. *J. Bone Miner. Metab.* **1999**, *17*(4), 233–244.

93 Loomer, P.M. *Crit. Rev. Oral Biol. Med.* **2001**, *12*(3), 252–261.

94 Zerath, E. *Adv. Space Res.* **1998**, *21*(8–9), 1049–1058.

95 Greenleaf, J.E., Bulbulian, R., Bernauer, E.M., Haskell, W.L., Moore, T. *J. Appl. Physiol.* **1989**, *67*(6), 2191–2204.

96 Norman, T.L., Bradley-Popovich, G., Clovis, N., Cutlip, R.G., Bryner, R.W. *Aviat. Space Environ. Med.* **2000**, *71*(6), 593–598.

97 Wimalawansa, S.M., Chapa, M.T., Wei, J.N., Westlund, K.N., Quast, M.J., Wimalawansa, S.J. *J. Appl. Physiol.* **1999**, *86*(6), 1841–1846.

98 Tabony, J., Job, D. *Proc. Natl. Acad. Sci. U.S.A.* **1992**, *89*(15), 6948–6952.

99 Glade, N., Demongeot, J., Tabony, J. *BMC Cell Biol.* **2004**, *5*(1), 23.

100 Papaseit, C., Pochon, N., Tabony, J. *Proc. Natl. Acad. Sci. U.S.A.* **2000**, *97*(15), 8364–8368.

101 Cowin, S.C. *Bone* **1998**, *22*(1), 119S–125S.

102 Lau, A.W., Hoffman, B.D., Davies, A., Crocker, J.C., Lubensky, T.C. *Phys. Rev. Lett.* **2003**, *91*(19), 198101-1–198101-4.

103 Tabony, J., Glade, N., Papaseit, C., Demongeot, J. *Adv. Space Biol. Med.* **2002**, *8*, 19–58.

104 Veldhuijzen, J.P., Van Loon, J.J.W.A., The Effect of Microgravity and Mechanical Stimulation on the In Vitro Mineralization and Resorption in Fetal Mouse Long Bones, in: *Biorack on Spacelab IML-1*, Mattok, C. (Ed.), pp. 129–137, ESA SP-1162, ESA Publications Division, ESTEC, Noordwijk, **1995**.

7
Bone Cell Biology in Microgravity

Geert Carmeliet, Lieve Coenegrachts, and Roger Bouillon

7.1
Overview

Space flight results in loss of bone mass, especially in weight-bearing bones, a condition resembling disuse osteoporosis. The general picture that emerged is that mainly bone formation is decreased, whereas bone resorption is unaltered or increased. This decrease in bone mass can be restored, but the time span for recovery exceeds the period of unloading. The pathway by which microgravity is transduced into biochemical signals has been progressively elucidated. *In vitro* studies using osteoblasts or their precursors show that their nuclear morphology, cytoskeletal structure and intracellular signalling cascades are altered during (simulated) microgravity, resulting in impaired differentiation of mesenchymal stem cells and osteoblasts.

7.2
Introduction

To carry out its function of mechanical integrity and its involvement in mineral homeostasis, bone is continuously remodelled. Bone remodelling is a highly complex process that is tightly regulated by local and endocrine factors and by mechanical usage. Mechanical loading plays a critical role in maintaining bone mass and strength at areas where increased force is sensed. Consequently, the architecture of bone is correlated with the mechanical stresses exerted on it, resulting in material with optimal functional design. Various models of physical exercise overloading have been shown to preserve or increase skeletal mass. Correspondingly, skeletal unloading or disuse – as in sedentary people or in diseases associated with paralysis or prolonged bed rest – are associated with bone loss and are likely to contribute in age-related osteopenia. An extreme example is the response of bone to weightlessness during space flight. This bone loss that develops under microgravity conditions represents the most significant hindrance for long-term space travel, e.g. a flight to Mars, or for lengthy stays under conditions of reduced

Biology in Space and Life on Earth. Effects of Spaceflight on Biological Systems. Edited by Enno Brinckmann

ISBN: 978-3-527-40668-5

gravity (i.e., stays on the Moon). In this chapter, we describe only briefly the alterations in bone metabolism in humans and small animals induced by space flight and skeletal unloading; in contrast, we focus mainly on the cellular mechanisms underlying this bone loss.

7.3
Bone Remodelling: An Equilibrium between Osteoblasts and Osteoclasts

Adult bone is continuously remodelled and this process consists of two phases, resorption of pre-existing bone tissues by osteoclasts, followed by *de novo* bone formation by the osteoblasts. Osteoblasts are derived from mesenchymal precursor cells that also give rise to chondrocytes, myoblasts, adipocytes and tendon cells [1]. Mesenchymal stem cells (MSC) are multipotent cells present in adult bone marrow and can replicate as undifferentiated cells. Thus, osteoblasts share a common precursor with adipocytes, and the expression of specific transcription factors directs these mesenchymal cells to either of these lineages: differentiation to adipocytes is controlled by C/EBPs and PPARγ whereas Runx2, also known as Cbfa1, is essential for the differentiation of mesenchymal precursors to osteoblasts (Fig. 7.1). Pre-osteoblasts expressing Runx2 then differentiate into osteoblasts expressing alkaline phosphatase and collagen 1, which is the major component of the bone matrix (osteoid). In a later stage, mature osteoblasts mineralize the osteoid and are typified by expression of osteocalcin. Ultimately, the cells become embedded within the bone matrix as osteocytes or are converted into bone lining cells.

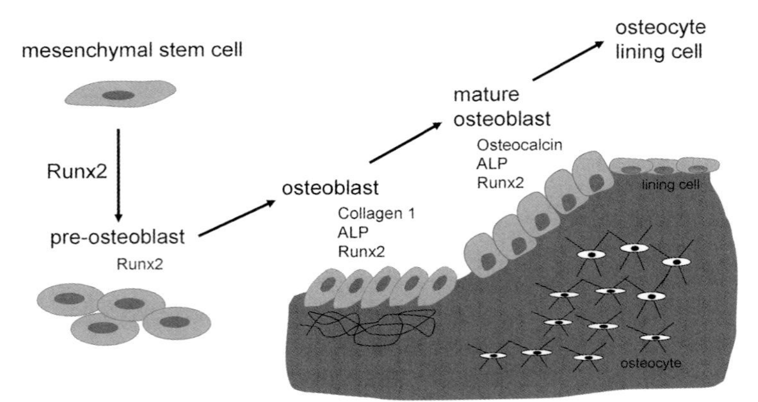

Fig. 7.1 Osteoblast differentiation. Osteoblasts derive from mesenchymal stem cells that are directed into the osteoblast lineage by Runx2. Subsequently, pre-osteoblasts differentiate, expressing alkaline phosphatase (ALP) and collagen 1, which is the major component of the bone matrix (osteoid). In a later stage, mature osteoblasts mineralize the osteoid and are typified by expression of osteocalcin. Ultimately, the cells become embedded within the bone matrix as osteocytes or are converted into bone lining cells.

The bone resorbing cells or osteoclasts are specialized cells derived from the monocyte/macrophage haematopoietic lineage that develop in close contact with osteoblasts/stromal cells that secrete factors promoting osteoclastogenesis [2]. Mature osteoclasts adhere to the bone matrix, and then secrete acid and lytic enzymes that degrade this bone matrix in a specialized, extracellular compartment. The degradation products of bone matrix (hydroxyproline, collagen crosslinks) can be measured in serum and urine.

Although resorption is much faster than formation (it takes at least three months to rebuild bone resorbed in 2–3 weeks), bone resorption and formation are "coupled" locally by mechanisms not fully understood. Imbalances of remodelling can result in gross perturbation in skeletal structure and function, and one of the regulators of bone mass, besides calcium availability and sex steroids, is mechanical usage.

7.4 Human Studies: Response of Bone to Space Flight

It has been estimated that, in a space flight environment, on average about 0.9% of the skeleton is mobilized and lost each month [3–5]. This general pattern emerged even though these data are derived from a small number of subjects studied and marked differences between flights existed with respect to duration, diet, and use of countermeasures. These changes in bone mass are, however, site-specific, with the tendency that weight-bearing bones (calcaneus, tibia, femur, and vertebra) are more affected by space flight than non-weight-bearing bones (radius, ulna) [3, 5]. In addition, the tibial trabecular bone showed more striking and earlier bone loss than the tibial cortex [6]. This bone loss was also observed in cosmonauts receiving some physical training as countermeasure. Moreover, after a recovery period on Earth of similar duration as the space mission, the tibial bone loss persisted, suggesting that the time needed to recover is longer than the mission duration [6, 7]. The recovery of skeletal density after long-duration space flights (4–6 months) may even exceed one year, as was recently shown [8]. Notably, bone size was increased by periosteal apposition, probably as a compensatory response for bone loss. Although no pathological fractures have yet occurred, space flight-related bone loss might have potentially serious consequences in long-term space flight, especially because the recovery is a long-lasting process, if it is possible at all.

In normal adult bone, equilibrium exists between bone formation and bone resorption (see above). The bone loss observed during space flight resulted mainly from decreased bone formation in association with increased or unaltered bone resorption, as evidenced by biochemical markers of bone turnover. The general tendency is that bone formation markers (the COOH-terminal propeptide of human type 1 procollagen, bone alkaline phosphatase, osteocalcin) are decreased and that bone resorption markers measured in urine are increased (hydroxyproline, collagen crosslinks) [9–12]. This description is a general picture and some studies have reported other findings in bone density and biochemical parameters. The apparent contradiction is most likely related to the large inter-individual variations found, which are due to previous environmental and lifestyle factors and genetic predisposition.

7.5
Space Flight and Unloading in the Rat Mimics Human Bone Loss

To examine the effects of space flight on skeletal structure and metabolism in detail, the rat has been used extensively as a model system. Histomorphometric analyses demonstrated a decrease in the periosteal bone formation rate [13–15] and a decrease in trabecular bone mass [16, 17] after space flight, especially in weight-bearing bones. In addition, the formed bone matrix was abnormal and hypomineralized [18] and osteoblast surface and number were decreased. Osteoclast number was, however, unchanged [16]. Accordingly, endosteal bone resorption was not markedly affected [14, 19], whereas trabecular bone resorption was transiently increased with longer duration of the space flight [20]. These alterations contribute to the observed reduction in bone strength [21–23]. On testing the reversibility of this process during a period equal to the duration of the flight, bone loss was still observed at certain, but not all, sites, although dynamic parameters indicated that recovery was started [24, 25]. These findings agree with observations in astronauts and cosmonauts, making rodents a suitable model.

Hindlimb unloading by tail suspension is an established ground model of mechanical unloading that occurs during space flight. In such a model, the animal bears the whole weight on its forelimbs while the movement of its hindlimbs is free without weight bearing. Histomorphometric and biochemical analyses have shown that skeletal adaptations in these rats are similar to those observed in rats exposed to space flight [26–30]. The deficit in bone mass induced by skeletal unloading can be replaced if normal weight bearing is restored, but again the recovery is a long-lasting process [28, 29, 31, 32].

Taken together, both human and rat studies have shown that the negative bone balance induced by space flight is mainly due to impaired bone formation and to a lesser degree to increased bone resorption.

7.6
Mechanisms of Decreased Bone Formation Induced by Unloading or Space Flight

The bone loss observed during space flight or skeletal unloading is considered to result mainly from reduced bone formation, suggesting that the osteoblast lineage is susceptible to altered gravity or loading. Zerath et al. [33] have demonstrated that the number of total and alkaline phosphatase-positive endosteal bone cells isolated from rats after space flight was reduced, suggesting that osteoblast precursor cell recruitment was decreased (Fig. 7.2). In addition, skeletal unloading

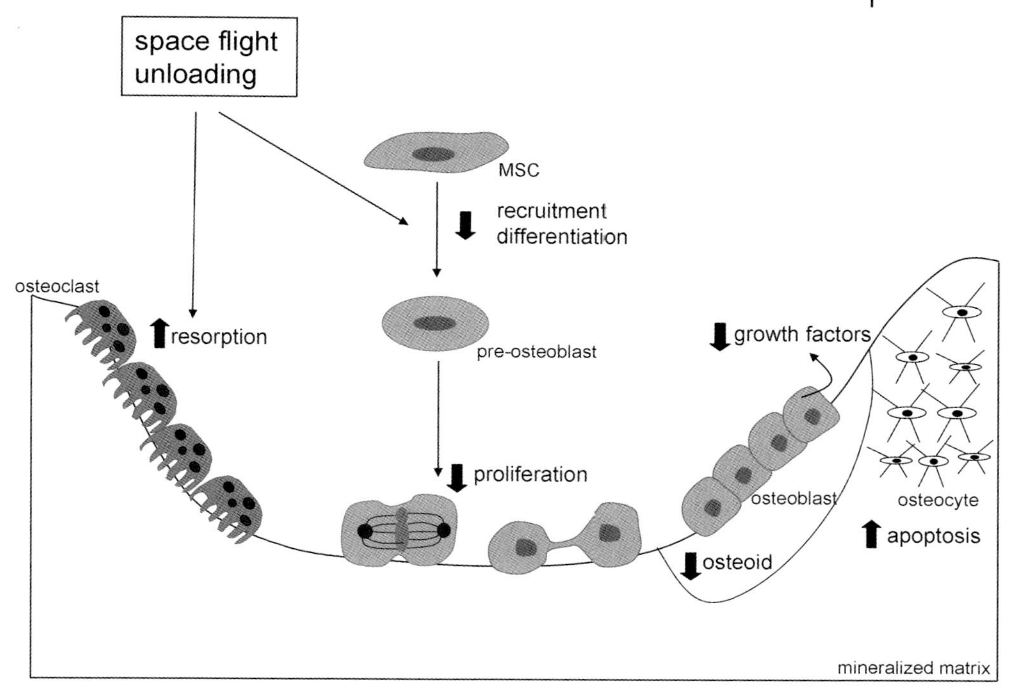

Fig. 7.2 Model of the cellular alterations underlying bone loss induced by space flight or unloading. Bone resorption is temporally increased or remains unaltered whereas several aspects of bone formation are decreased, resulting in an imbalance. The recruitment and differentiation of mesenchymal stem cells (MSC) to pre-osteoblasts is reduced. In addition, proliferation, and especially differentiation of osteoblasts, is impaired as the expression of matrix proteins is decreased and the release of growth factors is altered. Finally, more osteoblasts, and especially osteocytes, undergo apoptosis.

decreased the proliferation of periosteal and trabecular preosteoblasts as well as bone marrow cells assessed both *in vivo* [34] as *in vitro* [30, 35–37]. This decreased proliferation was (partly) due to unloading induced resistance to insulin-like growth factor 1 (IGF-1). More precisely, the activation of the IGF-1 signalling pathways was inhibited through downregulation of the integrin $\alpha_v\beta_3$ [38, 39].

In addition to reduced proliferation, a decrease in differentiation of bone marrow cells to osteoblasts was also observed during skeletal unloading, as evidenced by decreased expression of bone matrix proteins and reduced mineralized nodule formation [29, 35, 40, 41]. Accordingly, space flight resulted in reduced gene expression of bone matrix proteins [42], and these changes were site-specific [43]. Moreover, the message level of local growth factors was altered by space flight [26, 44]. In addition to the reduced proliferation and differentiation of (pre-) osteoblasts, hindlimb unloading increased the apoptosis of osteoblasts and osteocytes [45–47].

These findings support the hypothesis that bone loss induced by space flight or skeletal unloading in both humans and rats is partly based on decreased osteoblast proliferation, differentiation and function. The mechanism underlying this reduced bone formation is not yet fully understood. Changes in hormones, locally acting growth factors and neuronal signals, together with alteration in blood flow, are certainly contributing factors. Another possible mechanism suggests that the osteoblast in itself displays an altered cellular behaviour when exposed to microgravity.

7.7
Are Osteoblastic Cells *In Vitro* Responding to Altered Gravity Conditions?

The next question is whether microgravity conditions alter osteoblast function *in vitro*. To this end, several studies have been performed using osteoblastic cell cultures during short duration space flights (5–10 days). Ideally, an in-flight unit-gravity (1×g) centrifuge is used as an internal control, as described in certain experiments, although this opportunity is not always available. In addition, it remains difficult and impractical to conduct well-controlled *in vitro* studies in sufficient numbers in real microgravity conditions because of the limited and expensive nature of space flight missions and due to payload constraints. To meet these restrictions, several ground-based systems, including the two-dimensional (2D) and 3D clinostats and the rotating wall vessel (RCCS), have been developed to simulate microgravity conditions on cultured cells and tissues [48–51]. Simulated microgravity is based on the hypothesis that sensing no weight would have effects similar to those of weightlessness. The 3D clinostat simulates microgravity by continuously moving the gravity vector in three dimensions before the cell has enough time to sense it, which is a method called gravity-vector averaging. Although some disagreement exists regarding the validity of ground-based microgravity experiments, several cell types have been studied using the RCCS, and many of these studies reveal strikingly similar results to those obtained during space flight.

In this chapter we focus on studies using osteoblasts at different stages of development (mesenchymal cells, osteoblasts) derived from different species (human, rat and mouse) and describe the effect of space flight and simulated microgravity on proliferation, differentiation and the possible underlying mechanisms.

7.7.1
Proliferation and Apoptosis

In most rat and human osteoblastic cell cultures (simulated) microgravity did not affect cell proliferation [51–55], although a reduced cell number was noticed in a mouse osteoblastic cell line after space flight [56]. In agreement with the first observation, the proliferation of osteoblast precursor cells was not altered by simulated microgravity, as assessed in human MSC (hMSC) and murine pre-osteoblasts (2T3) [57, 58]. However, apoptosis was increased in rat osteoblastic cells during simulated microgravity [51].

7.7.2
Differentiation: Matrix Production

The production of matrix proteins is an essential feature in the differentiation of osteoblasts and is required for mineralization. We have shown that in human osteosarcoma cells, MG63, gene expression for collagen Ia1, alkaline phosphatase, and osteocalcin following treatment with 1,25-dihydroxyvitamin D_3 and transforming growth factor β (TGFβ) was reduced at microgravity (Fig. 7.3) [52, 53]. Similar results were obtained in simulated microgravity using the same cell line [59]. Moreover, other osteoblastic cells cultured in (simulated) microgravity show a comparable reduction in osteoblastic gene markers, including alkaline phosphatase, osteocalcin, and Runx2 [60, 61]. These data indicate that the differentiation of osteoblastic cells is impaired under microgravity conditions, a finding consistent with the *in vivo* observations (Section 7.6). In addition, also osteoblast precursor cells (hMSC) cultured in simulated microgravity failed to display detectable mRNA levels for major osteoblastic markers, including alkaline phosphatase, procollagen type I, osteonectin and Runx2, suggesting marked suppression of hMSC differentiation into osteoblasts [57]. Comparable findings were obtained using the pre-osteoblast cell line 2T3 [58]. At the same time, PPARγ2 became highly expressed in hMSC cultured in simulated microgravity, indicating that the development towards adipocytes was enhanced. Moreover, the changes observed in hMSC were not corrected after 35 days of culture under normal gravity. Taken together, (simulated) microgravity inhibits osteoblastic differentiation at several

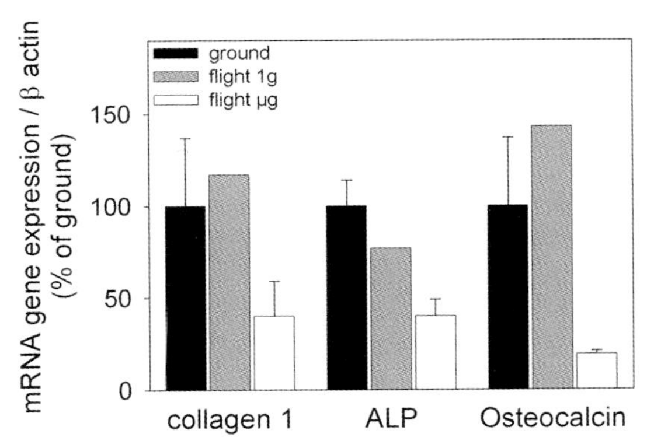

Fig. 7.3 Gene expression of differentiation-related genes in MG63 cells cultured at unit- and microgravity for 9 days. The mRNA level for collagen 1, alkaline phosphatase (ALP) and osteocalcin corrected for β-actin mRNA in 1,25-dihydroxyvitamin D_3 treated cells cultured at unit-gravity (ground), microgravity (flight μg) and in-flight unit-gravity (flight 1g) are shown. The mRNA level of treated cells at unit-gravity was taken as 100%.

stages of its development and may even switch the development of MSC toward the adipocytic lineage.

7.7.3
Differentiation: Growth Factors

Osteoblasts not only produce bone matrix while they differentiate but are also an important source of local growth factors that regulate the proliferation and differentiation of cells involved in bone homeostasis and remodelling. Under microgravity conditions the message level of interleukin-6, a cytokine involved in bone resorption, was increased in 1,25-dihydroxyvitamin D_3 treated rat stromal cells [62]. In contrast, several members belonging to the IGF binding protein family, which are modulators of IGF-1 action, were differentially regulated in these cells when cultured during space flight: IGF binding protein-3 message level was increased, whereas IGF binding protein-5 and -4 mRNA levels were decreased [63]. In addition, we observed that microgravity also altered the expression of TGFβ1 and its binding protein LTBP1 in human osteoblastic cells. TGFβ and fibroblast growth factor-2 mRNA levels were also decreased in microgravity exposed murine osteoblasts [64]. Moreover, distinct changes in the level of growth factor receptors were observed. The mRNA level for platelet-derived growth factor-β receptor was decreased in microgravity exposed rat osteoblastic cultures compared with ground controls, whereas the expression of epidermal growth factor receptor was not altered [65]. This changed behaviour of osteoblastic cells under microgravity can certainly contribute to the altered bone remodelling observed during space flight, although the exact mechanisms are not yet elucidated.

7.8
Potential Mechanisms of Altered Osteoblastic Behaviour

The molecular mechanisms responsible for the impaired osteoblastic differentiation and altered growth factor production are still poorly understood.

Since osteoblast differentiation following 1,25-dihydroxyvitamin D_3 treatment was decreased under microgravity we investigated the effect of microgravity on the intracellular signalling pathway of 1,25-dihydroxyvitamin D_3. The ligand 1,25-dihydroxyvitamin D_3 interacts with the vitamin D receptor (VDR) and this complex binds to vitamin D response elements (VDRE) in the promoter region of target genes to stimulate or suppress gene transcription (Fig. 7.4A). To investigate the interaction of liganded VDR with VDRE, the mouse osteoblastic cell line, MC3T3, was stable transfected with a construct containing multiple VDRE of the rat osteocalcin promoter fused to growth hormone as reporter gene. Treatment of these transfectants with 1,25-dihydroxyvitamin D_3 resulted in a time- and dose-dependent release of growth hormone in the culture medium. During space flight these cultures responded to 1,25-dihydroxyvitamin D_3 treatment with increased growth hormone production that was comparable to the induction observed in

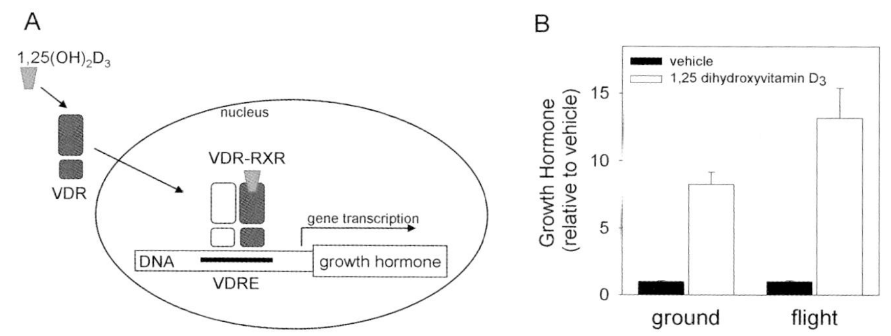

Fig. 7.4 (A) Scheme of investigated 1,25-dihydroxyvitamin D_3 signalling pathway. Mouse osteoblastic cells (MC3T3) were stably transfected with a construct containing multiple VDRE (vitamin D response elements) of the rat osteocalcin promoter fused to growth hormone as reporter gene (MC3T3-VDRE). 1,25-Dihydroxyvitamin D_3 will interact with VDR (vitamin D receptor), which will translocate to the nucleus and bind as a heterodimer with RXR to the osteocalcin VDRE, resulting in increased growth hormone release in the medium. (B) Growth hormone release after 1,25-dihydroxyvitamin D_3 treatment of MC3T3-VDRE cells cultured during space flight or on the ground and expressed relative to vehicle treated cells.

ground cultures (Fig. 7.4B). These data indicate that microgravity did not alter the interaction of VDR with the osteocalcin VDRE or the subsequent gene transcription. However, it has recently become evident that chromatin modifications outside the binding sites for transcription factors contribute to gene transcription, and this has also been observed for vitamin D regulated gene expression [66]. Moreover, gravity has lately been shown to affect the inner nuclear structure of murine osteoblasts [64]. These nuclear changes may be linked to cytoskeletal alterations observed in osteoblasts exposed to (simulated) microgravity.

Several experiments using different osteoblastic cell types have shown that cell morphology is altered under microgravity conditions [62, 63, 65, 67]. Microgravity resulted in less mature focal adhesion plaques in rat osteoblastic cells by disorganizing cytoskeletal actin and losing vinculin-positive focal contacts (Fig. 7.5) [55]. Disruption of F-actin stress fibres was also detected in hMSC exposed to simulated microgravity. A potential mechanism responsible for this cytoskeletal disruption of hMSC is the reduced activation of the small GTPase RhoA observed in simulated microgravity [68]. RhoA is required for tyrosine phosphorylation and full activation of FAK and also contributes to integrin-mediated ERK activation, which are both affected by simulated microgravity [69]. These data suggest that integrin interaction with extracellular matrix proteins is disrupted and integrin signalling is reduced in microgravity. Notably, simulated microgravity resulted in reduced expression of collagen I but increased $\alpha 2$ and $\beta 1$ integrin expression, pointing to the importance of the impaired downstream signalling cascade.

This altered cytoskeletal organization may explain the reduced osteoblastogenesis and enhanced adipogenesis of hMSCs (see above), it may control the signalling cascades induced by hormones and growth factors in osteoblasts and it may be

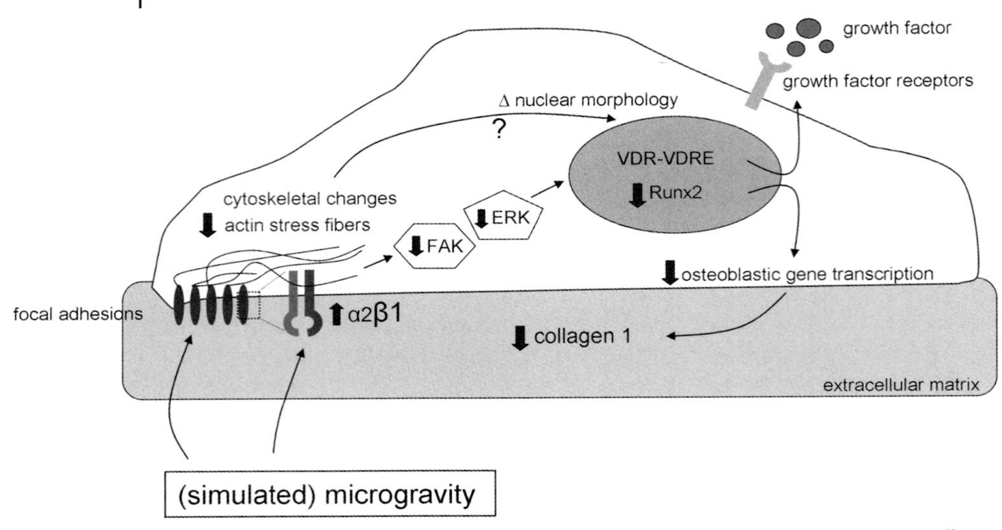

Fig. 7.5 Schematic representation of the effect of (simulated) microgravity on osteoblastic behaviour. Osteoblastic cells exposed to (simulated) microgravity show altered cell morphology, with less mature focal adhesion plaques and disorganized cytoskeletal structures. Moreover, downstream signalling cascade is impaired and, together with altered nuclear morphology, affects gene expression for matrix proteins and growth factors, thereby reducing osteoblast differentiation.

linked to increased apoptosis as observed in rat osteoblastic cells exposed to simulated microgravity [51].

7.9 Conclusion

During space flight, bone formation becomes impaired, resulting in decreased bone mass, primarily in bones most loaded in the 1×g environment. The limited number of studies performed in space and the restrictions inherent with such experiments do not yet allow us to drawing final conclusions on the precise, specific effects of space flight on bone. Nevertheless, it has gradually become apparent that, besides alterations in blood flow, systemic hormones and locally produced growth factors, the behaviour of osteoblasts may also be directly affected by microgravity. *In vitro* studies using osteoblasts or their precursors exposed to (simulated) microgravity indicate that their differentiation is impaired, probably due to changes in cytoskeletal structure and intracellular signalling pathways. Whether and how these isolated cells sense microgravity is, however, not yet defined. The answers to such questions are not only relevant to understanding the effects of space flight on living organisms and to preventing specific problems in astronauts but could also increase our insight into fundamental processes of cell biology and, possibly, provide links to the pathophysiology of certain diseases such as age-related bone loss.

Acknowledgments

The authors thank the European Space Agency and the personnel at the European Space Research and Technology Centre (ESTEC) in Noordwijk for their invaluable support and good collaboration. Our microgravity projects have been financially supported by Diensten van de Eerste Minister, Wetenschappelijke, Technische en Culturele Aangelegenheden (DWTC).

References

1 Harada, S., Rodan, G.A. *Nature* **2003**, *423*, 349–355.
2 Boyle, W.J., Simonet, W.S., Lacey, D.L. *Nature* **2003**, *423*, 337–342.
3 Stupakov, G.P., Kazeykin, V.S., Kozlovskiy, A.P., Korolev, V.V. *Space Biol. Med.* **1984**, *18*, 42–47.
4 LeBlanc, A.D., Schneider, V., Shackelford, L.L., West, S., Oganov, V., Bakulin, A., Veronin, L. *J. Bone Miner. Res.* **1996**, *1*, S323.
5 Collet, P., Uebelhart, D., Vico, L., Moro, L., Hartmann, D., Roth, M., Alexandre, C. *Bone* **1997**, *20*, 547–551.
6 Vico, L., Collet, P., Guignandon, A., Lafage-Proust, M.H., Thomas, T., Rehailia, M., Alexandre, C. *Lancet* **2000**, *355*, 1607–1611.
7 Tilton, F.E., Degioanni, J.J., Schneider, V.S. *Aviat. Space Environ. Med.* **1980**, *51*, 1209–1213.
8 Lang, T.F., Leblanc, A.D., Evans, H.J., Lu, Y. J. *Bone Miner. Res.* **2006**, *21(8)*, 1224–1230.
9 Smith, S.M., Nillen, J.L., LeBlanc, A., Lipton, A., Demers, L.M., Lane, H.W., Leach, C.S. *J. Clin. Endocrinol. Metab.* **1998**, *83*, 3584–3591.
10 Smith, S.M., Wastney, M.E., Morukov, B.V., Larina, I.M., Nyquist, L.E., Abrams, S.A., Taran, E.N., Shih, C.Y., Nillen, J.L., Davis-Street, J.E., Rice, B.L., Lane, H.W. *Am. J. Physiol.* **1999**, *277*, R1–R10.
11 Caillot-Augusseau, A., Lafage-Proust, M.H., Soler, C., Pernod, J., Dubois, F., Alexandre, C. *Clin. Chem.* **1998**, *44*, 578–585.
12 Caillot-Augusseau, A., Vico, L., Heer, M., Voroviev, D., Souberbielle, J.C., Zitterman, A., Alexandre, C., Lafage-Proust, M.H. *Clin. Chem.* **2000**, *46*, 1136–1143.
13 Morey, E.R., Baylink, D.J. *Science* **1978**, *201*, 1138–1141.
14 Wronski, T.J., Morey, E.R. *Am. J. Physiol.* **1983**, *244*, R305–R309.
15 Wronski, T.J., Morey-Holton, E.R., Doty, S.B., Maese, A.C., Walsh, C.C. *Am. J. Physiol.* **1987**, *252*, R252–R255.
16 Jee, W.S.S., Wronski, T.J., Morey, E.R., Kimmel, D.B. *Am. J. Physiol.* **1983**, *244*, R310–R314.
17 Vico, L., Chappard, D., Palle, S., Bakulin, A.V., Novikov, V.E., Alexandre, C. *Am. J. Physiol.* **1988**, *255*, R243–R247.
18 Turner, R.T., Bell, N.H., Duvall, P., Bobyn, J.D., Spector, M., Morey-Holton, E., Baylink, D.J. *Proc. Soc. Exp. Biol. Med.* **1985**, *180*, 544–549.
19 Cann, C.E., Adachi, R.R. *Am. J. Physiol.* **1983**, *244*, R327–331.
20 Vico, L., Bourrin, S., Genty, C., Palle, S., Alexandre, C. *J. Appl. Physiol.* **1993**, *75*, 2203–2208.
21 Spengler, D.M., Morey, E.R., Carter, D.R., Turner, R.T., Baylink, D.J. *Proc. Soc. Exp. Biol. Med.* **1983**, *174*, 224–228.
22 Shaw, S.R., Vailas, A.C., Grindeland, R.E., Zernicke, R.F. *Am. J. Physiol.* **1988**, *254*, R78–R83.
23 Vailas, A.C., Zernicke, R.F., Grindeland, R.E., Kaplansky, A., Durnova, G.N., Li, K.C., Martinez, D.A. *FASEB J.* **1990**, *4*, 47–54.

24 Zerath, E., Godet, D., Holy, X., Andre, C., Renault, S., Hott, M., Marie, P.J. *J. Appl. Physiol.* **1996**, *81*, 164–171.
25 Lafage-Proust, M.H., Collet, P., Dubost, J.M., Laroche, N., Alexandre, C., Vico, L. *Am. J. Physiol.* **1998**, *274*, R324–R334.
26 Bikle, D.D., Harris, J., Halloran, B.P., Morey-Holton, E. *Am. J. Physiol.* **1994**, *267*, E822–E827.
27 Globus, R.K., Bikle, D.D., Morey-Holton, E. *Endocrinology* **1986**, *118*, 733–742.
28 Halloran, B.P., Bikle, D.D., Harris, J., Foskett, H.C., Morey-Holton, E. *Am. J. Physiol.* **1993**, *264*, E712–E716.
29 Sakata, T., Sakai, A., Tsurukami, H., Okimoto, N., Okazaki, Y., Ikeda, S., Norimura, T., Nakamura, T. *J. Bone Miner. Res.* **1999**, *14*, 1596–1604.
30 Machwate, M., Zerath, E., Holy, X., Hott, M., Modrowski, D., Malouvier, A., Marie, P.J. *Am. J. Physiol.* **1993**, *264*, E790–E799.
31 Sessions, N.D.V., Halloran, B.P., Bikle, D.D., Wronski, T.J., Cone, C.M., Morey-Holton, E. *Am. J. Physiol.* **1989**, *257*, E606–E610.
32 Bourrin, S., Palle, S., Genty, C., Alexandre, C. *J. Bone Miner. Res.* **1995**, *10*, 820–828.
33 Zerath, E., Holy, X., Roberts, S.G., Andre, C., Renault, S., Hott, M., Marie, P.J. *J. Bone Miner. Res.* **2000**, *15*, 1310–1320.
34 Barou, O., Palle, S., Vico, L., Alexandre, C., Lafage-Proust, M.H. *Am. J. Physiol.* **1998**, *274*, E108–E114.
35 Zhang, R., Supowit, S.C., Klein, G.L., Lu, Z., Christensen, M.D., Lozano, R., Simmons, D.J. *J. Bone Miner. Res.* **1995**, *10*, 415–423.
36 Kostenuik, P.J., Halloran, B.P., Morey-Holton, E., Bikle, D.D. *Am. J. Physiol.* **1997**, *273*, E1133–E1139.
37 Basso, N., Jia, Y., Bellows, C.G., Heersche, J.N. *Bone* **2005**, *37(3)*, 370–378.
38 Bikle, D.D., Harris, J., Halloran, B.P., Turner, R.T., Morey-Holton, E.R. *J. Bone Miner. Res.* **1994**, *9*, 1789–1796.
39 Sakata, T., Yongmei, W., Halloran, B.P., Elalieh, H.Z., Cao, J., Bikle, D.D. *J. Bone Miner. Res.* **2004**, *19*, 436–446.
40 Bikle, D.D., Harris, J., Halloran, B.P., Currier, P.A., Tanner, S., Morey-Holton, E. *Endocrinology* **1995**, *136*, 2099–2109.
41 Grano, M., Mori, G., Minielli, V., Barou, O., Colucci, S., Giannelli, G., Alexandre, C., Zallone, A.Z., Vico, L. *Calc. Tissue Int.* **2002**, *70(3)*, 176–185.
42 Backup, P., Westerlind, K., Harris, S., Spelsberg, T., Kline, B., Turner, R. *Am. J. Physiol.* **1994**, *266*, E567–E573.
43 Evans, G.L., Morey-Holton, E., Turner, R.T. *J. Appl. Physiol.* **1998**, *84*, 2132–2137.
44 Westerlind, K.C., Turner, R.T. *J. Bone Miner. Res.* **1995**, *10*, 843–848.
45 Dufour, C., Holy, X., Marie, P.J. Skeletal unloading induces osteoblast apoptosis and targets alpha5beta1-PI3K-Bcl-2 signaling in rat bone. *Exp. Cell. Res.* **2007**, *313(2)*, 394–403.
46 Aguirre, J.I., Plotkin, L.I., Stewart, S.A., Weinstein, R.S., Parfitt, A.M., Manolagas, S.C., Bellido, T. *J. Bone Miner. Res.* **2006**, *21(4)*, 650–615.
47 Basso, N., Heersche, J.N. *Bone* **2006**, *39(4)*, 807–814.
48 Al-Ajmi, N., Braidman, I.P., Moore, D. *Adv. Space Res.* **1996**, *17(6–7)*, 189–192.
49 Kobayashi, K., Kambe, F., Kurokouchi, K., Sakai, T., Ishiguro, N., Iwata, H., Koga, K., Gruener, R., Seo, H. *Biochem. Biophys. Res. Commun.* **2000**, *279(1)*, 258–264.
50 Nakamura, H., Kumie, Y., Morita, S., Shimokawa, H., Ohya, K., Shinomiya, K. *Ann. New York Acad. Sci.* **2003**, *1010*, 143–147.
51 Sarkar, D., Nagaya, T., Koga, K., Nomura, Y., Gruener, R., Seo, H. *J. Bone Miner. Res.* **2000**, *15(3)*, 489–498.
52 Carmeliet, G., Nys, G., Bouillon, R. *J. Bone Miner. Res.* **1997**, *12(5)*, 786–794.
53 Carmeliet, G., Nys, G., Stockmans, I., Bouillon, R. *Bone* **1998**, *5*, 139S–143S.
54 Harris, S.A., Zhang, M., Kidder, L.S., Evans, G.L., Spelsberg, T.C., Turner, R.T. *Bone* **2000**, *26(4)*, 325–331.
55 Guignandon, A., Lafage-Proust, M.H., Usson, Y., Laroche, N., Caillot-Augusseau, A., Alexandre, C., Vico, L. *FASEB J.* **2001**, *15(11)*, 2036–2038.
56 Hughes-Fulford, M., Lewis, M.L. *Exp. Cell Res.* **1996**, *224(1)*, 103–109.
57 Zayzafoon, M., Gathings, W.E., McDonald, J.M. *Endocrinology* **2004**, *145(5)*, 2421–2432.
58 Pardo, S.J., Patel, M.J., Sykes, M.C., Platt, M.O., Boyd, N.L., Sorescu, G.P., Xu, M.,

van Loon, J.J., Wang, M.D., Jo, H. *Am. J. Physiol. Cell Physiol.* **2005**, *288*(*6*), C1211–C1221.

59 Narayanan, R., Smith, C.L., Weigel, N.L. *Bone* **2002**, *31*(*3*), 381–388.

60 Landis, W.J., Hodgens, K.J., Block, D., Toma, C.D., Gerstenfeld, L.C. *J. Bone Miner. Res.* **2000**, *15*(*6*), 1099–1112.

61 Ontiveros, C., McCabe, L.R. *J. Cell. Biochem.* **2003**, *88*(*3*), 427–437.

62 Kumei, Y., Shimokawa, H., Katano, H., Hara, E., Akiyama, H., Hirano, M., Mukai, C., Nagaoka, S., Whitson, P.A., Sams, C.F. *J. Biotechnol.* **1996**, *47*(*2–3*), 313–324.

63 Kumei, Y., Shimokawa, H., Katano, H., Akiyama, H., Hirano, M., Mukai, C., Nagaoka, S., Whitson, P.A., Sams, C.F. *J. Appl. Physiol.* **1998**, *85*(*1*), 139–147.

64 Hughes-Fulford, M., Rodenacker, K., Jutting, U. *J. Cell. Biochem.* **2006**, *99*(*2*), 435–449.

65 Akiyama, H., Kanai, S., Hirano, M., Shimokawa, H., Katano, H., Mukai, C., Nagaoka, S., Morita, S., Kumei, Y. *Mol. Cell. Biochem.* **1999**, *202*, 63–71.

66 Carlberg, C., Dunlop, T.W. *Anticancer Res.* **2006**, *26*(*4A*), 2637–2645.

67 Guignandon, A., Usson, Y., Laroche, N., Lafage-Proust, M.H., Sabido, O., Alexandre, C., Vico, L. *Exp. Cell Res.* **1997**, *236*, 66–75.

68 Meyers, V.E., Zayzafoon, M., Douglas, J.T., McDonald, J.M. *J. Bone Miner. Res.* **2005**, *20*(*10*), 1858–1866.

69 Meyers, V.E., Zayzafoon, M., Gonda, S.R., Gathings, W.E., McDonald, J.M. *J. Cell. Biochem.* **2004**, *93*(*4*), 697–707.

8
Cells of the Immune System in Space (Lymphocytes)

Augusto Cogoli and Marianne Cogoli-Greuter

8.1
Introduction

When, in 1976, we proposed our first experiment with lymphocytes in Space it had been known for decades that the immune response of humans undergoing physical and psychological stress was significantly depressed. Moreover, measurements conducted on blood samples from Space crew members before and after flight revealed that some immunological parameters were reduced after landing. In general, the values of such parameters returned to the pre-flight baseline within a few days.

The novelty was that we proposed to mimic in Space an infection by activating cultures of lymphocytes drawn from a donor on the ground by means of a protein called concanavalin A (Con A), which is known to activate one of the two major lymphocyte populations, namely the T cells that are responsible of the cellular immunity. The other major lymphocyte population is that of the B cells responsible for the production of antibodies. We were speculating that such *in vitro* studies could deliver important information on the immune system of humans exposed to the stress of space flight.

Our first experiment with T cells was conducted, after many delays, in 1983 in Spacelab-1 (the STS-9 flight of the US Space Shuttle). The data were quite surprising [1] and led to a series of investigations conducted by us and others in several Space Shuttle flights as well on the MIR and ISS Space stations, on sounding rockets and stratospheric balloons. At the same time, parallel studies were performed in ground-based devices such as clinostats and rotating vessels. We investigated also the response of T cells to Con A on marathon runners to compare the effect of physical stress on Earth with that of humans in Space.

Briefly, the activation of T cells *in vitro* is depressed by 80–93% in microgravity compared with control samples kept at 1×g, whereas the response of T cells from humans in Space as well as immediately after landing is depressed by 20–30%. Importantly, the data of the effect of space flight on cell cultures prepared on the

Biology in Space and Life on Earth. Effects of Spaceflight on Biological Systems. Edited by Enno Brinckmann

ISBN: 978-3-527-40668-5

ground and processed in Space have to be considered with great caution when compared with the status of the immune system of an astronaut. In principle, the Con A test is a valuable check of the status of the cellular immunity (that means of the T cells) of the respective donor. We call this test an *ex vivo* test. In addition, tests, like the skin test, conducted directly on the subject are reliable measurements of the cellular immune fitness. We call these tests *in vivo* tests. Conversely, experiments in Space with cells prepared on the ground provide important information on the effect of microgravity on signal transduction and cell differentiation. We refer to this approach as *in vitro* experiments.

In parallel with our investigations in Space and on ground, analytical technology and discoveries in cellular biology made great progresses. Therefore, we had to design our protocols accordingly. As often in science, each new experiment aiming to provide new information raised also new questions that were asked in subsequent experiments. We are still preparing new investigations to be performed on the International Space Station (ISS) and on sounding rockets in the next few years.

Since the early 1990s our studies on the cells of the immune system in Space have been conducted in close collaboration with colleagues of the University of Sassari in Italy and of the University of California in San Francisco. Further collaborative projects were carried out with the Universities of Rome, Berne and Zurich and with the Institute of Biomedical Problems (IBMP) in Moscow.

We have chosen to present the data in chronological order, thus giving the reader the opportunity to follow the progress of knowledge on the effect of space flight on T cells in particular and on single cells in general. Thereby, we discuss here also data obtained on the ground with devices capable of randomizing the gravity vector, such as the Random Positioning Machine (RPM). In fact, valuable data were obtained recently while the program was on hold due to the catastrophe of Columbia.

This chapter gives an updated overview of the work performed by our teams in the last 25 years. Several reviews on the behaviour of immune cells in Space and in ground-based facilities, including also work performed by other investigators in the USA and Europe, have been published by us and others; see, for instance, Refs. [2–5]. Important contributions to the understanding of the behaviour of immune cells in Space and simulated microgravity have been made by Didier Schmitt, Ben Hashemi, Neal Pellis, Clarence Sams, Marian Lewis, Laurence Schaffar and their collaborators.

Five kinds of immune cells were used in our studies: (1) T lymphocytes from donors in cultures either of whole blood, or of white blood cells purified by gradient centrifugation on Ficoll and containing all populations of leucocytes, or of pan T cells purified through affinity columns; (2) Jurkat cells, a cell line derived from human T lymphocytes; (3) S49 WT cells and their kin^- mutant are a subclone of murine S49 lymphoma cells; (4) hybridoma cells producing monoclonal antibodies; and (5) J-111 cells, a monocyte/macrophage cell line derived from human acute monocytic leukaemia.

8.2 Activation of T Cells

The mechanism of T cell activation is very complex and has for decades been the subject of extensive investigations worldwide. The fact that they can be transformed *in vitro* from a resting to an activated status offers an ideal system for the study of cell differentiation in addition to the study of how the immune system reacts to antigens (Fig. 8.1).

We know today that three signals are required for full T cell activation [6]. The mechanism is quite complex and still not yet fully understood. The first signal is delivered to the TCR/CD3 complex either by the antigen presenting cell (antigen fragment+major histocompatibility complex, MHC) or by anti-CD3, or by the mitogen Con A (Fig. 8.2). Con A does not bind to a specific receptor but rather to all glycoproteins on the membrane fitted with α-glycosides. The second signal is a costimulatory signal delivered either by the accessory cells – usually monocytes secreting interleukin-1 (IL-1) – or by anti-CD28. Upon interaction with T cells the monocytes secrete IL-1. After stimulation of the TCR/CD3 complex and CD28 the signal is transferred to the nucleus, resulting in the synthesis of interleukin-2 (IL-2) [7, 8]. IL-2, acting as third signal, is secreted and bound to its receptor (IL-2R). This induces further synthesis of IL-2 and its receptor, resulting finally in full activation. It takes three days to reach the endpoint of activation. Such process consists of several steps in which specific cell products, called cytokines, are secreted, cell locomotion is reduced and the structure of the cytoskeleton is altered before the onset of cell division.

Several pathways are activated in parallel and have been identified only recently (see below). Conceivably, gravitational forces may interact with cell organelles and structures like the cytoskeleton, having significant density differences. Theoretical considerations in this respect have been published recently [9].

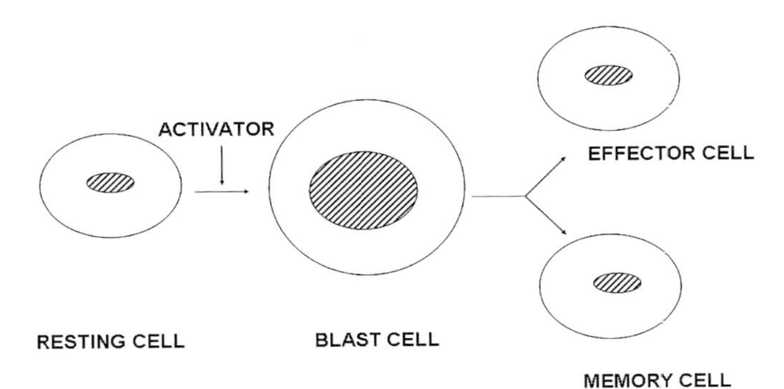

Fig. 8.1 Activation of T lymphocytes. Resting cells in the G_0 phase are activated *in vitro* with a mitogen, thus triggering the events occurring *in vivo* during antigenic activation, e.g. with a virus or a bacterium. Within two days the cells increase in volume and enter mitosis on the third day of activation.

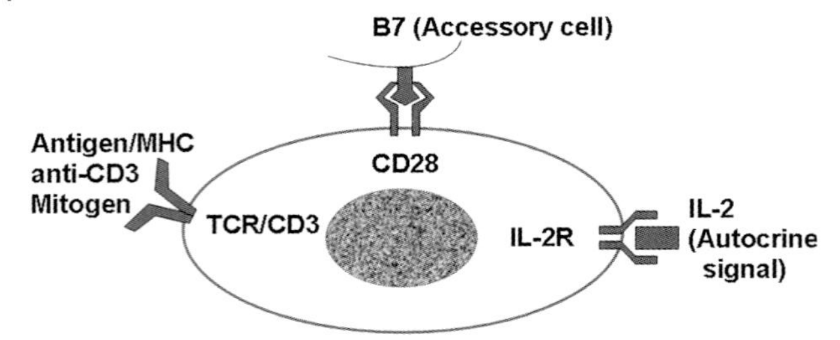

Fig. 8.2 Three signals are required for full T cell activation. The first signal is delivered *in vivo* by fragments of an antigen, first degraded and then "presented" by a monocyte via the major histocompatibility complex, to the T cell receptor/CD3 complex of the T lymphocyte. Such interaction is mimicked *in vitro* with a mitogen (e.g. Con A) or by anti-CD3 antibody. The second signal is delivered by an "accessory" cell (usually a monocyte) carrying a B7 ligand that is recognized by the CD28 receptor on the T cell. As the third signal, during the activation process, T cells produce α and β subunits of the IL-2 receptor that combine with the γ subunit constitutively present on the membrane and secrete IL-2 as autocrine.

8.3 Earliest Data

Three pioneering experiments that inspired our research were conducted in the early days of Space biology in the 1970s and are reviewed in Ref. [9]. One was a Soviet–Hungarian study by Talas, Batkay et al. on human lymphocytes that were activated with polynucleotides on board of the spaceship Salyut 7. Although the conditions of the experiment were not ideal, the results showed that lymphocyte function changed in 0×g. A five-fold increase of the interferon-α production was observed. The second investigation was performed by a US team led by Montgomery on board of Skylab with WI38 human embryonic lung cells. In what is probably the most sophisticated instrument for cell biology ever used in a Space laboratory, the cells were cultivated over weeks under controlled conditions. A microscope and a camera permitted cinematographic recording. However, cinematographic recording, phase contrast, electron and scanning microscopy revealed no observable differences in ultrastructure and in cell migration between flight and ground controls. The third study was conducted independently by US and Soviet scientists and was dedicated to the study of the immune system of humans in Space. Lymphocytes, taken from crew members of Skylab and of Salyut before and after flight, were activated with mitogens. Kimzey reported that the rate of RNA synthesis was significantly decreased after flight. Similar results were presented by Konstantinova and collaborators. Although these last investigations were not true cell biology experiments, they showed that it was possible to simulate an immune reaction *in vitro* and thus to study a very intriguing differentiation process.

8.4 Spacelab-1, 1983

Our first experiment with lymphocytes in Space was performed in November 1983 with an incubator manufactured in our workshop by our former collaborator, the late Alex Tschopp. We were offered the opportunity to test the proper function of such equipment on an earlier flight of the Space Shuttle in July of the same year. At that time we did not dispose of 1×g controls in flight. A parallel control experiment was conducted in the ground laboratory at the Kennedy Space Center with an identical incubator and with aliquots of the same culture as in Space. White blood cells were purified from peripheral blood donated 24 h before launch by centrifugation with the well-known procedure based on the Ficoll gradient. The incubator was carried on board of the Shuttle Columbia 5 h before launch. The cells were activated in flight by addition of the T cell mitogen concanavalin A (Con A); 72 h later they were treated for 3 h with tritium-labelled thymidine followed by freezing in liquid nitrogen after addition of a cryopreservative. Both the late access, 5 h before launch, and the availability of a liquid nitrogen container on board of the Space Shuttle were conditions never available on later flights. Safety regulations following the Challenger accident in 1986 worsened remarkably the experimental conditions.

The analysis was performed in our laboratory in Zurich a few days after landing by determining the mitotic index, expressed as the amount of radioactive thymidine incorporated into DNA. To our great surprise the activation of the three flight samples was depressed by 95% compared with the ground controls [1]. The fact that the cells consumed an amount of glucose similar to the controls during the 3-day incubation showed that the cells were not damaged by the flight conditions.

Why did microgravity have such an impact on the activation process? We tried to answer the question in the following experiments.

8.5 Spacelab D-1, 1985

The advent of the Biorack facility opened a new era and contributed to great progress in Space biology due to the availability of 1×g centrifuges, a freezer, a cooler and a glove box. Unlike Spacelab-1, in which the investigators had to provide their own facility, Biorack required that only the experiment-specific hardware, such as the cell culture flasks or the syringes, had to be provided by the investigators. We had the opportunity to perform two different experiments on the Spacelab D-1 mission managed by the German Space Agency DLR and flown with the shuttle Challenger in November 1985.

The protocol of one experiment was a repetition and an extension of that on Spacelab-1. In addition to the measurement of the mitotic index of Con A-activated

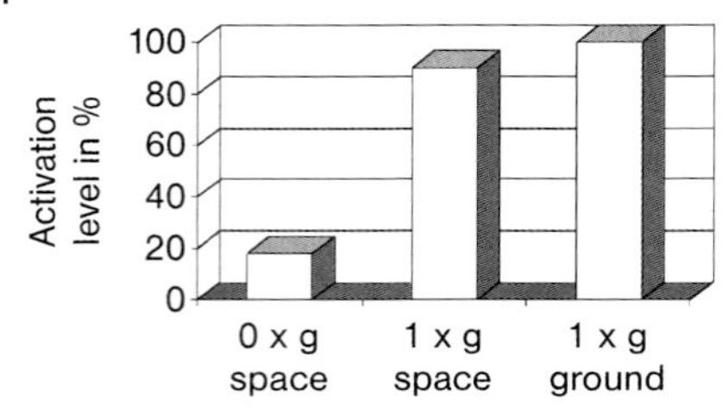

Fig. 8.3 Mitotic index, determined as amount of tritiated thymidine (2-h pulse) incorporated into DNA, of T lymphocytes activated for 72 h with Con A. White cells from the blood taken from a healthy donor were purified by gradient centrifugation 36 h before flight. The data are expressed as percent of the control processed on the ground. The slightly reduced activation of samples kept at 1×g in a reference centrifuge is most probably due to the stopping of the centrifuge to operate other experiments.

T cells, a few samples were fixed chemically in flight by the addition of glutaraldehyde, thus permitting electron-microscopic analysis. The other experiment consisted of the *ex vivo* determination of the mitotic index of T cells in whole-blood cultures, i.e. in cultures diluting peripheral blood with culture medium, obtained from donations of three crew members before, in and after flight. For the first time scientists had the opportunity to process in-flight blood samples from the astronauts.

The first experiment confirmed the results obtained on Spacelab-1, namely a remarkable depression of the mitotic index (Fig. 8.3). However, the credibility of the data was sustained by the 1×g control samples in flight, which showed an activation index only slightly lower than that of the samples incubated on ground in a Biorack unit identical to that in flight. The slight decrease was attributed to the short stops of the 1×g centrifuge due to the servicing of other experiments installed in the same facility.

Electron microscopy of the in flight 1×g samples showed the typical structure of activated lymphocytes with increased cytoplasmic volume together with mitotic cells (Fig. 8.4). The 0×g samples showed damaged cells and a structure typical of apoptotic cells. At that time, however, the phenomenon of apoptosis was still unknown [10, 11].

The probes from the astronauts confirmed the depression already observed by the Russian and US investigators on samples taken immediately after landing. We provided further support to such evidence with the data obtained at 1×g in flight (Fig. 8.5). Although both the *in vitro* and the *ex vivo* data showed a depression of mitogenic activation, the extent of the effect was much larger in the *in vitro* samples (between –80 and –90%) than in the samples taken from the crew in flight and immediately after landing (approx –50%). While the effect *in vitro* may be attributed to microgravity, the *ex vivo* effect may be caused by other effects such as psychological or physical stress or both.

Three months after D-1, Challenger exploded a few minutes after launch. This catastrophe caused a five-year delay in the Spacelab programme.

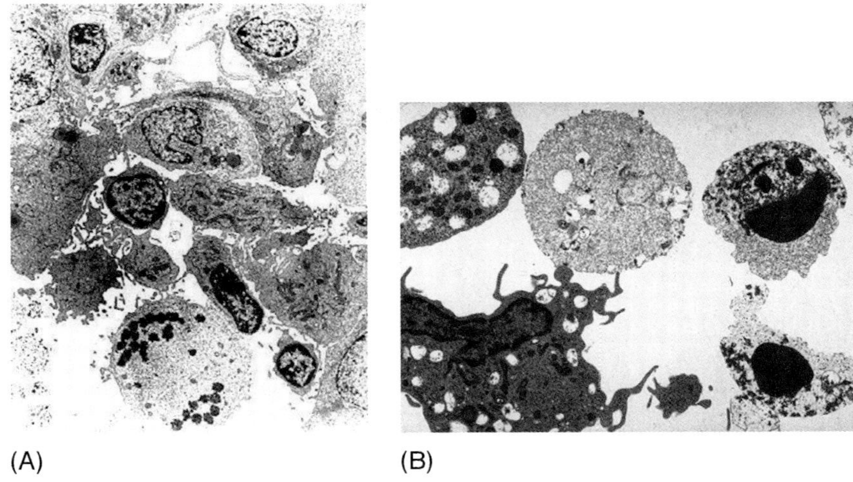

Fig. 8.4 Electron microscopy of white blood cells activated in flight with Con A. (A) 1×*g* in-flight control sample, showing the typical structure of activated T cells with enlarged cell volume and mitotic cells. (B) 0×*g* samples showing cell damage and apoptotic cells (11).

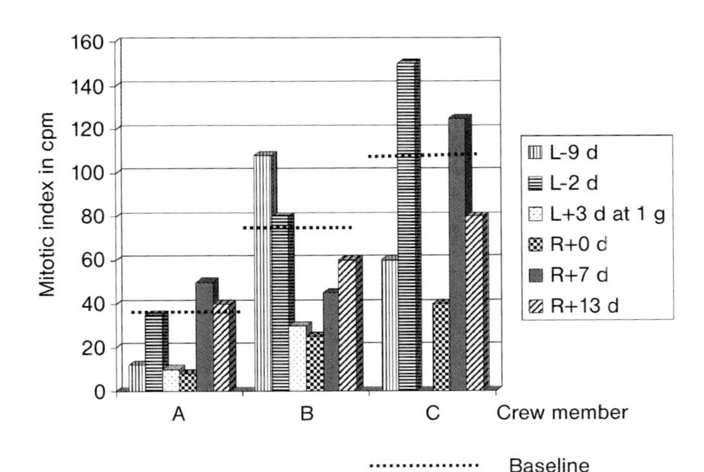

Fig. 8.5 Mitotic index of T cells in whole-blood cultures from three crew members on Spacelab D-1. Whole-blood samples taken 9 and 2 days before launch, in-flight on the 3rd mission day, and 0, 7, 13 days after landing where cultured for 3 days in the presence of Con A. The in-flight samples were incubated in a centrifuge at 1×g. Data are expressed as counts per minute of tritiated thymidine incorporated into DNA (9).

8.6
Stratospheric Balloon, 1986

Although the data from the 1×g in flight centrifuge were excluding such a hypothesis, at least part of the loss of activity *in vitro* of samples exposed to the Space environment could also be explained by damage of the cells by cosmic radiation. Therefore, we gladly accepted the invitation of the French Space Agency, CNES, to perform an experiment with lymphocytes on board of a stratospheric balloon flying from Sicily to Spain at an altitude of approximately 40 km. The level of radiation under such conditions is comparable to that on board an orbital vehicle while gravity is still 1×g. After a failure in 1985 due to a leak in the balloon envelope we had a successful flight in 1986. T cells were activated immediately before launch and kept in an incubator at 37 °C for the duration of the flight (20 h). The results were surprising. In fact, instead of a loss of activity, we observed an increase of 60% of the mitotic index compared with identical ground control samples [12]. There is a theory – not known to most biologists – called the theory of hormesis. The theory says that small doses of irradiation exert an activating rather than a deleterious action on biological objects. We still wonder whether our data are the result of hormesis on the activation process of T cells. Conversely, the loss of activity in Space is certainly due to the effect of microgravity.

8.7
Sounding Rockets Maser 3, 1989, and Maser 4, 1990

Sounding rockets are an ideal tool to study events of cell activation occurring within a short time. We have performed a series of investigations on Maser and Maxus rockets giving a microgravity time of 6–7 or 12.5 min, respectively.

Several experiments addressed the question of whether binding of Con A to the cell membrane, i.e. the delivery of the first signal of the *in vitro* activation of lymphocytes, is altered in microgravity. The mitogen Con A, a lectin carrying four binding sites specific for α-glucosides, is a specific activator of T-lymphocytes. Its binding to the glycoproteins of the cell membrane initiates first the formation of patches and, second, of caps of Con A receptors. Patching, i.e. the formation of clusters of Con A receptors on the cell membrane, is an energy-independent step. Capping, i.e. the formation of one cluster of Con A receptors at one pole of the cell, occurs only at 37 °C and is an energy-dependent step. Capping is also a necessary step towards lymphocyte activation. Patching and capping occur within a few minutes at 37 °C under 1×g conditions. Abnormalities in this process would obviously alter the transduction of the first activation signal.

On Maser 3 and 4, automated pre-programmed instruments permitted the injection of Con A and fixative to Ficoll purified lymphocytes incubated at 37 °C at preset times in flight.

In the first investigation (Maser 3), fluorescent-labelled Con A was added to the cells immediately after microgravity conditions were established. Fixation with paraformaldehyde followed after different preset times between 0.3 and 390 s. No significant differences in the rate of binding, patching and capping compared with the 1×g ground control were observed [13]. Binding was very fast and was completed after 30 s.

The second experiment was carried out on Maser 4, to test if the lack of differences between flight and 1×g control samples was due to a lack of influence of microgravity on this process or to the fact that the cells did not have sufficient exposure time to "adapt" to these conditions. Therefore, Con A was injected into the samples 3 and 5 min after the onset of microgravity. The data revealed that even a 5-minute exposure of the cells to microgravity prior to the addition of Con A does not affect the binding of the mitogen to the cell membrane compared with the ground control. Furthermore, there were no significant differences in patching after the cells had been exposed for 3 min to microgravity before addition of the mitogen. However, patching was significantly retarded, when Con A was added after 5 min of microgravity exposure. In the flight samples, fixed after 90 s incubation with Con A, $2.9 \pm 0.9\%$ of the cells showed patching compared with $7.4 \pm 3.4\%$ in the ground control. Thus, patching was clearly retarded after the cells had been exposed for 5 min to microgravity before the addition of Con A [14]. The question of differences in capping could not be answered with statistically significant data as the number of cells showing capping was too low in this experiment. However, there seemed to be a tendency that capping is also retarded in the flight samples.

8.8
Spacelab Life Sciences SLS-1, 1991

This mission was scheduled shortly after Spacelab D-1. Unfortunately the Challenger catastrophe delayed the flight by a few years. Spacelab Life Sciences mission 1 (SLS-1) was one of two NASA missions dedicated to Life Sciences. Ours was one of the two non-US experiments on board. The instrumentation was composed of two incubators identical to those used in Spacelab-1 and of a multi-g centrifuge providing 0.6, 1, 1.36 and 1.75×g (Fig. 8.6). For the first time a g-level between 0 and 1×g was available in Space.

The investigation consisted of two parts: in the first we used lymphocytes Ficoll-purified pre-flight from the blood of a donor. A novelty was that in aliquots of the cultures the cells were attached to microcarrier beads coated with ligands capable of binding lymphocytes (Fig. 8.7). In fact, lymphocytes are adhesion-independent cells and, in contrast to monocytes, remain in suspension when cultured *in vitro*.

The cells were activated with Con A as in the previous experiments and frozen with the cryopreservative dimethyl sulfoxide 46 and 65 h after activation. Another novelty was that, in addition to the mitotic index, we also determined the concentrations of IL-1, IL-2 TNF-α, IL-2R and IFN-γ in the supernatants, thus permitting the detection of changes of the so-called genetic expression of specific T cells

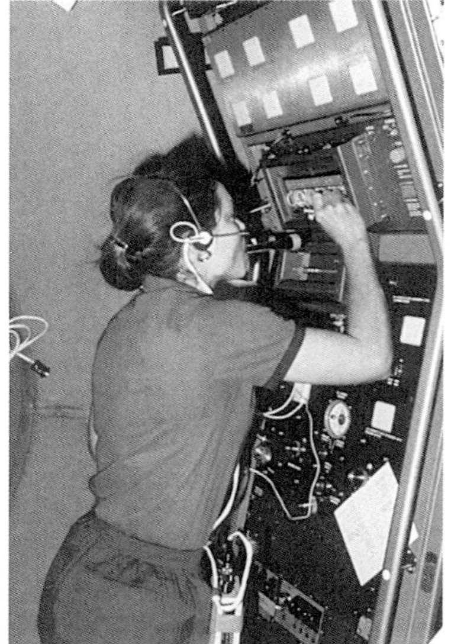

Fig. 8.6 Instruments used in Spacelab SLS-1. Left: rack showing one closed incubator (top), an open incubator (centre), and the multi-g centrifuge (bottom). Right: mission specialist Tamara Jernigan injecting the activator into the cultures in the central incubator. (Photograph NASA.)

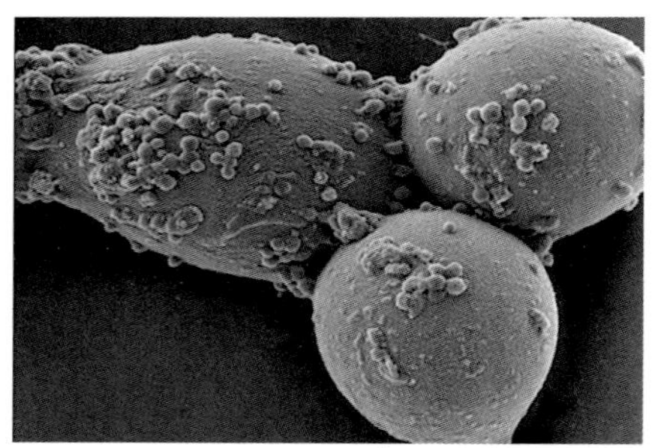

Fig. 8.7 Cells attached to microcarrier beads flown in SLS-1. Left-hand side: Thin section electron micrograph showing one microcarrier. Right-hand side: Scanning electron micrograph showing three microcarrier beads with attached cells (16).

products, the cytokines. At that time, the RT-PCR as well as the gene chips technologies were not yet developed, so that with the term "genetic expression" we refer improperly to the synthesis and secretion of proteins rather than to the expression of specific genes (see below). The 0×g samples with re-suspended cells confirmed the previous results: in addition to the mitotic index, the secretion of IL-2, TNF-α and IL-2R were clearly depressed. For the first time the data suggested

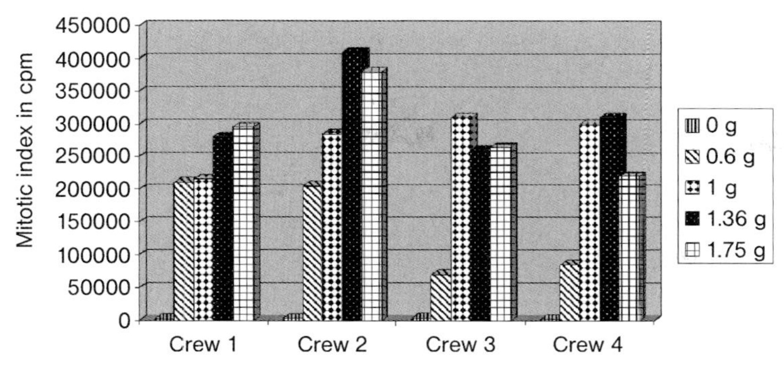

Fig. 8.8 Mitotic index of whole-blood cultures from four crew members of SLS-1 incubated at five different g levels.

that microgravity may influence genetic expression. Conversely, to our great surprise the cells attached to microcarrier beads displayed an activation that was remarkably higher at 0×g than the corresponding 1×g ground and flight controls [15, 16]. We are still seeking an explanation. One hypothesis is that the anchorage of the cells to a substratum stabilizes the cytoskeleton, in contrast to what is occurring in re-suspended free cells.

In the second part of the investigation we used blood samples donated in flight by the crew and kept at 0×g in the incubator and at 0.6, 1, 1.36 and 1.75×g in the multi-g centrifuge. Again, the data confirmed the dramatic depression of the mitotic index (Fig. 8.8). But the striking result was that at 0.6×g there is already a significant activation of the T cells. This means that there must be a threshold of sensitivity between 0 and 0.6×g. We are still waiting to further investigate this phenomenon. This was the subject of our experiment in Biopack that was lost with the Columbia accident in 2003 and has been re-flown in the KUBIK facility on ISS in September 2006. The data are not yet available. Due to a number of contingencies occurred in flight further ground-based experiments are required before the analysis of the flight samples is carried out.

8.9 Russian MIR Station, Missions 7, 8, 9, 1988–1990

A Swiss physician, who heard about our experiments on the immune system in Space, encouraged us to use a commercially available "skin" test to assess the fitness of the cellular immune system. The test consists of an applicator bearing seven antigens (tetanus, diphtheria, *Streptococcus*, tuberculin, *Proteus*, *Trychophyton* and *Candida*) and glycerine as control that are applied on the forearm. The physiological reaction after three days is the appearance of red indurations, similar to mosquito's bites, of approx. 10 mm diameter, where the antigens were applied. Colleagues at the Institute of Biomedical Problems (IBMP) in Moscow were enthusiastic about our proposal to conduct the skin test on cosmonauts spending a few months on the MIR Space station. Consequently, the test was performed on five subjects a few weeks before launch, after a few weeks in flight and shortly after

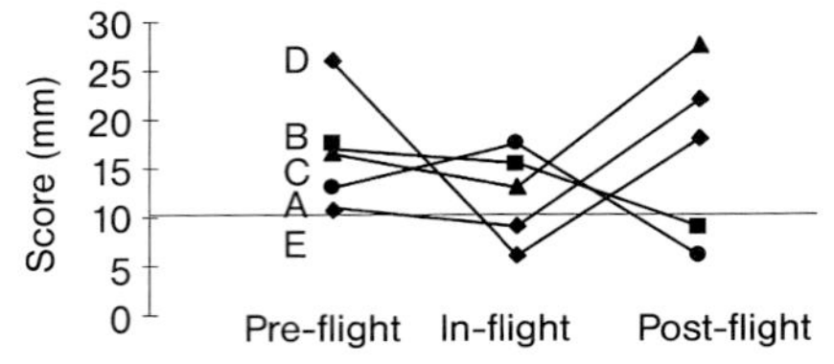

Fig. 8.9 Skin test on MIR. Seven antigens were applied on the forearm of five crew members of the Russian MIR space station a few weeks before, 3 months in-flight, and a few days after landing. The score is determined by averaging the diameter of the indurations caused by the seven antigens. The physiological value of the population is 10 mm (17).

landing. All five showed a depression of the response either in-flight or after flight compared with the response before flight (Fig. 8.9).

Four had a score below the physiological threshold, two in-flight and two after flight respectively. Interestingly, the lower response in-flight coincided with a life threatening extravehicular activity to repair the station [17].

8.10 Spacelab IML-1, 1992

The second flight of Biorack took place in Spacelab with the first International Microgravity Laboratory (IML-1). Our team participated with three experiments, of which only one was related to the cells of the immune system. One was with Hybridoma cells, which are the product of the hybridization of B lymphocytes with myeloma cancer cells and are capable of producing large amounts of monoclonal antibodies. The second was with Friend leukaemia virus-transformed cells, a cell line that differentiates in the presence of dimethyl sulfoxide to a curious type of red blood cells capable of producing haemoglobin. The third experiment, not discussed here, was a test flight of a miniaturized bioreactor filled with hamster embryonic kidney cells. All three investigations had a biotechnological background, asking the question of whether microgravity could be used as a tool to improve the production of specific cell products of medical and commercial relevance. Today, based on the experience gained with other experiments on the ISS, we can say that such an aspect has been overestimated. The uncertainty of the Space programme – the Space Shuttle is suffering several launch delays and the Soyuz spaceship can not carry large payloads – and the severe (maybe too severe) safety regulations will probably never permit the development of commercial applications on the ISS. In addition, the spectacular success of terrestrial technologies like bio- and tissue engineering will have a negative impact on the competition of parallel endeavours in Space. Furthermore, earth-bound devices like clinostats and rotating vessels capable of producing "functional weightlessness" or "vectorless gravity" may be used in future for interesting bioprocesses.

Nevertheless, Space bioreactors could serve in the next few decades as life support systems on Space stations and Space vehicles travelling in the Solar System and beyond.

Concerning the two experiments of the two mammalian cell lines, the data were interesting but not as spectacular as those with T cells reported above. The parameters determined were cell proliferation, biosynthesis of specific cell products, consumption of glucose, glutamine and production of ammonia and lactate. Murine Friend leukaemia virus-transformed cells (Friend cells) were induced to differentiate and express haemoglobin (Hg) genes upon induction with dimethyl sulfoxide (DMSO). No change was observed in all metabolic parameters, including the production of Hg and the number of Hg-positive cells. Electron microscopy analysis showed no difference in morphology, mean cell volume and mitotic index between the different cell samples. Murine hybridoma cells revealed an increase (+30–40%) of cell proliferation rate in microgravity, whereas the metabolic parameters, production of monoclonal antibodies included, were lower in 0×g than in the 1×g controls. The results clearly show that not all mammalian cells undergo dramatic changes in microgravity and that the effects reported on human T lymphocytes are unique [18].

8.11
Sounding Rockets Maxus 1B, 1992, and Maxus 2, 1995

Cell–cell interactions and aggregate formation are important means of cell communication and signal delivery in the mitogenic T lymphocyte activation. From earlier experiments we had indirect evidence that cells are moving and interacting also in Space, as aggregates were observed in samples fixed in Space and analysed on ground [1, 10].

On Maxus 1B we studied the movements of free floating non-activated peripheral blood lymphocytes incubated at 37 °C in real time under microgravity conditions. A microscope, tele-manipulated from the ground and connected to a CCD camera, permitted us to record cell motility and interactions in microgravity [19].

The images recorded under microgravity clearly showed that the free-floating non-activated cells were able to display autonomous motion in random directions [14].

A detailed analysis revealed that the movements are much more complex. The cells often change direction, move back and forwards and sometimes cross the same point several times (Fig. 8.10). The average velocity, calculated from the displacement in 13 s increment, is $8.4 \pm 1.2\,\mu m\,min^{-1}$, with a range of $0–29.4\,\mu m\,min^{-1}$. Also of interest is the observation that the cells in low gravity were not all round. Very often they exhibited longitudinal forms, rotated around their axis and also showed contraction waves similar to those described in the literature for lymphocytes that move in collagen gels under 1×g conditions [20]. All eleven cells in the observation field showed motion capability under low gravity conditions. This is in contradiction to the behaviour of lymphocytes under 1×g conditions, where

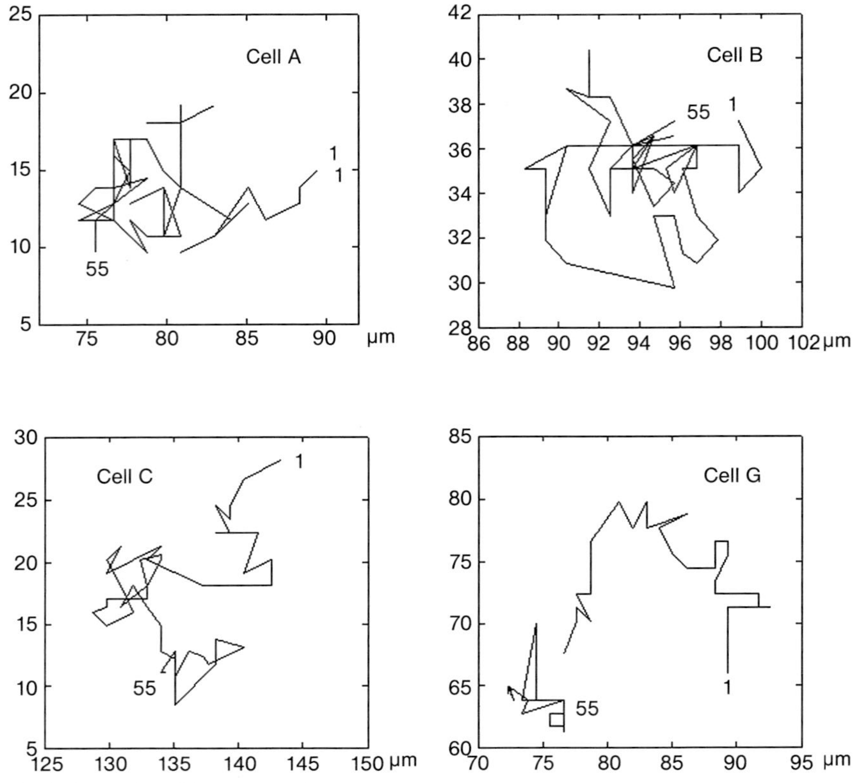

Fig. 8.10 Motility of human lymphocytes in microgravity during the Maxus 1B flight. Detailed motion behaviour of four cells (one measuring point every 13 s). The numbers on the axis, in μm, correspond to the position of the cell in the viewing field (14).

mainly activated cells or cells in the presence of a chemo attractant show this capability [21].

The existence of active movements and cell–cell contacts supports – but does not prove – the notion that T cells can communicate and transmit signals in microgravity. Thus, another experiment was performed on IML-2 to obtain more information concerning this question (Section 8.12).

The cytoskeleton plays an important role in the aggregation of the receptors on the cell membrane and signal transduction (for reviews see Refs. [22–24]), but also in cell motility. In two different experiments we studied the influence of gravity changes on the cytoskeleton. In particular, we have analysed the pattern of the intermediate filaments of vimentin and of microtubule in Jurkat cells. The higher cytoplasmic volume and size of these cells, compared with resting lymphocytes, render them more suitable for this type of analysis.

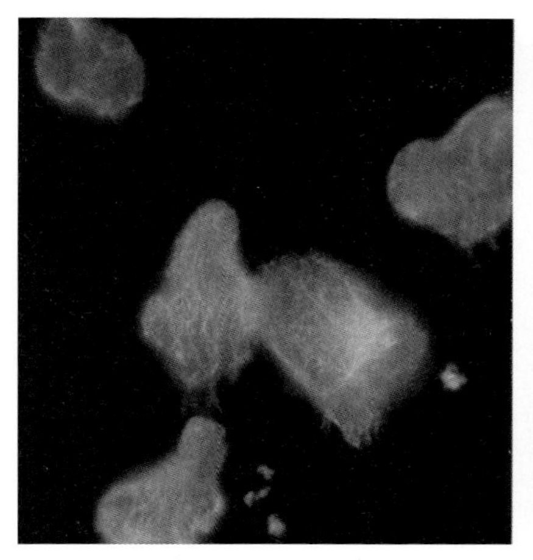
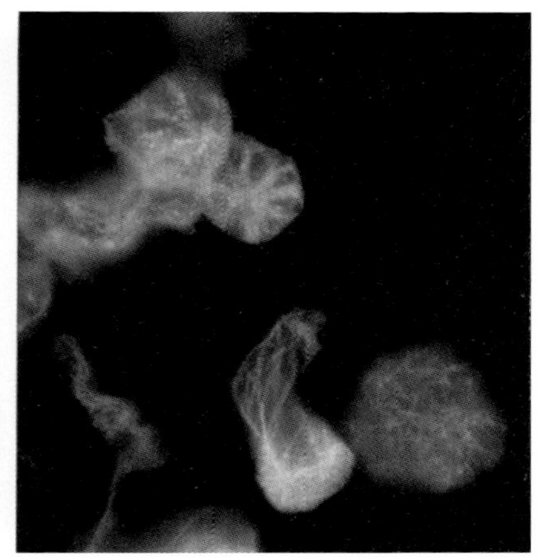

Fig. 8.11 Cytofluorographs of the intermediate filaments of vimentin in Jurkat cells. Left-hand side: ground control; right-hand side: Maxus 1B flight sample fixed after 30 s in microgravity (14). (Photograph by L. Sciola, University of Sassari.)

In a first experiment on Maxus 1B, the cells, incubated at 37 °C, were fixed at different times after the onset of microgravity. After recovery, the cytoskeleton was marked with the monoclonal antibodies anti-vimentin and anti-tubulin. Morphological analysis revealed that structural changes in the cytoskeleton occur in microgravity compared with ground controls and that they are most evident in vimentin (Fig. 8.11). Under microgravity conditions, the intermediate filaments of vimentin assemble into thick bundles in 20.9 ± 1.7% of the cells compared with 9.9 ± 1.3% in the ground controls. This is observed after only 30 s of microgravity. The number of cells showing large bundles is slightly but significantly decreased at microgravity times of 420 s (17.8 ± 0.4%) and 720 s (18.8 ± 1.6%). The other statistically significant alterations observed in microgravity consist of the formation of aggregates of protein, appearing as fluorescent points and suggesting depolymerization processes, as well as of discontinuities of the filamentous network (11.9 ± 1.3% at 30 s, 12.1 ± 2.3% at 420 s and 11.0 ± 1.3% at 720 s compared with 8.8 ± 1.6% of the ground control).

A significant difference in the appearance of large bundles has also been observed in the microtubules network, where 13.2 ± 1.2% of the cells show these alterations after 30 s of microgravity compared with 5.9 ± 2.8% in the ground control. Other changes are less evident [14].

During the launch of the sounding rocket, the cells were exposed to hyper-gravitational forces. Ground controls conducted at hypergravity (centrifuge) showed that the observed alterations in the structure of vimentin and tubulin are not due to the acceleration peak of 13–15×g following the launch of the rocket. But there where still doubts as to whether observed changes were due to microgravity conditions. Therefore, it was quite important to repeat this experiment in a module with an in-flight 1×g centrifuge, as we had on Maxus 2.

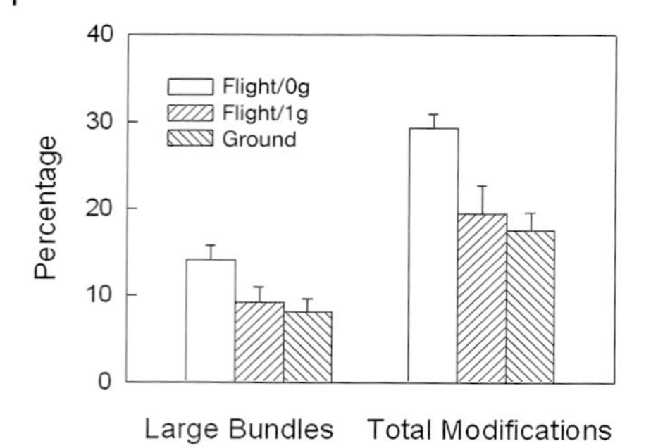

Fig. 8.12 Influence of gravity changes on the structure of vimentin in Jurkat cells (Maxus 2 flight). The bars represent the percentage of cells showing either the appearance of large bundles or the total of modifications. Standard errors of the means are given (14).

On Maxus 2 we not only studied the influence of microgravity on vimentin in Jurkat cells, but also possible changes in the structure of these intermediate filaments after binding of Con A to the cell. Jurkat cells were incubated at 37 °C. Part of the samples was fixed at the onset of microgravity. To the other samples Con A was added at the beginning of the microgravity phase and the cells were fixed before the end of 0×g. The analysis clearly showed that the changes in the structure of vimentin observed on Maxus 1B can be confirmed. Significantly more cells showed changes in the structure of vimentin in microgravity (29.3 ± 1.6%) than in the in-flight 1×g control (19.5 ± 3.2%) and in the ground control (17.5 ± 2.1%). Once more, the appearance of large bundles significantly increased in the microgravity samples (in 14.2 ± 1.6% of the cells) as compared with the on-board 1×g control (9.2 ± 1.8%) and the ground control (8.2 ± 1.4%) (Fig. 8.12). The other changes observed in microgravity are similar to the ones found in the experiment on Maxus 1B. In the samples fixed at the onset of microgravity, 20.9 ± 3.1% of the cells showed modifications and 9.9 ± 1.7% large bundles. This and the results of the samples of the in-flight 1×g centrifuge clearly demonstrate that the changes in the structure of vimentin are due to microgravity and not to the g-stress caused by the launch. Binding of the mitogen Con A has no influence on the observed microgravity effect.

The results of the experiments on Maser 3 and 4 (Section 8.7) indicated that the percentage of cells showing patching and capping increases with incubation time after the addition of Con A. Therefore, the experiment on binding of the mitogen to the cell membrane followed by patching and capping was repeated on Maxus 2, taking advantage of the longer microgravity time (12.5 min) compared with the Maser flights (7 min). An in-flight 1×g centrifuge allowed discrimination between the effects of low gravity and the effects of launch. Instead of lymphocytes we used Jurkat cells – a human leukemic T cell line. Fluorescent-labelled Con A was added to Jurkat cells incubated at 37 °C at the onset of the microgravity phase. The

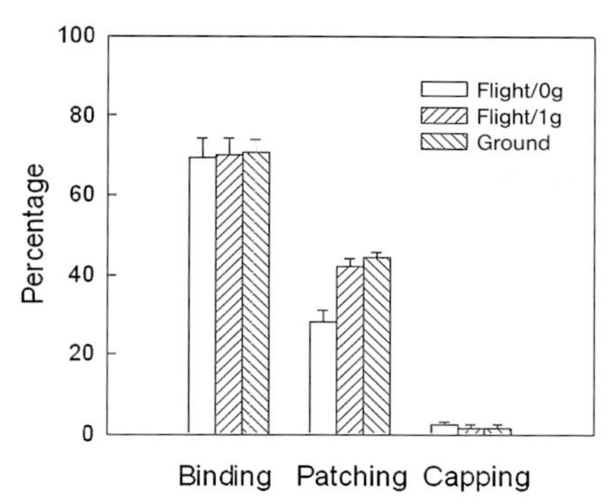

Fig. 8.13 Binding of Con A to Jurkat cells followed by patching and capping of the Con A receptors (Maxus 2 flight). The bars represent the percentage of cells showing binding, patching or capping. Standard errors of the means are given (14).

samples were fixed before the end of microgravity after 12 min incubation by the addition of paraformaldehyde.

Analysis showed that binding of Con A to Jurkat cells is not influenced by gravity changes (Fig. 8.13). Patching of the Con A receptors, however, is slightly but significantly lower in microgravity compared with the in-flight 1×g control as well as the ground control [14]. The results confirmed, thus, the previous findings obtained on Maser 3 and 4. They showed that patching is a dynamic process also under microgravity conditions that might be completed with longer incubation. Despite the longer microgravity time, the number of cells showing capping was too low for a statistical evaluation. From these observations we concluded that the influence of low gravity on the delivery of the first signal of activation is rather small. Rapid processes like binding of the mitogen, patching (although slightly retarded) and probably also capping are not involved in the depression of the *in vitro* activation of T lymphocytes observed in several Spacelab experiments. The delayed patching and capping processes are certainly due to the observed changes in the cytoskeletal structures.

8.12
Spacelab IML-2, 1994

During this mission we resumed with two experiments our work with lymphocytes Ficoll-purified from human peripheral blood. Importantly, the purification of white blood cells on Ficoll gradient delivers a mixture of T and B cells and approximately 5% monocytes, which are required for full activation as accessory cells (see above). In the first experiment, T cells were activated with Con A with or without exogenous recombinant interleukin-1 (IL-1) alone or IL-1 + interleukin-2 (IL-2) under microgravity conditions to test the hypothesis that lack of production of IL-1

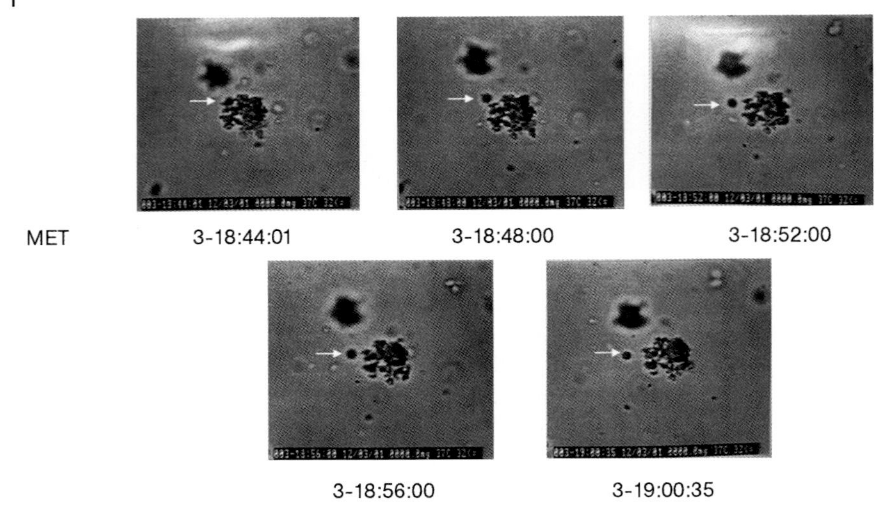

Fig. 8.14 "Space" aggregate 78 h after the addition of Con A observed in NIZEMI. A single cell (arrow) is moving out of the aggregate. MET: Mission elapse time; the total sequence covers 16 min 34 s (26).

by monocytes is the cause of the near total loss of activation observed earlier on several Spacelab flights. The 60-min failure of the on-board 1×g reference centrifuge at the time of the addition of the activator renders the in-flight data at 1×g unreliable. However, data from a previous experiment on SLS-1 show that there is no difference between the results from the in-flight 1×g centrifuge and 1×g on ground. Comparison between the data of the cultures at 0×g in Space and the synchronous control at 1×g on ground show that exogenous IL-1 and IL-2 do not prevent the loss of activity (measured as the mitotic index) at 0×g; production of interferon-gamma, however, is partially restored. In contrast to a previous experiment in Space, the production of IL-1 is not inhibited [25].

As cell–cell contacts are one of the elements essential for activation, the behaviour of human leukocytes (mainly lymphocytes and monocytes as accessory cells) in the presence of the mitogen Con A was studied in the centrifuge microscope NIZEMI at 0×g (Fig. 8.14). Aggregates (formed by intercellular bindings of membrane glycoproteins via the tetravalent α-glucoside ligand Con A) were found at 0×g as well as at 1×g already 12 h after the addition of the mitogen. In general, the aggregates observed at 0×g after an incubation time of 46 and 78 h were smaller than the corresponding aggregates in the ground control. The findings are of primary importance since they confirm the indirect evidence we had from earlier Spacelab experiments and demonstrate that cell–cell contacts are also occurring in microgravity. In addition, single cells in 0×g show a significant higher locomotion velocity than the cells at 1×g. The fact that the locomotion capability is not decreased during the 78 h incubation with Con A provides further evidence that the cells are not proceeding through the cell cycle [26].

8.13
Sounding Rocket Maser 9, 2002

Simulated microgravity (RPM) has previously been shown to modulate gene expression in T lymphocytes [27]. In an experiment flown on Maser 9 we performed a wide range screening of early transcriptional modulation on T lymphocytes. Peripheral blood lymphocytes kept at 37 °C were exposed to microgravity for 5.5 min immediately after activation with Con A. The cells were then separated from the culture medium and lysed in RNA-preserving conditions. The control experiment was performed in an in-flight 1×g centrifuge to discriminate launch effects known to trigger gene expression. The transcriptional response was monitored by the cDNA microarray hybridization technology. The high-density microarray filter-based cDNA microarrays, constituted by single membranes, contained approximately 4000 human genes, all of which are of known function. We found that in T lymphocytes from two different donors about 1–2% of the genes monitored show significant modulation in response to microgravity as compared with the in-flight 1×g control [28].

Most of the modulations observed appear to be donor-dependent, but the genes coding for the small cytokine A5 (RANTES) and the IL-10Rβ were found similarly modulated in the two different donors examined. Furthermore, the expression of the IL-2Rα gene is clearly depressed, whereas the IL-1 gene is induced (Table 8.1). Clearly, IL-2 production and secretion and their inhibition observed in microgravity conditions have a crucial role in regulating the expression of IL-2 receptors [16].

Cytokine A5 (RANTES) has been shown to be early transcriptionally induced after T lymphocytes stimulation. Its prompt repression by microgravity conditions could, therefore, be meaningful. Down-regulation of IL-10Rβ could also be relevant in the early stages of microgravity interference on T lymphocytes activation program. The cytoplasmic domain of the receptor for interleukin-10 (IL-10R) contains two box 3 sequence motifs that have been identified in the signal-transducing receptor subunits for IL-6-type cytokines and noted to be required for activating STAT3 and inducing transcription through IL-6-responsive elements.

Table 8.1 Early transcriptional response of T lymphocytes to microgravity conditions (Maser 9, 2002).

Gene modified[a]	*Type of modulation*
ILβ1	Induced
IL-4Rα, IL-13Rα1, IL-13Rα2	Induced
Cytokine A13	Induced
IL-8	Depressed
IL-2Rα, IL-7Rα, IL-10Rβ	Depressed
Cytokine A2	Depressed
Cytokine A5 (RANTES)	Depressed

a Compared with in-flight 1×g control.

8.14
Shuttle Flight STS-107, Biopack, 2003

Based on the knowledge gained so far on the behaviour of T cells in microgravity, we proposed an experiment called "LEUKIN" to investigate two main aspects: first, to identify the sensitivity threshold between 0 and 1×*g* and, second, to determine, in the first 4 h, changes of genetic expression of the main activation markers by RT-PCR technology. Moreover, we improved the quality of the cell cultures by separating the T cells (both CD^{4+} and CD^{4+} T subsets) from the B cells and the monocytes by newly developed and commercially available affinity columns. The second activation signal usually delivered by the monocytes was replaced by anti-CD28 monoclonal antibodies. Biopack was the name of the new facility provided by the ESA, installed in Spacehab, a smaller version of Spacelab. As with the multi-*g* centrifuge on SL-1 in 1991, Biopack could provide *g*-levels below 1×*g*. The telemetry software permitted us to follow on the ground the main operations like start and stopping of the centrifuges (operating for our experiment at 0.2, 0.6 and 1×*g*), opening and closing of the incubator, cooler and freezer and to check the respective temperatures. Everything went smoothly with our experiment until the catastrophic re-entry, with the loss of the crew and of the experiments (Fig. 8.15).

As mentioned above, the LEUKIN experiment was launched again in September 2006 with Soyuz TMA-9 from Baikonur and was installed in the KUBIK facility on the ISS.

8.15
Ground Simulations

There are two stringent reasons to conduct ground based investigations either with humans or with biological specimens in parallel to the flight experiments. One

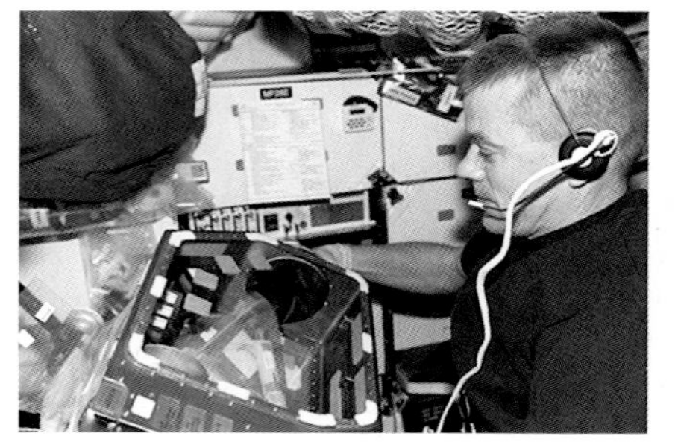

Fig. 8.15 Willie McCool operating the LEUKIN experiment in a portable glove box (left-hand side) in Columbia on STS-107 and the recovered, but unreleased, experiment containers (right-hand side). (Photographs NASA.)

reason is to optimize the conditions of a unique and expensive opportunity such as an experiment in Space by testing the protocols, incubation times and to gain useful indications of the effects of real microgravity. The second reason is that the time interval between two experiments in Space is in general 2–3 years or even longer.

The head-down tilt test is recognized by the scientific community as a valuable simulation of the effects of microgravity conditions on human test subjects. A marathon run is a good model to understand the effect of physical stress on the immune system.

The three-dimensional clinostat, called also Random Positioning Machine (RPM), is an important device used to generate vectorless gravity. In this section we summarize our ground-based work.

8.15.1
Marathon Run and Head-down Tilt Test

As stated above, data obtained with cultures of lymphocytes from *in vitro* experiments cannot be extrapolated to what is happening *in vivo*. In fact, it has been known for decades that stress provokes a depression of the immune system via the neuroendocrine system and independently of microgravity.

To better understand whether the effect seen in the astronauts is due to microgravity *per se* or to psychological and physical stress we conducted two studies, one on marathon runners, the other on subjects undergoing the head-down tilt test (HDT). This is a typical example of how microgravity research can contribute to the understanding of certain physiological functions, in particular of the immune system.

The response of critical immunological parameters in seven athletes to the sustained physical stress of marathon running was assessed. Variables analysed were the responsiveness of lymphocytes (measured as mitogenic response to Con A), the numbers of lymphocytes, their subsets, and leukocyte numbers. In addition, blood levels of cortisol, epinephrine, and norepinephrine were determined. After the run, lymphocyte responsiveness was severely depressed to 1–70% of the resting values, even though the lymphocyte counts did not change. Leukocyte counts were elevated 2.8-fold. No dramatic changes were found within the lymphocyte subsets, although an increase in pan T cells and the helper/inducer subset two days after the run was significant. In addition, the numbers of B cells decreased significantly. No change was observed within the suppressor/cytotoxic subset. Cortisol increased 2.1-fold, epinephrine 3.2-fold and norepinephrine 2.7-fold. All these parameters returned to baseline values within two days. These data were compared with data obtained during and after space flight. We concluded that the prolonged physical stress of marathon running induces changes in immunological responsiveness that are strikingly similar to those arising from the stress of space flight [29].

Immunological responses of six healthy males to ten days of head-down tilt bed rest (HDT) were assessed. Lymphocyte responsiveness was severely reduced immediately before(!), during, and immediately after the HDT, even though the lymphocyte numbers did not change. By contrast, delayed-type hypersensitivity was not affected. No dramatic changes were found in white blood cells counts and lymphocyte subpopulations, with the only exception of natural killer (NK) cells, which transiently decreased immediately after HDT. Plasma cortisol levels were elevated above normal immediately before(!) and during the HDT. The data suggest that the mitogenic response of lymphocytes was affected by psychological and fluid shift stress. These results have been compared with data obtained during and after space flight. Again, we concluded that the stress of HDT induces changes in immunological responsiveness that are strikingly similar to those arising from the stress of space flight [30].

8.15.2 Clinostats

Although we started to work with the fast rotating clinostat in the 1970s, reviewed in Ref. [2], in this section we present the most recent data obtained with the Random Positioning Machine (RPM). The RPM was introduced in gravitational biology in the 1990s by Hoson at the University of Osaka [31]. To qualify this instrument as a good device for microgravity simulation, we performed a study to compare the data from Space with those from the RPM and other devices. The results with T cells confirmed that the RPM can reproduce, at least qualitatively, the results from our previous Space experiments [32].

The data obtained from cultures of T cells in Space indicated that a failure of the expression of the interleukin-2 receptor (measured as protein secreted in the supernatant) is responsible for the loss of activity. To test such a hypothesis we have studied the genetic expression of interleukin-2 and of its receptor in Con A-activated lymphocytes kept on the RPM with the RT-PCR technology. The data clearly showed that the expression of both IL-2 and IL-2Rα genes is significantly inhibited in simulated 0×g. Thus, full activation is prevented [27].

The hypothesis that apoptosis is also triggered in 0×g conditions was confirmed in a study conducted with Mauro Maccarrone, then at the University of Rome [33].

We reported in Section 8.8 that the inhibition of the activation was overcome by attaching the cells to a substratum. Another way to bypass the inhibition is to use anti-CD3/IL-2, PHA/IL-2 or anti-CD3/anti-CD28 instead of Con A/anti-CD28 as activators. This suggests that microgravity has a discriminatory effect on different activation pathways [34].

While preparing the LEUKIN experiment for STS-107, we started the study of the signal transduction pathways involved in T cell activation on cells activated at 1×g. Affymetrix oligonucleotide arrays were used to globally monitor expression of 8500 genes in activated T cells; analysis revealed that 217 genes were significantly upregulated at 1×g within 4 h. Forty-eight of the 217 induced genes are known or predicted to be regulated by a CRE promoter/enhancer. Commercially available chemicals were used to specifically inhibit/activate components of the activation pathways: forskolin as activator and H89 as inhibitor of PKA, GF109203X as inhibitor of PKC, and UO126 as inhibitor of MEK1. Inhibition of PKA significantly reduced IFN-γ, IL-2 and IL-2Rα gene expression by approximately 40%. Induced genes included transcription factors, cytokines and their receptor genes. Analysis by semi-quantitative RT-PCR technology confirmed the significant induction of IL-2, IL-2Rγ and IL-2Rα. PKC is known to play a major role in T cell activation: inhibition of PKC reduced the expression of IFN-γ, IL-2 and IL-2Rα. To study potential cross-talk between the PKC and the PKA/MAPK we used the T lymphocyte-derived Jurkat cell line. Phorbol ester PMA-stimulated Jurkat cells were studied with specific signal pathway inhibitors at 1×g. The extracellular signal-regulated kinase-2 (ERK-2) pathway was found to be significantly activated (sevenfold); however, there was no activation of ERK-1, JNK or p38 MAPK. U0126 significantly blocked expression of IL-2 and IL-2Rα. Gene expression of IL-2Rα and IFN-γ was dependent on PKA in S49 wt cells but not in kin$^-$ mutants. Taken together, these results suggest that the PKA pathway, in addition to PKC and MAPK, plays a role in T cell activation and induction of IL-2, IL-2Rα and IFN-γ gene expression [35].

Based on the data obtained at 1×g we continued this line of work in the RPM. We analysed differential gene expression in the first 4 h after activation to find gravity-dependent genes and pathways (Fig. 8.16). We found inhibited induction of 91 genes in simulated microgravity when compared with cells activated in normal gravity. Altered induction of 10 genes regulated by key signalling pathways was verified using RT-PCR technology. We discovered that impaired induction of early genes regulated primarily by transcription factors NF-KB, CREB, ELK, AP-1, and STAT contribute to T cell inhibition in microgravity (Table 8.2). Mitogenic T cell activation leads at 1×g to phosphorylation of CREB and LAT (linker of activation in T cells, an upstream regulator of PLC). While CREB phosphorylation is suppressed in microgravity, LAT phosphorylation is not downregulated. Such findings rule out the involvement of lipid rafts in the inhibition of activation [36].

In an investigation successfully flown in the KUBIK BIO#1 mission on the ISS in March/April 2006, we studied the motility of adherent monocytes and their interaction with T lymphocytes. The interaction between T cells and monocytes is important for the delivery of the second signal in the activation of lymphocytes. In earlier experiments we had found that lymphocytes are highly motile in microgravity (Sections 8.11 and 8.12) and form aggregates after activation with Con A. Cell motility, and with it a continuous rearrangement of the cytoskeletal network within the cell, is essential for cell–cell contacts.

In preparation for this experiment we have studied the motility of monocytes and changes in the cytoskeletal structures in these cells, specifically F-actin, β-tubulin and vinculin, in simulated microgravity conditions achieved on the RPM. The J-111 cells, an adherent monocyte cell line, were attached to cover slides coated with colloidal gold particles and exposed for 1 and 24 h, respectively, at 37 °C to simulated microgravity conditions in the RPM.

Microscopic analysis revealed migration tracks of monocytes at 1×g similar to those described in the literature [37]. A normal pattern of cell migration was

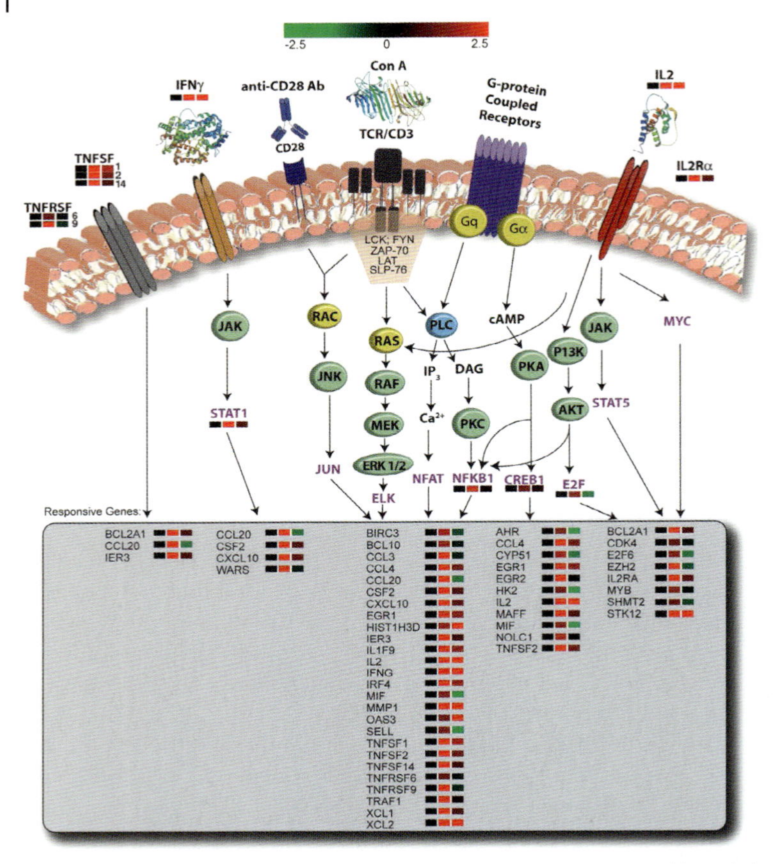

Fig. 8.16 Outline of signal transduction pathways initiated and genes significantly upregulated after T cell activation with Con A and anti-CD28 antibody. Transcription factors shown induce specific gene transcription that lead to proliferation, differentiation, apoptosis, or chemotaxis. Coloured blocks correspond to average normalized gene expression ratios at 1×g 0 h, 1×g 4 h, or variable g 4 h from left to right, respectively. The colour key represents the degree of up- or down-regulation in $\log_2$ scale. Yellow, green, and blue disks represent G-proteins, kinases, and lipases, respectively. Purple text refers to transcription factors (36).

observed on gold particle-coated cover slides of 1×g samples (Fig. 8.17). Areas around cells appeared completely cleaned out of gold particles partly internalized inside the cells or accumulated to the surface of the cells [38]. Conversely, very short migration tracks were observed both after 1 and 24 h of exposure to modelled low gravity; thus, the monocytes were less motile under this condition. In fact, the cell shape appeared more contracted with shorter protrusions reaching the neighbouring cell, whereas the cells of the 1×g control showed the typical morphology of migrating monocytes.

Table 8.2 Key genes and transcription factors of human lymphocytes: expression ratios under different gravity conditions. (From Ref. [36].)

Gene symbol	*1×g 0 h*	*1×g 4 h*	*0×g 4 h*	*Pathways involved*
IFNG	1.1	2147.2	38.6	NFKB (AKT, PKC); STAT (JAK)
XCL2	0.9	300.4	13.7	NFAT/Ca^{2+}
IL-2Rα	0.9	52.5	2.4	STAT (JAK)
IL-2	0.9	12.3	1.1	AP1 (JNK); CREB (MEK, PKA, PKC); NFAT (Ca^{2+}); NFKB (AKT, PKC); STAT (JAK)
CSF2	1.0	6.4	3.0	NFKB (AKT, PKC); STAT (JAK)
STAT1	1.1	3.6	1.6	STAT (JAK)
LTA	0.8	3.1	1.5	NFKB (AKT, PKC)
TNFα	0.9	2.7	1.4	AP1 (JNK); CREB (MEK, PKA, PKC); NFAT (Ca^{2+}); NFKB (PKC, AKT)
MIF	0.5	2.5	0.5	AP1 (JNK); CREB
NFKB1	0.8	2.2	0.9	NFKB (AKT, PKC)

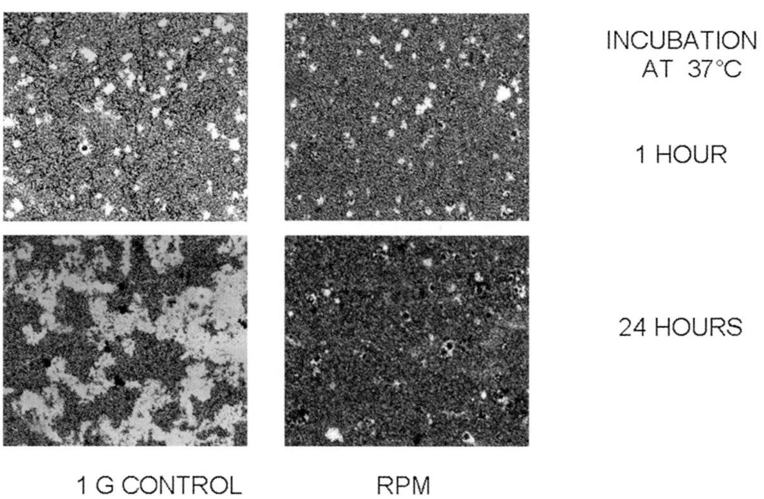

Fig. 8.17 Locomotion tracks of J-111 on gold-particle coated cover slides incubated for 1 h and 24 h at 37 °C. Left-hand panels: 1×g control; right-hand panels: exposure to simulated microgravity on the Random Positioning Machine (RPM) (38).

A quantitative analysis demonstrated a remarkable difference in the locomotion ability of the cells even after 1 h simulated microgravity in the RPM, when compared with 1×g controls. Cells exposed to simulated microgravity for 1 h moved on average 6 μm, and the more frequent displacement was 0–4 μm, whereas the cells in the 1×g controls showed a displacement of 14.9 μm on average and the more

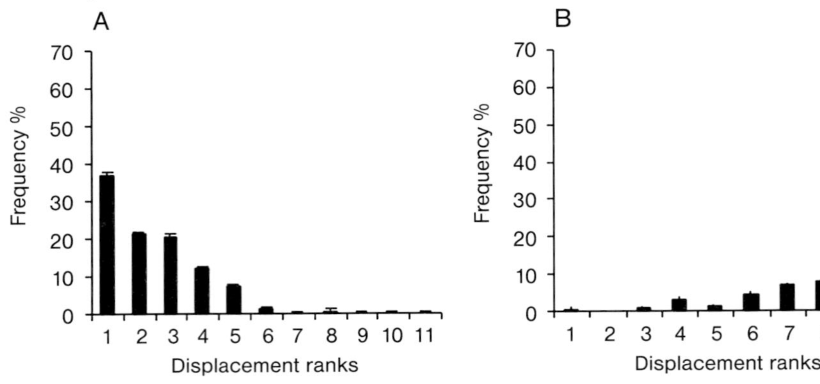

Fig. 8.18 Displacement frequencies (%) of J-111 cells on gold particle-coated cover slides (*Y*-axis) after 24 h of exposure to simulated microgravity (Random Positioning Machine) (A) and to 1×g (B). Displacement rank (μm): (1) 0–4, (2) 5–9, (3) 10–14, (4) 15–19, (5) 20–24, (6) 25–29, (7) 30–34, (8) 35–39, (9) 40–44, (10) 45–49, (11) ≥50. The results are the average of at least three independent experiments with 50 cells for each experiment (38).

frequent displacement was 10–14 μm. Cells exposed to 24 h of simulated microgravity showed a total displacement of 8.7 μm on average, but the more frequent displacement was still 0–4 μm (Fig. 8.18). Therefore, the total locomotion of monocytes during 24 h of simulated microgravity appeared to be not much different than during 1 h. In contrast, the control cells showed a total average displacement of 59.7 μm, and within 24 h most of the cells moved >50 μm.

The severely reduced locomotion ability of the monocytes found in simulated microgravity could hinder the delivery of the second signal necessary for a full activation of lymphocytes *in vitro*.

The importance of an intact and dynamic cytoskeletal network for cell movement has been pointed out by Horwitz and Parson [37]. We have found severe alterations in the cytoskeletal structures of F-actin, β-tubulin and vinculin in J-111 cells exposed to 1 and 24 h of simulated microgravity. F-actin filaments appeared abundant and well organized into cytosolic bundles and in the elongated and extended filopodia in the control samples. Conversely, the F-actin network of J-111 cells exposed to simulated microgravity showed a remarkable decrease in the filamentous biopolymers density and the actin stress fibres appeared localized like continuous sub-plasmatic bundles, both after 1 and 24 h (Fig. 8.19). β-Tubulin showed a perinuclear position with reduction in arborization, losing the radial disposition, even after 1 h of simulated microgravity in the RPM, in contrast to 1×g controls where they appeared radiated from the microtubule organizing centre to the plasma membrane.

The anchor protein vinculin appeared as focal contacts linking actin filaments to the plasma membrane in the 1×g controls, whereas samples exposed to 1 h as well as 24 h of modelled low gravity showed vinculin proteins not evenly spread but thickened close to the cell membrane as globular clusters.

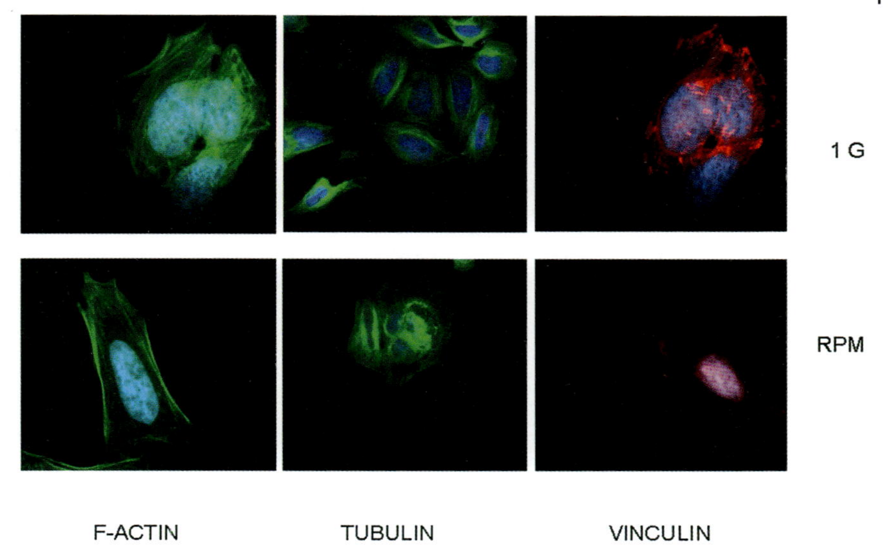

Fig. 8.19 Immunofluorescence images of F-actin, β-tubulin and vinculin in cultured J-111 cells exposed for 24 h to simulated microgravity in the Random Positioning Machine (bottom panels). The 1×g controls are shown in the upper panels (38).

Based on these results it can thus be speculated that the impaired motility of adherent monocytes on a surface in low gravity might be due to the disruption of the cytoskeletal network. Conversely, this is not valid for lymphocytes in suspension, showing a high motility, despite the fact that the structures of vimentin and tubulin were found to be affected by exposure to microgravity (Sections 8.11 and 8.12).

8.16 Conclusion

In conclusion, experiments *in vitro* with T cells in Space and in clinostats have shown that:

- mitogenic activation with Con A (concanavalin A) or Con A/anti-CD28 is strongly reduced;
- the structure of the cytoskeleton in T cells and monocytes is altered;
- cell movements, aggregation and contacts are occurring;
- binding, patching and capping of the mitogen are occurring, patching and capping are delayed;
- motility of monocytes is reduced, the cytoskeletal structures of F-actin, β-tubulin and vinculin are affected;

- the genetic expression of IL-2 and IL-2R is depressed;
- there is a threshold of g-sensing between 0 and 0.6×g;
- the cells undergo apoptosis;
- NF-KB, CREB, ELK and STAT contribute to T cell inhibition in microgravity;
- lipid rafts are not involved in the dysfunction;
- the depression can be bypassed by attaching the cells to a substratum or by using another activator;
- the PKA pathway is also involved in T cells activation.

Most probably, the alteration of the structure of the cytoskeleton is the primary cause of the phenomena observed.

Concerning the response of the immune system of humans in Space, we have a strong indication that physical and psychological stress rather than microgravity *per se* are responsible for the reduced immune response in-flight and immediately after flight.

The lines of research outlined here will continue with experiments on ISS, sounding rockets and with the RPM.

In conclusion, the findings clearly show that new and important knowledge can be achieved with experiments in gravitational biology and physiology. However, the safety constraints on manned Space laboratories strongly hamper the execution of experiments involving transfer of harmful liquids commonly used in biological research. In our opinion, the easiest approach should be with automated sounding rockets and satellites that can stay in orbit for 2–3 weeks and then bring the samples back to Earth.

For the same reasons mentioned above, it will be extremely difficult to perform experiments with biotechnological applications on the ISS. The processing of dozens of samples as required by such technology will be impossible also on satellites. Much more potential is offered by terrestrial devices like rotating wall vessels and clinostats.

Acknowledgments

We would like to express our warmest thanks to all colleagues who contributed to the success of our experiments and shared with us the ups and downs, the joys and the sad moments of thirty years of research and development: Alex Tschopp (who passed away in 2006), Miriam Valluchi, Pia Fuchs-Bislin, Birgitt Bechler, Felix Gmünder, Giovanna Lorenzi, Thomas Schopper, Nadine Conza, ETHZ; Isabelle Walther, ***Zero-g LifeTec*** GmbH; Millie Hughes-Fulford and her team, University of California, San Francisco; Proto Pippia, M.A. Meloni, G. Galleri, L. Sciola, A. Spano, University of Sassari; Mauro Macarrone and his team, University of Teramo; Otfried Müller, University of Bern; Rodolfo Negri, University of Rome; the teams of ESA (in particular of Biorack), NASA (in particular of SLS-1), IBMP Moscow; the astronauts of STS-8, SL-1, D-1, SLS-1, IML-1, IML-2; the cosmonauts of MIR.

Financial support was provided by: ETHZ, ESA/PRODEX, ASI, NASA, SNSF, and Contraves Space.

References

1 Cogoli, A., Tschopp, A., Fuchs-Bislin, P. *Science* **1984**, *225*, 228–230.
2 Cogoli, A. *J. Grav. Physiol.* **1996**, *3*, 1–9.
3 Cogoli, A., Cogoli-Greuter, M., Activation and Proliferation of Lymphocytes and other Mammalian Cells in Microgravity, in: *Advances in Space Biology and Medicine*, Bonting, S. (Ed.), pp. 33–79, JAI Press Inc., Vol. 6, **1997**.
4 Cogoli, A. *Gravit. Space Biol. Bull.* **1997**, *10*, 5–16.
5 Lewis, M.L. *Adv. Space Biol. Med.* **2002**, *8*, 77–128.
6 Crabtree, G.R., Clipstone, N.A. *Annu. Rev. Biochem.* **1994**, *63*, 1045–1083.
7 Foletta, V.C., Segal, D.R., Cohen, D.R. *J. Leukocyte Biol.* **1997**, *63*, 139–152.
8 Leonard, W.J., O'Shea, J.J. *Annu. Rev. Immunol.* **1998**, *16*, 292–322.
9 Cogoli, A. Cell Biology, in: *Fundamental of Space Biology*, Clement, G., Slenzka, K. (Eds.), pp. 121–170, Springer, Heidelberg, **2006**.
10 Bechler, B., Cogoli, A., Mesland, D. *Naturwissenschaften* **1986**, *73*, 400–403.
11 Cogoli, A., Bechler, B., Müller, O., Hunzinger E., Effect of Microgravity on Lymphocyte Activation, in: *Biorack on Spacelab D1*, Longdon, N., David, V. (Eds.), pp. 89–100, ESA SP-1091, ESA Publications Division, ESTEC, Noordwijk, **1988**.
12 Cogoli, A. *ASGSB Bull.* (now *Gravitat. Space Biol. Bull.*) **1991**, *4*, 107–115.
13 Cogoli, M., Bechler, B., Cogoli, A., Arena, N., Barni, S., Pippia, P., Sechi, G., Valora, N., Monti, R. Lymphocytes On Sounding Rockets, in: *Proceedings of the 4th European Symposium on Life Sciences Research in Space, Trieste*, David, V. (Ed.), pp. 229–234, ESA SP-307, ESA Publications Division, ESTEC, Noordwijk, **1990**.
14 Cogoli-Greuter, M., Spano, A., Sciola, L., Pippia, P., Cogoli, A. *Jpn. J. Aerospace Med.* **1998**, *35*, 27–39.
15 Bechler, B., Cogoli, A., Cogoli-Greuter, M., Müller, O., Hunzinger, F., Criswell, S.B. *Biotechnol. Bioeng.* **1992**, *40*, 991–996.
16 Cogoli, A., Bechler, B., Cogoli-Greuter, M., Criswell, S.B., Joller, H., Joller, P., Hunzinger, E., Müller, O. *J. Leukoc. Biol.* **1993**, *53*, 569–575.
17 Gmünder, F.K., Konstantinova, I., Cogoli, A., Lesnyak, A., Bogomolov, W., Grachov, A.W. *Aviat. Space Environ. Med.* **1994**, *65*, 419–423.
18 Bechler, B., Hunzinger, E., Müller, O., Cogoli, A. *Biol. Cell* **1993**, *79*, 45–50.
19 Cogoli, A. *J. Leukocyte Biol.* **1993**, *54*, 259–268.
20 Haston, W.S., Shields, J.M. *J. Cell Sci.* **1984**, *68*, 227–241.
21 Wilkinson, P.C. *J. Cell Sci.* **1987**, *8*(Suppl.), 104–119.
22 Penninger, J.M., Crabtree, G.R. *Cell* **1999**, *96*, 9–12.
23 Bauch, A., Alt, F.W., Crabtree, G.R., Snapper, S.B. *Adv. Immunol.* **2000**, *75*, 89–114.
24 Acuto, O., Cantrell, D. *Annu. Rev. Immunol.* **2000**, *18*, 165–184.
25 Pippia, P., Sciola, L., Cogoli-Greuter, M., Meloni, M.A., Spano, A., Cogoli, A. *J. Biotechnol.* **1996**, *47*, 215–222.
26 Cogoli-Greuter, M., Meloni, M.A., Sciola, L., Spano, A., Pippia, P., Monaco, G., Cogoli, A. *J. Biotechnol.* **1996**, *47*, 279–287.
27 Walther, I., Pippia, P., Meloni, M.A., Turrini, F., Mannu, F., Cogoli, A. *FEBS Lett.* **1998**, *436*, 115–118.
28 Negri, R., Costanzo, G., Galleri, G., Meloni, M.A., Pippia, P., Schopper, T., Cogoli-Greuter, M., Early Transcriptional Response of Human T-lymphocytes to Microgravity conditions during MASER 9 Vector Flight, in: *16th ESA Symposium on European Rocket and Balloon Programmes and Related Research*, St.Gallen, Switzerland, 2–5 June 2003, pp. 243–247, ESA SP 530, ESA Publications Division, ESTEC, Noordwijk, **2003**.

29 Gmünder, F.K., Lorenzi, G., Bechler, B., Joller, P., Muller, J., Ziegler, W.H., Cogoli, A. *Aviat. Space Environ. Med.* **1988**, *59*, 146–151.

30 Gmünder, F.K., Baisch, F., Bechler, B., Cogoli, A., Cogoli, M., Joller, P.W., Maass, H., Muller, J., Ziegler, W.H. *Acta Physiol. Scand.* **1992**, *604*(Suppl.), 131–141.

31 Hoson, T., Kamisaka, S., Masuda, Y., Yamashita, M., Buchen, B. *Planta* **1997**, *203*, S187–S197.

32 Schwarzenberg, M., Pippia, P., Meloni, M.A., Cossu, G., Cogoli-Greuter, M., Cogoli, A. *Adv. Space Res.* **1999**, *24*, 793–800.

33 Maccarrone, M., Battista, N., Meloni, M.A., Bari, M., Galleri, G., Pippia, P., Cogoli, A., Finazzi-Agrò, A. *J. Leukoc. Biol.* **2003**, *73*, 472–481.

34 Vadrucci, S., Lovis, P., Henggeler, D., Lambers, B., Cogoli, A. *J. Gravit. Physiol.* **2005**, *12*(*1*), P177–P178.

35 Hughes-Fulford, M., Sugano, E., Schopper, T., Li, C.-F., Boonyratanakornkit, J.B., Cogoli, A. *Cell. Signalling* **2005**, *17*, 1111–1124.

36 Boonyaratanakornkit, J.B., Cogoli, A., Li, C.-F., Schopper, T., Pippia, P., Galleri, G., Meloni, M.A., Hughes-Fulford, M. *FASEB J.* **2005**, *19*, 2020–2022.

37 Horwitz, A.R., Parson, J.T. *Science* **1999**, *286*, 1172–1174.

38 Meloni, M.A., Galleri, G., Pippia, P., Cogoli-Greuter, M. *Protoplasma*, **2006**, *229*, 243–249.

9
Evaluation of Environmental Radiation Effects at the Single Cell Level in Space and on Earth

Patrick Van Oostveldt, Geert Meesen, Philippe Baert, and André Poffijn

9.1
Introduction

Leaving the Earth's atmosphere to explore the ultimate frontier, Space, is possible through the tremendous advances in rocket and spaceship technologies developed during the last 50 years. Although much information has been obtained by unmanned Space exploration, the curiosity of the human species is an ever driving force not only to set up new missions to ever deeper Space, but also to allow man the feeling and sensation of zero gravity and contact with other planets.

Although the realization of satellite communication and the setting up of a global positioning system convinced even the most sceptical people of the possible benefits of Space research and technology, every Space-related accident that costs the life of a man rose again the question about the necessity of human exposure to the dangers of Space travels. The impact of the loss of the Space shuttle Columbia and its crew in February 2003 has a more important effect on the further development of Space policy than the assessment of risks associated with microgravity, radiation or other physiological effects on man. Nevertheless, humans leaving Earth's orbit for extended periods should be aware of different risks having short- and long-term effects on their optimal functioning during prolonged exposure to the Space environment. In addition, having gained the sensation of interplanetary Space travel, the safe return of the crew to Earth does not exclude any long-term health effects. If Space travel is extended to non-professionals, the risk evaluation of the complex exposure should be well defined to exclude any legal steps against flight organizers.

Assessments of radiation risks associated with personnel in Space have proven not to be amenable to simple approaches. The problem is exacerbated by the fact that almost every known type of radiation is present with varying intensities and varying environments. The primary particles, for example, produce a wealth of secondary particles, including neutrons and pions, when they interact with a spacecraft or even with its occupants [1]. These effects are expressed in cells, tissues or biological systems that are exposed to zero gravity, which has severe health implications for astronauts. If we plan long-term Space flights, the risk

Biology in Space and Life on Earth. Effects of Spaceflight on Biological Systems. Edited by Enno Brinckmann

ISBN: 978-3-527-40668-5

assessment of radiation not only concerns the well-documented effects of radiation on the genetic stability of different tissues but also has to take into account the risks of loss of valuable function, acute bodily injury or infection, cancer etc. There are three potential threats to the immune system on manned space flight according to Todd et al. [2]: (a) the over-proliferation and potential lack of control of microbial populations in combination with allergenic or inflammatory atmospheric contaminants; (b) the compromising of specific components of the immune response due to low gravity; and (c) the killing of progenitor cells by radiation.

To evaluate the combination of risks for these situations, specific evaluations of the different components, macromolecules, single cells, tissues, organs and organisms should be studied. Once the risk factors and their relative biological effects (RBE) have been evaluated in specific conditions by ground experiments in combination with specific short-term Space experiments, we can advance to more complex experiments with small animals in combination with specific countermeasures. Having a good estimate of the risk, we can then inform the motivated astronaut and agree on the specific balance between risks and costs to optimize the long-term Space travel. In this chapter we discuss different strategies that can be adopted to reach the goal of interplanetary travelling. We also define some border conditions that are to be fulfilled to evaluate risks of extraterrestrial missions. The idea is that a long journey in Space should take into account the spacecraft as a small autonomous unit with specific ecological constrains that will have an impact on any aspect of the global functioning of the system. As such the spacecraft is the integration of different physical and biological systems that need to be under perfect control. Space is an unforgiving environment that does not tolerate human errors or technical failure [3]. The failing of a single element can impact on any other compartment of this ecosystem, which is the manned spacecraft. The success of manned interplanetary Space exploration can only be guaranteed if the crew returns in a healthy status back to Earth.

9.2 The Space Radiation Environment

If we define the spacecraft as an autonomous unit, we can not suppose it is isolated. To survive and continue to work will require a regular delivery of basic supplies (food, water, air, building material, and so on). The supply of these materials is only possible at high costs and should, therefore, be minimized. In addition, the spacecraft and its inhabitants are continuously irradiated by a wide spectrum of electromagnetic radiation and charged particles. This electromagnetic radiation can be used to deliver the necessary energy to support the working of the station but part of this radiation is known to be injurious. The complex radiation in low Earth orbit is composed of electromagnetic radiation and charged particles of solar and galactic origin. Among the galactic cosmic rays (GCR) and the solar particle radiation we find the high-energy, ionizing cosmic ray (HZE) particles with a nuclear charge $Z>2$. GCR originate from outside our solar system and contain particles of all charges from protons (87%) to heavy ions (1%) with energies in the range from a few MeV per nucleus (nuc) to $>10^{15}$ MeV per nuc. The flux of GCR particles is modulated by the solar activity, whereby the flux is lowest during solar maximum and highest during solar minimum. The cosmic ray flux varies, depending on the particle energy, up to a factor of 10 within the solar cycle [4].

In low Earth orbits between 200 and 600 km altitude, the main absorbed dose inside a spacecraft is contributed by trapped protons from the South Atlantic Anomaly, an area where the radiation belt reaches lowest altitude due to a displacement of the magnetic dipole from the Earth's centre. Only a fraction of the incident flux, depending on the rigidity of the incident particles, can reach the spacecraft in low Earth orbits. However, leaving the protective magnetic field of the Earth, considerable higher numbers of HZE nuclei – debris from collapsing stars and supernova explosions that were thrown in Space – are present. Curtis and Letaw [5] estimated that on a three-year Mars mission, about 30% of cells in the body will be traversed by HZE nuclei with Z values between 10 and 28, and that virtually all cells will be traversed by nuclei with values between 3 and 9. Although we can produce HZE nuclei on Earth and study their effects on biological material, simultaneous studies in extended periods of zero gravity exposure in combination with HZE traversals in cells, tissues or organs are only possible on the International Space Station (ISS).

9.3 HZE Track Detection

One of the most commonly used nuclear track detectors is based on the poly(allyldiglycol carbonate) (PADC) CR-39® detector, which was discovered by Cartwright [6]. Another most commonly used nuclear track material is cellulose nitrate. The most well-known detector in this group is sold under the commercial name LR 115. Other kinds of detectors are also in use, such as the Makrofol® or Lexan® detector, which is based on polycarbonate. Some natural materials that show the track effect, such as apatite, mica, olivine, and others are used for fission or fossil track studies.

The primary process of charged-particle interaction with the detector material is ionization and excitation of the molecules in the detector. The initial charged particle loses its energy through the many interaction processes. Most interactions occur with electrons, and only a small number of interactions are with nuclei. Since the initial heavy charged particle (only such particles can produce tracks) is much heavier than electrons, the direction of the particle effectively does not change and the path is almost completely a straight line. This may not be true if the particle interacts with an atomic nucleus, where a significant deviation from the initial direction may occur. However, such interactions are relatively rare.

Some deviations from a straight line can happen close to the end of the particle range, when the energy of a particle becomes very low. The particle loses its energy in many small interaction processes, so the energy loss each time is usually very small when compared with its energy. For example, ionization of one molecule in air needs on average about 32 eV, which is 10^{-5} to 10^{-6} of the particle energy (assuming that the particle energy is in the MeV region). As a result of these many small interaction processes, the particle will continuously slow down in the detector material. The energy lost by a particle in the distance dx is the energy transferred to the material, and so this quantity is also called the linear energy transfer (LET).

9.3.1 Confocal Scanning Laser Microscopy for Track Analysis in PADC

For nuclear track analysis the shape, size and orientation of the track are important parameters. We have developed a confocal visualization method based on the negative staining of the tracks created after etching the PADC with 6.25 M NaOH at 70 °C for 7 h [7]. After etching, the detector is covered with a solution of FITC (1 ppm fluorescein isothiocyanate dissolved in glycerine at pH 8) and the tracks are visualized by recording different optical sections of the stained track. The use of long working distance (WD) objectives with high numerical aperture (NIKON, planapochromatic 60× NA 1.4, WD: 200 μm) allows both high resolution optical section and visualization of the track cone to analyse the shape. However, the working distance of this objective limits the detection of etched cones at both sides of the detector. To see the full-length track through the detector, lower numerical aperture objectives with longer working distances are used (Nikon 10× NA 0.3, WD: 17.3 mm). As an alternative to visualizing the shape of the track, the detector surface can be stained after etching by immersing the PADC detector in a 1% Nile Blue solution at 50 °C for 20 min. After staining, the unbound dye is washed away with distilled water.

The use of a precise computer controlled *X-Y* table on the confocal microscope allows precise relocation of every track on the detector. At low magnification it is possible to discriminate both sides of the detector and reconstruct the track path through the detector by using the coordinates of the cones from incoming and outgoing particles. This analysis gives the number of particles that passed through the detector and their orientation relative to the detector. Combining a stack of detectors with a cell culture on top allows precise location of the hit cell by reconstruction of the path through the whole experimental unit. Eventually, further characterization of the track is possible at high resolution by using a separate analysis of the top and the bottom side of the detector by high objectives with high numerical aperture. Because every detector is fully digitized and every cone is unequivocally localized with its *X-Y* coordinates, a full 3D track visualization is possible after computer reconstruction of all tracks together with the evaluated cell culture.

9.3.2 Time-resolved Track Detection

The use of a stack of detectors for recording HZE particles in combination with 3D recording of their path was further improved to discriminate tracks passing through the cell culture before and after fixation. This was realized by using a specific shifting mechanism in the experimental container (Fig. 9.1). After changing the cell culture medium by a formalin-based fixative, parts of the detectors were shifted for about 150 μm. This shift could be used to identify tracks passing through the cell culture after fixation. After aligning every detector back to its specific position (Fig. 9.1, number 8) the track coordinates per detector are recorded

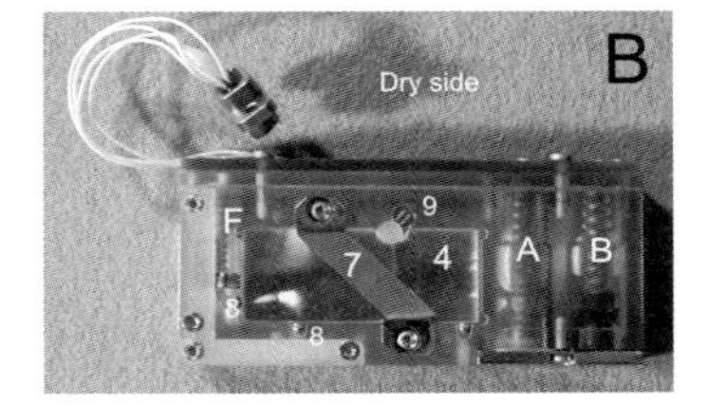

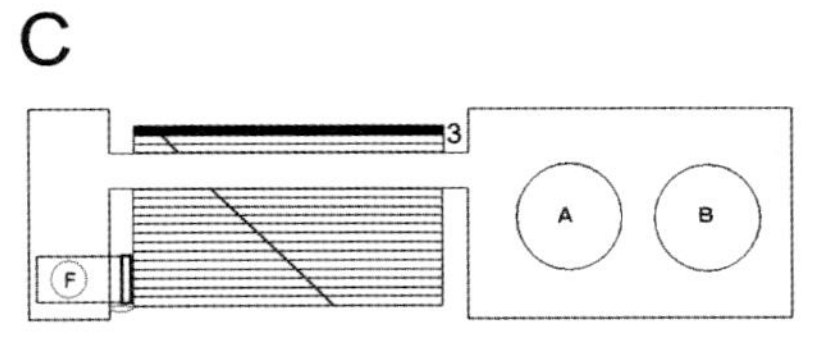

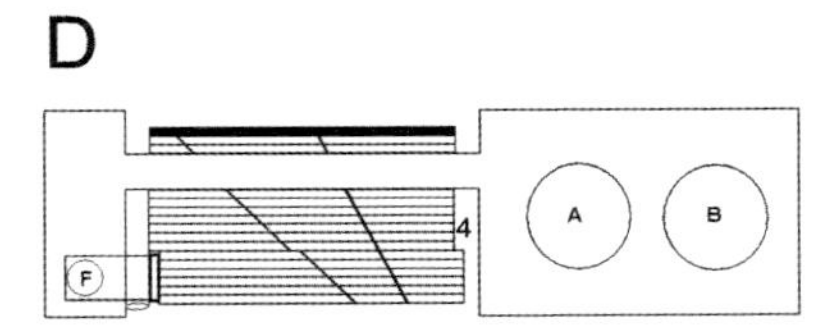

Fig. 9.1 Overview of the Radcells/Ramiros plungerbox units. The wet (3) and dry compartments (4) are shown. (A) The wet compartment is a compartment in which the cell culture is placed and that is filled with cell culture medium for maintenance. The cell culture is covered by a gas permeable polyethylene foil enabling O_2 exchange. Structural rigidity is provided by a metal grid (not shown). (B) The dry compartment is filled with the PADC stack. A total of three plungers are required for activation of cell culture refreshment (labelled A), cell fixation (labelled B) and partial shift of the PADC detector stack (labelled F) at the time of fixation. The cell culture support and the PADC detectors are kept in place by a positioning mechanism consisting of an eccentric shaft (9), a silicone ring and three positioning pins (8). Once the cell substrate or PADC detectors are put in place, the eccentric shaft is turned 180°, resulting in a directional force that pushes the substrate or detectors against the pins. For flight, a printed circuit board with RS2 connector is attached to the unit in replacement of a side plate, allowing for a hand-operated plunger activation. The automated plunger activation relies on a voltage given at a predefined moment. This voltage heats a wire, resulting in the release of the plunger and the compression of the flexible fluid bags present in the space in front of the releasing plunger. Fresh medium enters the cell culture compartment, while old medium flows into the empty plunger compartments via dedicated channels. Plungerbox dimensions: 8×4×2 cm^3. (C) Schematic top view of the plungerbox, showing the detector stack before fixation: uninterrupted track of particle. (D) Detector stack after fixation: due to the shift of the lower stack at time of the fixation, original tracks before fixation are interrupted, whilst particles hitting the detector after fixation leave uninterrupted tracks.

and the full track record is reconstructed. These reconstructed tracks show only one continuous path along the different detector strips, without a specific shift: they were created by a particle passing the cell culture before killing the cells by fixation. Tracks representing a specific shift at the appropriate detector are caused by HZE particles hitting cells after fixation because shifting part of the detector's stack after the fixation results in a discontinuity in the *reconstructed* track.

Within the experiment container used during the cell culture experiments on the Belgian Soyuz taxi flight to the ISS (November 2002), this system proved its performance and showed good stability. Shifted tracks could be well reconstructed and the cells hit by the particle during their live status could be well localized. This is an essential step in the analysis of HZE effects on live cells during Space exploration.

Figure 9.2 illustrates the results of computer tracking. This detector contains 22 tracks with a clear shift and 38 without a shift. This should indicate that the studied cell culture was hit 38 times by a HZE particle during its life and that after fixation 22 particles passed through the container. This corresponds well with the time schedule of the Soyuz flight. Sixty hours after launch the temperature of the cell container was raised to 37 °C and the cell cultures were all kept at this temperature until fixation 96 hours (h) later, also at 37 °C. After fixation it was another 105 h

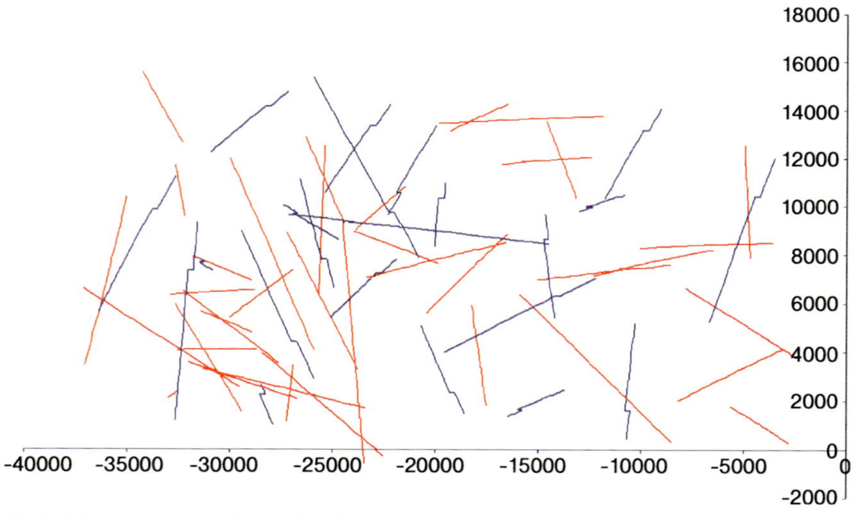

Fig. 9.2 Representation of a graphic Excel output summarizing the high-energy, ionizing cosmic ray (HZE) tracks as reconstructed from the microscopic analysis of the detector stacks. Because all tracks are recorded relative to the position pins at the microscope, only the interrupted tracks originate from particles passed after fixation and shifting of the detectors. Dimensions of the axis are in μm. The red tracks are straight lines and represent tracks passed through the culture compartment when the cells were alive. The blue lines represent tracks reconstructed after part of the stacks was shifted, i.e. after fixation. The coordinates of the tracks were used to calculate the point where the cell culture was hit by the detected HZE.

before the Soyuz landed. Consequently, we postulate that the live cells of this experiment were exposed to HZE particles for 156 h.

9.4 Results of the RAMIROS Experiment on board of the Soyuz Taxi Flight to the ISS

To show the feasibility of the approach, we present here some results of the Belgian Soyuz taxi flight to the ISS in November 2002, named "Odissea". This flight was joined by Frank De Winne as a Belgian ESA crew member. The experiment was titled "Cosmic *ra*diation and *mi*crogravity *r*elated *o*xidative *s*tress" (RAMIROS).

9.4.1 Methods

Since a more detailed description of the materials and methods will be published elsewhere, we describe only the basic experiment set-up. Primary bone marrow cells were isolated from mice femora three weeks before hardware integration. A total of 2.6×10^6 cells were inoculated in 4 mL of Iscove's cell medium (Invitrogen) supplemented with 10% calf serum (Invitrogen), 0.6 g per 100 mL β-glycerophosphate (Sigma-Aldrich, St-Louis, MO), 100 μg mL^{-1} ascorbic acid (Sigma-Aldrich) and 1% (v/v) of a combined L-glutamine/penicillin/streptomycin solution. The cells were grown on Aclar plastic supports (Tadpella Inc. Redding, CA), cut in the customized 8 cm^2 shape (Pedeo Technics, Oudenaarde, Belgium) to fit the Space hardware. Aclar supports were placed in chamber slides (Nalge Nunc International, Rochester, NY), inoculated with cells and placed in a humidified 5% CO_2/37 °C incubator. Half of the cell culture volume was refreshed twice a week. Half a week before the hardware integration, Iscove's medium was completely exchanged by α-MEM (Invitrogen) supplemented with 10% foetal bovine serum, 10% horse serum, 1% (v/v) of a combined L-glutamine/penicillin/streptomycin solution and 0.4% (v/v) Fungizone (Invitrogen). Integration of the whole plungerbox units and cell culture took place in Belgium, and cells were transported to Baikonur by ESA carrier.

The experiment used the experimental hardware designed by CCM (Neunen, The Netherlands), containing a plungerbox with three plungers each (Fig. 9.1). The firing of the plungers was controlled by electronics attached to the hardware and worked fully automatically. The crew had only to plug the experiment container into the incubator. One plunger contained fresh culture medium that was fed to the cells 48 h after the culture was exposed to 37 °C. The second plunger, containing the fixative (0.5% paraformaldehyde) was fired 96 h after the cell culture started growing at 37 °C. The third plunger was used to fire the shifting mechanism of the radiation detectors.

After return to ground, the samples were collected from the disassembled hardware, washed in PBS (phosphate buffered saline), dehydrated in ethanol series and stored at −20 °C until ground controls were obtained. Ground controls were carried out with freshly prepared cell cultures (cultured in the same conditions as the Space flight cell cultures) one month later based on the retrieved data of a temperature logger included in the experiment container.

After 9 h etching of the PADC plastics in 6.25 M NaOH at 70 °C, HZE particle hits were visualized by light transmission microscopy. The corresponding stage co-ordinates were stored in a database as well as the extrapolated stage co-ordinates of the cell culture support. Extrapolation took into account the direction of the track and the thickness of the Aclar support (200 μm). The corresponding cell culture support was calibrated based on the same predefined calibration points (a total of six) of the PADC and Aclar supports having the same shape. After calibration, the microscope stage moved to the points of interest where a HZE hit had traversed the cell culture. Cells were labelled by an in-house developed protocol for staining DNA double strand breaks and single strand breaks or oxidative stress [8]. Negative controls for Fpg enzyme could not be performed due to the scarcity of samples. Instead, the activity of the enzymes was checked on samples exposed to 200 μM H_2O_2.

9.4.2 Results and Discussion

9.4.2.1 Retrieval of Cell Cultures

The "Odissea" mission was performed under nominal conditions. RAMIROS plungerbox units were returned to the investigator two days after landing. Morphological inspection of the bone marrow stroma (BMS) cell cultures showed similar morphology compared with the cell cultures integrated in the hardware at the start of the experiment. The initial option of using glass slides for cell growth was abandoned in favour of the Aclar support slides because of a potential breaking of the glass at the moment of the rather hard landing of the Soyuz capsule. A Soyuz landing contrasts with the relatively soft landing of a Space Shuttle. The cell density on Aclar supports was, however, lower than on glass. The use of Aclar was a compromise of cell growth characteristics versus material reliability.

9.4.2.2 Number of HZE Tracks

An average of 150 tracks was found in the PADC nuclear track detector stack on a surface of 7 cm^2 after a stay of 11 days in Space. This number does not include proton hits and corresponds to a fluency of 1 HZE hit per 4.7 mm^2. This figure equals 21.2 tracks cm^{-2} or an HZE fluency rate of 1.94 HZE-hits per cm^2 per day. This fluency rate is significantly higher than the reported 0.96 HZE-hits per cm^2 per day measured in Biobox-2 on Foton-10 [9]. When assuming a confluent cell culture and a cell and nucleus size of 250 and 100 μm^2, respectively, the number of 1 HZE hit per 4.7 mm^2 should correspond with 0.4 nucleus particle traversals every 4.72 mm^2 or 1 nucleus traversal every 11.8 mm^2. However, this theoretical assumption overestimated the real coverage of cell nuclei by a factor of 4: it was indeed calculated that the actual coverage of cell nuclei in the cell culture was only

10%. This can be explained by the fact that the size of the cell measured in fixed condition underestimates its real size and by a smaller cell density on the Aclar growth substrate. A 10% coverage of the cell support by cell nuclei results in 1 nuclear HZE particle traversal every 47.2 mm^2 or 15 nuclear HZE particle hits per 7 cm^2 cell culture area. If this reasoning is continued, then 1.36 cell nuclei were traversed by a HZE particle per day per plunger-box unit. This fluency rate corresponds with 0.19 nuclear HZE particle traversals per day per cm^2. The indicated numbers obviate the need for single cell analysis and the link with positional information from the nuclear track detectors.

HZE particles passing through the cell culture and not through the plane where the shift of the detector was realized, create a track that does not give information about the status of the cell (live or dead) at the moment of exposure. These uninformative tracks (40%) were eliminated. Of the remaining 60% tracks, 55% could be unambiguously assigned as being incident before fixation. From this ratio, the total duration of the experiment was calculated in a retrospective way and was obtained by multiplying the total length of the mission by 0.55. This number, being 6.1 days, corresponds well with the predefined timeline of 6.5 days. The difference may be explained by the 2 days chasing time of the Soyuz to reach the ISS. During these two days the exposure to cosmic radiation may be lower than when Soyuz is docked to the ISS. Chasing of the ISS by the Soyuz spacecraft indeed proceeds from lower to higher orbits. Considering the 55% figure, it can be calculated that, of the 15 HZE hits traversing cell nuclei during space flight per cell culture, 8 were traversing nuclei before fixation. Table 9.1 summarizes these figures.

9.4.2.3 Correlation between HZE Hit and DNA Damage at Single Cell Level

From all cell culture spots coinciding with an HZE hit location, 99% did not show cells with a steep increase in the DNA damage endpoints. Only in four cases spread over three 7 cm^2 cell cultures could a significant increase in FPGSS

Table 9.1 Overview of characteristics relevant to HZE impact as measured after an 11-day exposure during the "Odissea" mission to the ISS at the level of the plungerbox unit (cell culture area 7 cm^2).

Features per plungerbox unit	*Value*
Cell culture area	700 mm^2
Number of tracks	~150
HZE fluency	1 track per 4.7 mm^{-2} = 21.3 tracks cm^{-2}
HZE fluency rate	1.94 tracks cm^{-2} day^{-1}
Ratio of HZE tracks produced before fixation	55%
Chance of all HZE impacts traversing a nucleus	10%
Calculated nuclear hits before fixation	~8
Number of DNA damage clusters	1–2

(formamidopyridine sensitive sites) [8] be seen as compared with the average cell. These four cases were characterized by a cluster of cells with increased FPGSS staining and divided over the three 7 cm^2 cell cultures. When the ratio of FPGSS positive clusters to the total amount – defined by probability – of the nuclear HZE particle traversals in three 7 cm^2 cell cultures, i.e. 45, is calculated, the figure is 9%. If the ratio is based on the number of HZE traversals through a nucleus before fixation, i.e. 24.8, the number becomes 16%. However, from two of the associated tracks no data were available, whether or not these tracks were produced before or after fixation. Intense TUNEL (staining indicative for chromatin fragmentation in apoptotic cells) could not be seen. Interestingly, two cell nuclei with the appearance of DNA synthesis and intensely stained for the FPGSS assay were located at the site of a HZE traversal, but not surrounded by a cluster of intensely stained nuclei for FPGSS. Figure 9.3 illustrates these observations.

Stained nuclei with typical track features based on the end labelling techniques were not seen. This finding contrasts with immunochemical methods developed by others [10]. The inability to visualize track structures is, however, in line with preliminary HZE studies performed at the heavy ion beam facility of GSI (Darmstadt, Germany), where, at doses of one carbon ion traversal per nucleus, only uniformly stained nuclei were seen with our assay. This may be explained by the fact that the direct DNA damage produced at the HZE traversal path is relatively small and somehow masked by the indirect damaging effect homogeneously spread throughout the nucleus. Preliminary studies carried out with the GSI heavy ion beam facility indicate that the response to HZE particles is first marked by a steep increase in DSB (double strand break) that significantly drops within 15 min post exposure. The decrease of DSB is followed by an increase of FPGSS staining that is prominent 24 h post irradiation. At that stage, some cells again began to intensely stain for double strand breaks, indicating the onset of late programmed cell death (apoptosis) [8]. Even though it was not possible to detect TUNEL positive cells in the Space-flown samples, the cells intensely staining for FPGSS may, however, still be linked with the process of programmed cell death. The argument for this is the finding of chromatin condensation at the nuclear periphery, first described by Kerr et al. [11], in several FPGSS positive cell clusters. Chromatin condensation associated with apoptosis is not an active process, but instead arises as a consequence of the loss of structural integrity of the euchromatin, nuclear matrix and lamina [12]. It was further reported that chromatin condensation occurs independently of, as well as prior to, DNA cleavage and involves a specific conformational change at the nucleosomal level [13]. The finding of chromatin condensation at the nuclear periphery may, therefore, indicate a hint for the onset of these early apoptotic events. Single strand nicks that are detected with the FPGSS assay may then primarily be produced during heat permeabilization procedures inherent to the described *in situ* end labelling assay. DNA strand nicking by heat permeabilization may become important on nuclear places marked by loss of structural integrity. This phenomenon may, therefore, also be seen in mitotic cells. The inability to find TUNEL positive nuclei may reflect the smaller

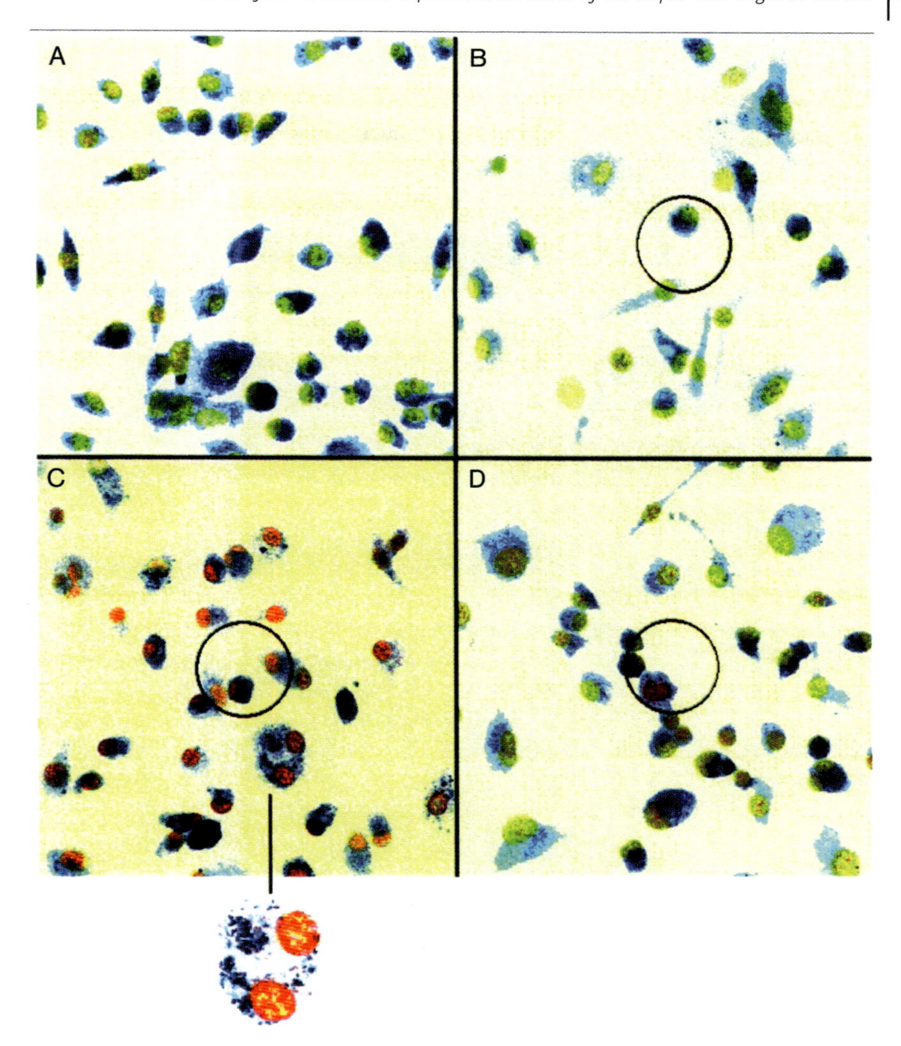

Fig. 9.3 Ground versus Space-flown cell cultures stained for FPGSS (formamidopyridine sensitive sites) and TUNEL (staining indicative for chromatin fragmentation in apoptotic cells). Images are inverted from the original. The blue cytoplasmic signal reflects autofluorescence inherent to bone crystals. Nuclei are counterstained with DAPI (represented in green). HZE-hits are indicated by a circle. (A) No extra nuclear staining referring to FPGSS or TUNEL was noticed in the ground controls. (B) A cases where a high-energy, ionizing cosmic ray (HZE) hit had traversed the cell culture but did not show a steep increase in DNA damage compared with neighbouring cells. This is 99% of the cases. (C) A representative image (1 out of 4) showing an increased FPGSS (red) seen in a number of cells at the location of the HZE hit. (D) A cell with DNA-synthesis appearance positively staining for FPGSS (red) within the spot of the HZE hit. Baseline of the different panes is 189 pm. Circles represent areas in which a single HZE particle has traversed the cell culture. Inset [below (C)]: enlarged image of two adjacent cells and nuclei in which FPGSS (red) are heterogeneously distributed and mainly seen at the nuclear periphery.

probability of finding late apoptotic cells because of the smaller time period, in which this late apoptotic phase is occurring.

The presence of a bystander effect may be evidenced by the four clusters of damaged cells, which ranged from 8 to 40 cells at the site of a single HZE traversal, and the fact that this phenomenon was not seen elsewhere. The first report of a bystander effect measured an increase in sister chromatid exchange in 20–40% of Chinese hamster ovary cells in cultures exposed to fluencies, by which only 0.1–1% of the cells' nuclei were actually traversed by a particle track [14]. Given the different endpoint used in this study, a comparison of the extent of this presumed bystander is difficult.

9.4.2.4 Space versus Ground Samples

Comparison of ground and Space cell culture samples revealed that the average level of DNA damage was low but differed significantly in both cases: image processing, evaluating whole cell fluorescence revealed, however, a significant increase of DNA damage in Space versus ground samples ($P < 0.001$). This finding was valid for both FPGSS and TUNEL. Figure 9.4 shows the bar charts of the different experimental groups for the two DNA damage endpoints investigated. FPGSS was 4.2 times increased as compared with the ground samples. TUNEL staining was 3.2-fold greater in Space samples than with ground samples. In theory, this overall increase in FPGSS could be due to the low LET radiation combined with microgravity. However, the very small dose of a few hundred $\mathrm{pGy\,d^{-1}}$ in low Earth orbit [15] may be too low to explain these differences. It was further reported that a stay in Space could result in increased oxidative stress [16]. However, an in-flight

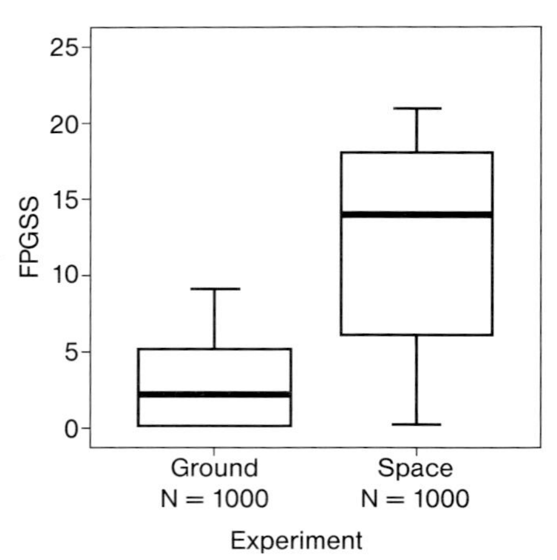

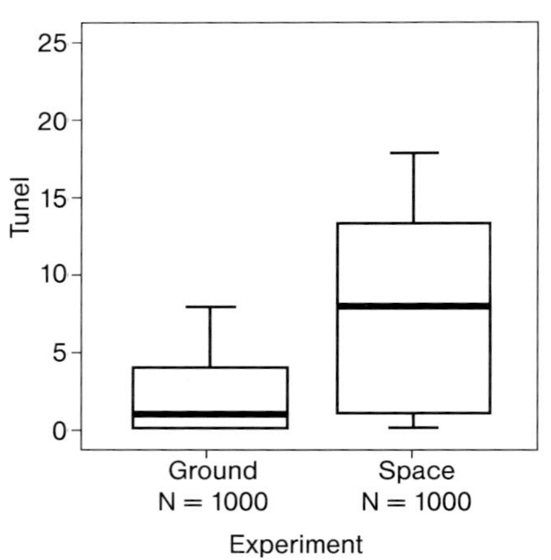

Fig. 9.4 Comparison between flight and ground samples at the level of the global cell culture after FPGSS labelling and TUNEL was applied. The intensity value of FPGSS or TUNEL signals in cells flown in Space is plotted as bar-chart. FPGSS intensity was significantly increased. The median value is increased by a factor of 4.2 as compared with the ground samples; TUNEL was also significantly increased 3.2-fold (right-hand side). N = number of cells, ordinate values represent the intensity in arbitrary units of the selected endpoints.

increase in oxidative stress is related to a deficit in energy intake while controls and space flight samples were equally provided with nutrients. In addition, temperature profiles were reproducible for Space and ground samples. Imaging parameters were equally set for Space and ground samples based on the cytoplasmic autofluorescence of cells in the bone marrow stroma. Significant bias may come from the fact that Space and ground cell cultures were not grown simultaneously. Another possible option is in line with the suggestion that heat pretreatment may induce strand breaks specifically at the site where structural loss of integrity is found in the nucleus. The differences seen between ground and Space samples at population level may accordingly reflect a difference in nuclear organization, not induced by apoptosis but by microgravity.

9.4.2.5 HZE Hits Before and After Fixation

A comparison was made of the degree of DNA damage at sites of HZE traversal before and after fixation. Intensely FPGSS stained nuclei, as seen at the spots of HZE traversal before fixation, were not seen at HZE spots that were produced after fixation. This finding may indicate that a high level of DNA damage induced by a HZE may be due to active processes, such as DNA repair and bystander effects extending DNA damage from a single track volume to a status that is characterized by a staining of multiple cells, as was demonstrated by previous experiments at the GSI accelerator in Darmstadt, Germany. The heavy ion beam study also revealed that DNA-DSBs precede single strand breaks linked to oxidative stress directly after exposure to particle irradiation.

9.4.2.6 Conclusions

The aim of this space flight experiment was to correlate HZE impact with DNA damage. A single cell approach was applied. No sound correlation could be drawn between the physical and biological measurements in Space-flown cell cultures. Nuclei intensely stained for DNA damage were only found in 1% of HZE cell culture traversals. This staining was only obtained for FPGSS (formamidopyridine sensitive sites) and not for TUNEL (indicative for chromatin fragmentation in apoptotic cells). It was calculated that the ratio of FPGSS positive cell culture spots to the number of nuclei traversed by a HZE particle before fixation may be up to 16%, dependent on the fate of two HZE traversals that were stopped in the stack, and from which it was not possible to determine if these hits were produced before or after fixation. It is argued that these four cases all represent the incidence of HZEs before fixation, because clustering of DNA damage is an active process [17]. Might this low ratio of 16% reflect the proficiency of repair? The four cases may then reflect HZE hits passing several hours before fixation. However, precise time information on HZE hits was not available. The experiment with live cells was run for 156 h (6.5 days); 16% of this time corresponds to 25 h, which is a reasonable time needed for repair of DNA damage. DNA condensation at the nuclear periphery is indicative for apoptosis. The occurrence of clusters of damaged cells around the HZE spot points to the presence of bystander effects in Space-flown cell cultures.

At the entire cell culture level, quantification of the fluorescence intensity revealed unexpected differences between ground and Space-flown cultures. The impact of microgravity in this finding could not be unravelled because no in-flight 1×g control was available. It may be that microgravity changes nuclear architecture, resulting in an increased sensitivity to end labelling procedures. However, the fact that the ground versus in-flight control originated from two different cell batches may compromise this Space-related significance. Future radiobiological experiments covering HZE impact should, therefore, include a simultaneous 1×g reference on an in-flight centrifuge for unravelling microgravity effects and study the impact of the bystander effect. The Biobox facility with its 1×g references is an opportunity to validate the hypothesized HZE-linked DNA damage and to study the extent of the bystander effect induced by HZE traversal in microgravity. Comparing wild-type cells with DNA repair or other relevant mutants is meaningful for unravelling possible sensitizing effects induced by microgravity.

9.5 Combination of Radiation with other Biological Stress in Space Travel

The physical characterization of the radiation in Space is fairly well defined by the constant collection of data all over the different Space missions. However, extrapolation to the relative biological effectiveness (RBE) is much more complex. Risk estimates are based on epidemiological data obtained on Earth for cohorts exposed predominantly to acute doses of γ rays, and extrapolation to the Space environment is highly problematic and error-prone. The classical approach is the use of well-defined biomarkers of exposure, sensitivity and disease [18, 19]. Biomarkers of *exposure* are biological parameters for which a dose–response relationship can be established – they can be broadly referred to as bio-dosimeters. Biomarkers of *sensitivity* are genetic markers, associated with an increase in individual susceptibility to radiation damage. Gender differences are good illustrations of sensitivity markers. The breast and ovaries are fairly sensitive to radiation, increasing the risk for females relative to males. Finally, biomarkers of *disease* are those biological events that can be used to anticipate the clinical diagnosis of a specific illness. In addition to these intrinsic bioindicators, Horneck [20] has defined extrinsic biomarkers such as those biological assays that can be used for genotoxic assessment in the environment.

A main problem in this approach is the large statistical uncertainty for dose or risks estimates with these models. This could be improved by drastically expanding the number of observations, but this is not feasible because of the elevated costs of large experimental set-ups. Phenomenological approaches, as often used in early experimental approaches, are too uncertain. One can project risks that way, but the error bars in the projection are just too high. Therefore, we need an approach based on fundamental biology, especially molecular biology in combination with advanced single-cell analysis techniques.

The combination of specific probes and advanced optical microscopy now allows quantitative probing of biochemical reactions in single living cells. Automated microscopy has the potential to allow fast, cheap collection of data describing protein behaviour and biological pathways within individual cells [21]. Accessing these data will produce useful profiles of cell phenotype, by analysis techniques that can provide massive and reliable information. The analysis of multidimensional single-cell phenotypic information is simple and inexpensive enough to allow extensive dose–response profiles for many disturbing factors. Since the very beginning of Space exploration, image recording and analysis has been a profound know-how of many participants, but, up till now, this expertise was not often used for the evaluation of biological phenomena. Miniaturization and validation of sensitive multiparametric cell-based assays can now be realized, even outside the molecular biology laboratory. We are convinced that an extensive, multidisciplinary collaboration program, within the frame of a Space project around these subjects, will also be of benefit for more performable diagnostic techniques in medicine were most of the image interpretation is done by a human observer.

Flow cytometry allows large-scale quantitative single-cell proteomics [22]. The application of this technique to a green fluorescent protein (GFP)-tagged yeast library, in which each protein is expressed as a C-terminal GFP fusion from its endogenous promoter at its natural chromosomal position, revealed cell-to-cell variation, or "noise", in proteins. The proteins with the most variable or noisiest expressions were the "first responders" to changing environmental conditions, such as those related to stress response and heat shock. This allows us to consider that cell populations generate diversity under one condition, so that if conditions change there will be at least a subset of cells that will be better optimized to that change. In terms of radiation protection, this can be described as an induced fit model, not for the cell but for the tissue. A low stress situation favours and protects tissues to higher doses. The expansion of this technology to image cytometry will reveal not only single-cell phenotypes but also can give information on cell–cell or cell–substrate interactions. The importance of cell–cell and cell–substrate interaction is becoming more and more prominent, mainly through the presentation of results obtained with advanced microscopical techniques [23, 24]. Bone development and mineralization is a typical example of cell–substrate interactions studied extensively in Space research (Chapters 6 and 7).

9.6 Interactions between Radiation and Gravity

Bone demineralization during residence at microgravity is a well-known phenomenon during Space travel. Just as with the radiation risk, it is an import problem for long-term Space travel. On Earth, gravity is an undeniable force that interacts with every organism. In a spacecraft and in extraterrestrial settlements, this force is absent or much reduced. Therefore, bone mineralization or development experiments have frequently been set up with a reference culture in a 1×g centrifuge. It is supposed that in this way unwanted interactions, e.g. with electromagnetic radiation on the cell culture, can be identified. A summary report of performance and results of 31 Biorack experiments flown on three Shuttle-to-Mir missions (STS-76, STS-81, STS-84) indicated that only about 50% of the results obtained with samples exposed in-flight to 1×g were identical to those obtained with control samples that had remained on ground [25]. Effects of different kinds of mechanical culture conditions on cells, especially the osteoblastic cell-line MC3T3-E1a, have been reported [26–28]. Work done on suspension cultures revealed a large set of genes responsive to mechanical culture conditions, including shear and turbulence [29]. Osteoblast proliferation can be achieved by mechanical stimulation, and this process is shown to be dependent on microtubular and microfilament components [30]. Owing to lack of experiment space, radiation-sensitive experiments on the "Odissea" mission were not repeated in a 1×g reference centrifuge and, therefore, we investigated whether a vibration load of the order those received during a Space Shuttle launch induced differential gene expression. Vibration load was applied in different directions on primary bone marrow stroma (BMS) cells to investigate a possible influence on gene expressions involved in DNA damage response. The effect of linear acceleration was not tested. We focussed on Gadd45, Mdm2, Bax, p21, Hsp70 and c-Fos mRNA known to be differentially expressed following exposure to biologically relevant doses of ionizing radiation [31–34]. The experiments, which used five housekeeping genes as reference and also checked the transcription of β-actin, showed that mechanical stimulation indeed activated the expression of different genes involved in DNA repair together with a stimulation of β-actin transcription [35]. The stimulation of actin was only significant if the direction of vibration paralleled the plane of the cell culture (Fig. 9.5). From

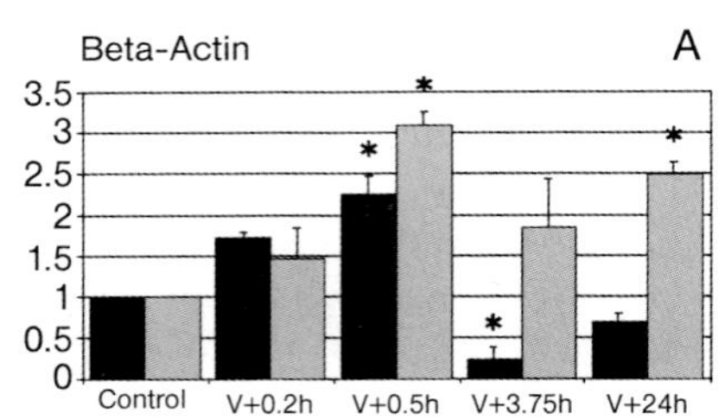

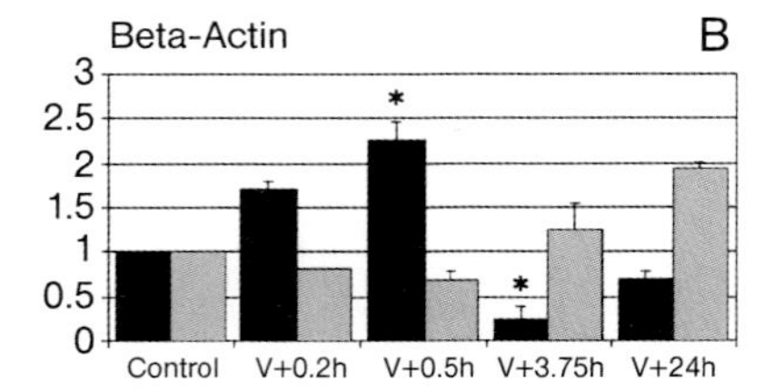

Fig. 9.5 (A) Gene expression profiles of β-actin at different time points following 150 s vibrations at 11.3×g rms (rms: root mean square, the energy in a vibrating medium) of primary bone marrow cells (BMS, black bars) and Ltk⁻ cells (immortal mouse L cells lacking thymidine kinase) (grey bars). Results for each sampling point post vibration (V) (V+0.2 h, V+0.5 h, V+3.75 h, V+24 h) are the average of real-time PCR data of two independent cDNA samples. Non-vibrated control samples were normalized to 1, and mean values are represented with the SEM. The expression profile of β-actin reflects the instability of the housekeeping gene over the set of vibrated and non-vibrated samples as first demonstrated by the geNorm algorithm. Asterisks indicate biologically and statistically significant differences to the control.
(B) Relative gene expression profiles of β-actin at different time points following the vibration of primary bone marrow cells (BMS) in *X*- (black bars) and *Z*- (grey bars) directions, measuring 11.3× and 4.5×g rms respectively. Time periods as in part (A).

the genes linked to DNA repair, c-Fos was the first and mostly affected gene. It reached a peak after 0.2 h, together with p21. Gadd45 is also significantly activated but reaches a maximum only after 0.5 h. Mdm2, Hspa4 and Bax were not significantly stimulated or repressed. Similar experiments performed with Ltk^- cells (immortal mouse L cells lacking thymidine kinase) did show a significant stimulation of β-actin, but the c-Fos or the other proteins studied did not show stimulation to the same extent as primary BMS cells. These facts led to the conclusion that mechanical stimulation can interfere with radiation response. It also shows that immortal cell lines do not necessarily react in the same way as primary cells. Consequently, we have to re-evaluate different experimental protocols before we can proceed with cell biology research in microgravity and Space.

9.7 General Conclusions and Perspectives

Estimation of the relative biological effect of different electromagnetic and ionizing cosmic radiation is an essential problem that needs to be under control before we can travel and survive on the way to Mars or any other extraterrestrial place. In 1996 the US National Research Council (NRC) produced a document entitled *Radiation Hazards to Crews on Interplanetary Missions: Biological Issues and Research Strategies* [36]. One of the conclusions was that the overall estimated uncertainty in the risk of radiation-induced biological effects ranges from a factor of 4- to 15-fold greater to a factor of 4- to 15-fold smaller than our present estimates because of uncertainties both in the way HZE particles and their spallation products penetrate shielding and the quantitative way in which these types of radiation affect biological functions. Another conclusion was that, to reduce these large uncertainties in particle transport behaviour and in the biological response function, a reliable source of HZE particles is necessary on Earth. New experiments on different Space travels to ISS reveal that the biological response to HZE is more complex than was then expected, and it is concluded that the relative biological effect (RBE) of Space radiation is far more complex and linked to many other signalling networks. Although very useful ground-based experiments can be performed at installations such as the Brookhaven National Laboratory, USA, or GSI in Darmstadt, Germany, the complexity of a real Space environment is difficult to mimic. In this complex environment, the absence of gravity is certainly a very important factor. But all other stress situations should also be considered. These can be physical (mechanical, thermal), chemical (toxic residues) or biological (hormone balance, bone demineralization). We, therefore, need to continue experiments on the ISS or any other spacecraft.

Executing experiments in the ISS is costly and difficult and thus restricted. Presently, the main priority of the crew occupying the ISS is the technical maintenance of the station and the material. Fully autonomous experiments that can be executed by telecommunication should therefore be set up. Failure of the experiments should be near 0% and fundamental errors in the protocol can not be allowed. Experimental protocols should be simple to increase the success rate of the experiment. Modern technologies make it now possible to perform experiments on single cells. This approach should be stimulated because it can reduce the need to manipulate large volumes of cells and growing media. The inclusion of high content single cell analysis techniques should be advised in combination with large data logging and storage facilities. The recording of high-resolution 3D images is advised. If extensive image storage facilities are available, it is always possible to re-evaluate specific cells or cell cultures at a later moment when new image algorithms are available. This implies the presence of a good image archiving and data management system.

References

1. Dicello, J.F. *Radiat. Prot. Dosim.* **1992**, *44*, 253–257.
2. Todd, P., Peccaut, M.J., Fleshner, M. *Mutat. Res.* **1999**, *430*, 211–219.
3. Setlow, R.B. *EMBO Rep.* **2003**, *4*, 1013–1016.
4. Kopp, J., Beaujean, R., Reitz, G., Enge, W. *Radiat. Meas.* **1999**, *31*, 573–578.
5. Curtis, S.B., Letaw, J.W. *Adv. Space Res.* **1989**, *9*, 293–298.
6. Cartwright, B.G., Shirk, E.K., Price, P.B. *Nucl. Instrum. Methods* **1978**, *153*, 457.
7. Meesen, G., Van Oostveldt, P. *Radiat. Meas.* **1997**, *28*, 845–848.
8. Baert, P., Meesen, G., De Schynkel, S., Poffijn, A., Van Oostveldt, P. *Micron* **2005**, *36*, 321–330.
9. Adams, L., Demets, R., Harboe-Sørensen, R., Nickson, R., Gmür, K., Heinrich, W., Radiation Dosimetry for Recoverable Payloads, in: *Proceedings of the Sixth European Symposium on Life Science Research in Space*, pp. 165–169, ESA SP-390, ESA Publications Division, ESTEC, Noordwijk, **1996**.
10. Jakob, B., Scholtz, M., Taucher-Scholz, G. *Int. J. Radiat. Biol.* **2002**, *78*, 75–88.
11. Kerr, J.F., Wyllie, A.H., Currie, A.R. *Br. J. Cancer*, **1972**, *26*, 239–257.
12. Hendzel, M.J., Nishioka, W.K., Raymond, Y., Allis, C.D., Bazett-Jones, D.P., Th'ng, J.P. *J. Biol. Chem.* **1998**, *273*, 24470–24478.
13. Allera, C., Lazzarini, G., Patrone, E., Alberti, I., Barboro, P., Sanna, P., Melchiori, A., Parodi, S., Balbi, C. *J. Biol. Chem.* **1997**, *272*, 10817–10822.
14. Nagasawa, H., Little, J.B. *Radiat. Res.* **1999**, *152*, 552–557.
15. Reitz, G., Beaujean, R., Heilmann, C., Kopp, J., Leicher, M., Stauch, K. *Radiat. Meas.* **1996**, *26*, 979–986.
16. Stein, T.P. *Nutrition*, **2002**, *18*, 867–871.
17. Sutherland, B.M., Bennett, P.V., Cintron-Torres, N., Hada, M., Trunk, J., Monteleone, D., Sutherland, J.C., Laval, J., Stanislaus, M., Gewirtz, A. *J. Radiat. Res.* **2002**, *43*, S149–S152.
18. Brooks, A.J. *J. Radiat. Biol.* **1999**, *75*, 1481–1503.
19. Durante, M. *Radiat. Res.* **2005**, *164*, 467–473.
20. Horneck, G. *Adv. Space Res.* **1998**, *22*, 1631–1641.
21. Perlman, Z.E., Slack, M.D., Feng, Y., Mitchison, T.J., Wu, L.F., Altschuler, S.J. *Science* **2004**, *306*, 1194–1198.
22. Newman, J.R.S., Ghaemmaghami, S., Ihmels, J., Breslow, D.K., Noble, M., DeRisi, J.L., Weisman, J.S. *Nature* **2006**, *441*, 840–846.
23. Xie, X.S., Yu, J., Yang, Y. *Science* **2006**, *312*, 228–230.
24. Giepmans, B.N.G., Adams, S.R., Ellisman, M.H., Tsien, R.Y. *Science* **2006**, *312*, 217–224.
25. Brillouet, C., Brinckmann, E., Biorack Facility Performance and Experiment Operations on three Spacehab Shuttle-to-Mir Missions, in: *Biorack on Spacehab*, Perry, M. (Ed.), pp. 3–21, ESA SP-1222, ESA Publications Division, ESTEC, Noordwijk, **1999**.
26. Nose, K., Shibanuma, M. *Exp. Cell Res.* **1994**, *211*, 168–170.

27 Fitzgerald, J., Hughes-Fulford, M. *Exp. Cell Res.*, **1996**, *228*, 168–171.

28 Tjandrawinata, R.R., Vincent, V.L., Hughes-Fulford, M. *FASEB J.* **1997**, *11*, 493–497.

29 Hammond, T.G., Bennes, E., O'Reilly, K.C., Wolf, D.A., Linneham, R.M., Taher, A., Kaysen, J.H., Allen, P.L., Goodwin, T.J. *Physiol. Genomics* **2000**, *3*, 163–173.

30 Rosenberg, N. *Hum. Exp. Toxicol.* **2003**, *22*, 271–274.

31 Prasad, A.V., Mohan, N., Chandrasekar, B., Meltz, M.L. *Radiat. Res.* **1995**, *143*, 263–272.

32 Ibuki, Y., Hayashi, A., Suzuki, A., Goto, R. *Biol. Pharm. Bull.* **1998**, *21*, 434–439.

33 Amundson, J.S.. Do, K.T., Fornace Jr., A.J. *Radiat. Res.* **1999**, *152*, 225–231.

34 Daino, K., Ichimura, S., Nenoi, M. *Radiat. Res.* **2002**, *157*, 478–482.

35 Baert, P., Van Cleynenbreugel, T., Vandesompele, J., De Schynkel, S., Vander Sloten, J., Van Oostveldt, P. *Acta Astron.* **2006**, *58*, 456–463.

36 US National Research Council (NRC), *Radiation Hazards to Crews on Interplanetary Missions: Biological Issues and Research Strategies*, National Academy Press, Washington D.C., **1996**.

10
Space Radiation Biology

Gerda Horneck

10.1
Radiation Scenario in Space

Radiation from Space, of solar as well as of galactic origin, is one of the external sources of energy of significant impact on our biosphere. During the first two billion years of Earth's history, solar UV radiation of wavelengths as short as 200 nm penetrated the atmosphere and reached the surface of the Earth. This is the time span when the decisive processes of prebiotic organic chemical evolution, appearance of life and diversification of microorganisms took place on Earth. In these processes, energetic solar UV radiation played a major role [1]. Today, the surface of the Earth is void of this energetic UV portion due to the effective screening by the stratospheric ozone layer. The role of cosmic radiation in the evolution of our biosphere is less clear, because the surface of the Earth is largely spared from such radiation due to the deflecting effect of the geomagnetic field and the huge shield of 1000 g matter per square meter provided by the atmosphere. However, in Space, the natural radiation encountered is a complex mixture of charged particles of galactic and solar origin and those particles trapped by the geomagnetic field.

Since the advent of space flight and the establishment of long-duration Space stations in Earth orbit, such as Skylab, Salyut, MIR and the International Space Station (ISS), the upper boundary of our biosphere has extended into Space. Such Space missions expose humans and any other biological systems to a radiation environment of a composition and intensity not encountered on Earth. To prevent detrimental health effects caused by the radiations of Space, radiation protection guidelines have been elaborated for humans in Space, based on (a) dosimetry and modelling of the radiation field in Space; (b) studies of the biological effects of the heavy ions of cosmic radiation encountered in Space or produced at heavy ion accelerators on ground; and (c) studies on potential interactions of cosmic radiation and other parameters of space flight, above all microgravity.

During space flight the biological systems, including humans, are protected from most of the hostile parameters of Space either by containment within a Space capsule, which is a pressurized module with an efficient life support system (LSS),

Biology in Space and Life on Earth. Effects of Spaceflight on Biological Systems. Edited by Enno Brinckmann

ISBN: 978-3-527-40668-5

or at least by a Space suit during extravehicular activity (EVA). In these cases, the dominant Space parameters of interest, or concern, are microgravity and radiation.

Alternatively, living beings arriving in Space without any protection, e.g. by natural processes, are confronted with an extremely hostile environment, characterized by a high vacuum, an intense radiation field of solar and galactic origin and extreme temperatures. This environment or selected parameters of it are the test bed for astrobiological investigations, thereby exposing chemical or biological systems to selected parameters of outer Space or defined combinations of them.

10.1.1
Cosmic Ionizing Radiation

In interplanetary Space, the radiation field is composed mainly of two groups: the solar cosmic radiations (SCR) and the galactic cosmic radiation (GCR). Near the Earth, a third radiation component is present: the radiation trapped by the Earth's magnetosphere, the so-called van Allen belts [2, 3]. Figure 10.1 shows typical integral energy spectra for these radiation components experienced in low Earth orbit (LEO).

SCR consist of the low energy solar wind particles that flow constantly from the sun, and the so-called solar particles events (SPEs) that originate from magneti-

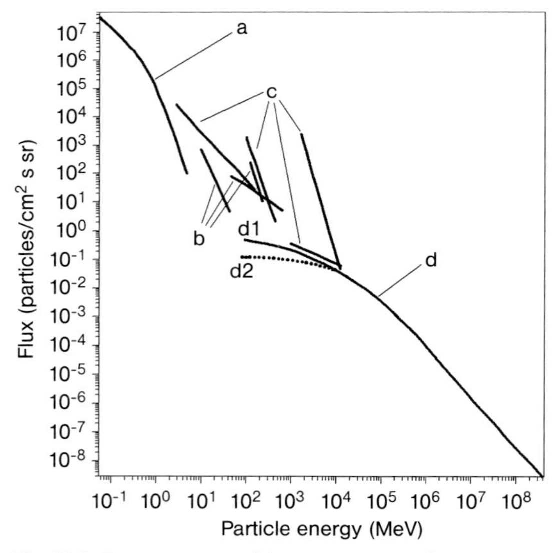

Fig. 10.1 Energy spectra of the components of cosmic radiation in low Earth orbit: (a) electrons (belts); (b) protons (belts); (c) solar particle events; (d) heavy ions of galactic cosmic radiation; (d1) during solar minimum; (d2) during solar maximum.

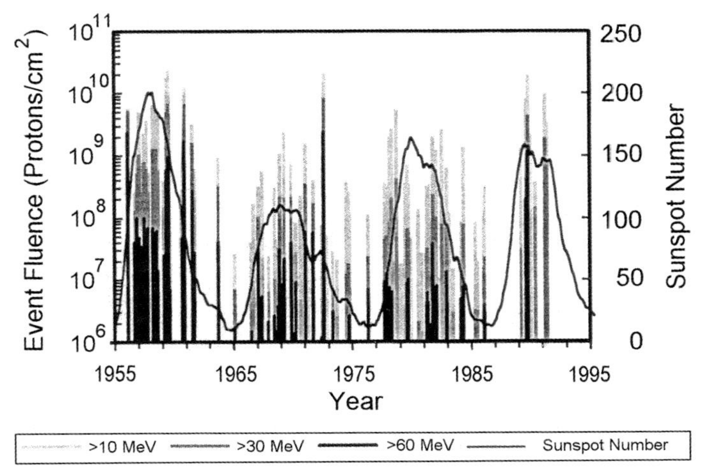

Fig. 10.2 Eleven-year cycle of solar activity; the solar particle events (SPEs, bars) are superimposed to the sunspot numbers (lines). The energy of each SPE is indicated by the grey bar (defined in the bottom line).

cally disturbed regions of the sun, which sporadically emit bursts of charged particles with high energies. These events are composed primarily of protons with a minor component (5–10%) being helium nuclei (alpha particles) and an even smaller part (1%) heavy ions and electrons. SPEs develop rapidly and generally last for no more than a few hours; however, some proton events observed near Earth may continue over several days. The emitted particles can reach energies up to several GeV (Fig. 10.1). In a worst case scenario, doses as high as 10 Gy could be received within a short time. Such strong events are very rare, typically about one event during the 11 year solar cycle (Fig. 10.2). Concerning the less energetic, though still quite intensive events, in cycle 22 (1986–1996), for example, there were at least eight events with proton energies greater than 30 MeV. For low Earth orbit, the Earth's magnetic field provides a latitude-dependent shielding against SPE particles. Only in high inclination orbits and in interplanetary missions do SPEs create a hazard to humans in Space, especially during extravehicular activities.

Galactic cosmic radiations (GCR) originate outside the solar system in cataclysmic astronomical events, such as supernova explosions. Detected particles consist of 98% baryons and 2% electrons. The baryonic component is composed of 85% protons (hydrogen nuclei), with the remainder being alpha particles (helium nuclei) (14%) and heavier nuclei (about 1%). The latter component comprises the so-called HZE particles (particles of High charge Z and high Energy), which are defined as cosmic ray primaries of charges $Z > 2$ and of energies high enough to penetrate at least 1 mm of spacecraft or of Space suit shielding. Though they only contribute roughly 1% of the flux of GCR, they are considered as a potential major concern to living beings in Space, especially for long-term missions at high altitudes or in high inclination orbits, or for missions beyond the Earth's magnetosphere. Reasons for this concern are based, on one hand, on the inefficiency of adequate shielding and, on the other hand, on the special nature of HZE particle-produced lesions. If the particle flux is weighted according to the energy deposition, iron nuclei will become the most important component (Fig. 10.3), although their relative abundance is comparatively small. The fluency of GCR is isotropic and energies up to 10^{20} eV can be present (Fig. 10.1). However, ultra-high energy nuclei of energies >10^{16} eV are very rare and are probably of extra galactic origin. When GCR enter our Solar System, they must overcome the magnetic fields carried along with the outward-flowing solar wind, the intensity of which varies according to the about 11-year cycle of solar activity. With increasing solar activity the interplanetary magnetic field increases, resulting in a decrease of the intensity of GCR of low energies. This modulation is effective for particles below some GeV per nucleon. Hence the GCR fluxes vary with the solar cycle and differ by a factor of approximately five between solar minimum and solar maximum, with a peak level during minimum solar activity and the lowest level during maximal solar activity (Fig. 10.1). At peak energies of about 200–700 MeV u^{-1} (u = unified atomic mass unit) during solar minimum, particle fluxes (flow rates) reach 2×10^3 protons per 100 μm^2 per year and 0.6 Fe-ions per 100 μm^2 per year (100 μm^2 being the typical cross section of a mammalian cell nucleus). Although iron ions are one-tenth as abundant as carbon or oxygen, their contribution to the GCR dose is substantial, since the dose is proportional to the square of the charge (Fig. 10.3).

The fluxes of GCR are further modified by the geomagnetic field. Only particles of very high energy have access to low inclination orbits, whereas towards higher

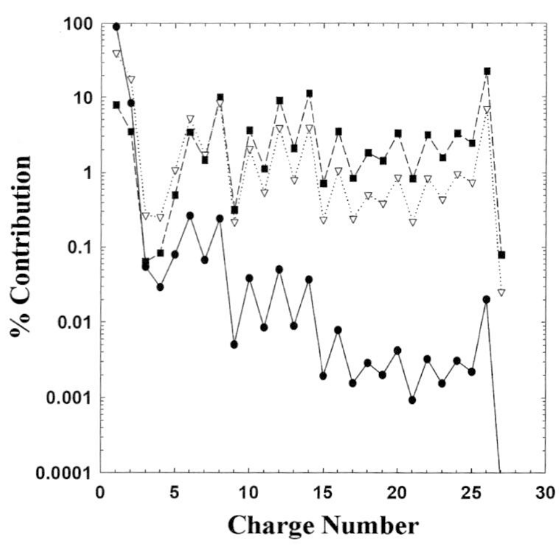

Fig. 10.3 Percent contributions from individual GCR (Galactic Cosmic Rays) nuclei to the particle flux (filled circles), dose (weighted by the square of the charge Z of the particle to give a measure of the "ionizing power", the radiation dose) (open triangles) and dose equivalent at solar minimum (filled squares). (Modified from Ref. [4].)

inclination particles of lower energies are also encountered. At the pole, particles of all energies can impinge in the direction of the magnetic field axes. Owing to this inclination-dependent shielding, the number of particles increases from lower to higher inclination.

The van Allen belts in the vicinity of the Earth are a result of the interaction of GCR and SCR with the Earth's magnetic field and the atmosphere. Two belts of radiation are formed, consisting of electrons and protons and some heavier particles trapped in closed orbits by the Earth's magnetic field. The main production process for the inner belt particles is the decay of neutrons produced in cosmic particle interactions with the atmosphere. The outer belt consists mainly of trapped solar particles. In each zone, the charged particles spiral around the geomagnetic field lines and are reflected back between the magnetic poles, acting as mirrors. Electrons reach energies of up to 7 MeV and protons up to about 200 MeV. The energy of trapped heavy ions is less than 50 MeV (Fig. 10.1), and their radiobiological impact is very small. The trapped radiation is modulated by the solar cycle: proton intensity decreases with high solar activity, while electron intensity increases, and vice versa.

Of special importance for LEO missions is the so-called "South Atlantic Anomaly" (SAA), a region over the coast of Brazil where the radiation belt reaches down to altitudes of 200 km. This behaviour is due to an 11° offset of the Earth's magnetic dipole axis from its axis of rotation and a 500 km displacement towards the Western pacific, with corresponding significantly reduced field strength values. The inner fringes of the inner radiation belts come down to the altitude of LEO, which results in a 1000× higher proton flux than in other parts of the orbit. Almost all radiation received by the crew in LEO at low inclinations is due to passages through the SAA.

10.1.2
Solar Electromagnetic Radiation

The spectrum of solar electromagnetic radiation spans over several orders of magnitude, from short wavelength X-rays to radio frequencies. At the distance of the Earth (1 Astronomical Unit = 1 AU), solar irradiance amounts to 1360 W m^{-2}, the solar constant. Of this radiation, a large fraction is attributed to the infrared fraction (IR: >800 nm) and to the visible fraction (VIS: 400–800 nm) and only about 8% to the ultraviolet range (UV: 100–400 nm).

On its way through the Earth's atmosphere, solar electromagnetic radiation is modified by scattering and absorption processes. Of special interest is the solar UV radiation, because it has a high impact on the health of our biosphere. During the first 2.5 billion years of Earth's history, UV-radiation of wavelengths >200 nm reached the surface of the Earth – due to lack of an effective ozone shield. Following the rapid oxidation of the Earth's atmosphere about 2.1 billion years ago – as a consequence of oxygenic photosynthesis – photochemical processes in the upper atmosphere resulted in the build up of a UV-absorbing ozone layer in the

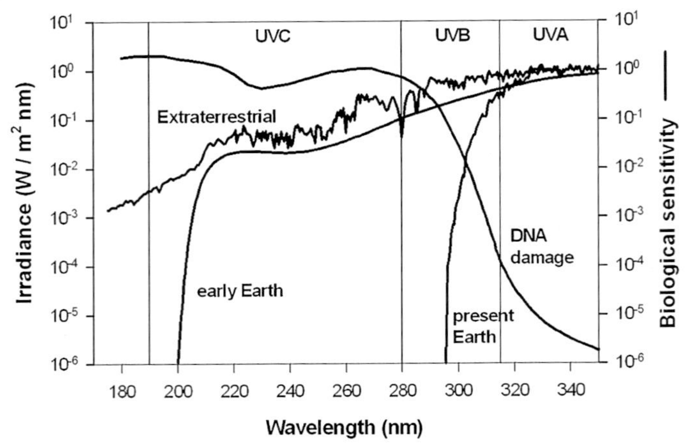

Fig. 10.4 Spectrum of solar UV radiation: extraterrestrial, at the surface of the early Earth (the first 2.5 billion years), at the surface of the present Earth, and spectral UV sensitivity of genetic material DNA.

stratosphere. Today, the stratospheric ozone layer effectively absorbs UV radiation at wavelengths shorter than 290 nm (Fig. 10.4).

The spectrum of extraterrestrial solar UV radiation has been measured during several Space missions, such as Spacelab 1 [5] and the *Eu*ropean *Re*trievable *Car*rier mission (EURECA) (Fig. 10.4). It can be divided into the following spectral ranges: UVC (200–280 nm) contributing 0.5%, UVB (280–315 nm) contributing 1.5%, and UVA (315–400 nm) contributing 6.3% to the whole solar electromagnetic spectrum, respectively. UV of wavelengths shorter than 200 nm, which is absorbed by the molecules of the atmosphere, belongs to the vacuum-UV. Although the UVC and UVB regions make up only 2% of the entire solar irradiance prior to attenuation by the atmosphere they are mainly responsible for the high lethality of extraterrestrial solar radiation to living organisms. The reason for this high lethality of extraterrestrial solar UV radiation – compared with conditions on Earth – lies in the absorption characteristics of the DNA, which is the decisive target for inactivation and mutation induction by UV (Fig. 10.4).

10.2 Questions Tackled in Space Radiation Biology

Space technology has opened a new opportunity for life sciences by providing the vehicle to transport terrestrial organic matter or organisms beyond the protective blanket of our atmosphere to study *in situ* their responses to selected conditions of Space or space flight. Since the 1980s, with the accessibility of Spacelab, the free flying satellite EURECA, the Russian MIR station and Cosmos or Foton satellites, and finally the ISS, the European Space Agency (ESA) has provided various opportunities for radiation biology experiments in Earth orbit.

Since the first supposition of the existence of ionizing radiation originating from outer Space at the beginning of the last century, much work has been done to characterize the radiation field surrounding the Earth. A series of experiments has been initiated on different components of cosmic radiation, on their radiobiological importance, and on the role of the factors of Space and of space flight in radiobiological processes. Besides offering opportunities for radiobiological experiments in Space, the current Space exploration program even requires the collection of radiobiological data in Space as baseline information for estimating radiation risks to human beings in future Space missions and for establishing radiation standards for man in Space.

The various fields of radiation biology in Space consist of:

- radiation detection and measurement
- studies on the biological response to radiation in Space
- studies on the impact of space flight environment on radiation effects
- radiation protection efforts for astronauts.

Of special concern are the heavy ions of cosmic radiation, the so-called HZE particles. To understand the ways by which HZE particles interact with biological systems, methods need to be developed to precisely localize the trajectory of an HZE particle relative to the biologically sensitive site and to correlate the physical data with the biological effects observed along the trajectory of the particle.

Such radiobiological data are even more important when approaching the next frontier of space flight, namely human missions beyond the Earth orbit, to the Moon or to Mars. Risk assessment, surveillance and countermeasures for the crew need to be optimized. Robotic precursor missions are required to precede the human missions, to provide data on the radiation climate on the surface and modes and efficiency of natural (regolith) and artificial (habitat) shielding.

Outer Space serves also as tool to tackle specific questions of astrobiology [6]. These studies include, but are not limited to, research on:

- the role of solar extraterrestrial UV radiation and cosmic radiation in the evolution of potential precursors of life in Space;
- the likelihood of interplanetary transfer and the limiting factors of Panspermia, the hypothesis of the distribution of life beyond its planet of origin;
- the simulation of the UV radiation climates of the early Earth or present Mars to assess their habitability;
- the role of the stratospheric ozone layer in protecting our biosphere;
- the upper boundary of the biosphere.

10.3
Results of Radiobiological Experiments in Space

10.3.1
Life and Cosmic Radiation

The accessibility of this unique radiation environment in Space and the increasing involvement of human beings in Space missions have initiated several Space activities in fields of radiobiological research. The Results of these studies are discussed below.

10.3.1.1 Radiation Dosimetry

Knowledge of the radiation situation inside a spacecraft is mandatory for each mission under consideration and is based on in-flight dosimetry data. Such measurements of radiation exposures have been performed during manned space flights at various altitudes, orbital inclinations, durations, periods during the solar cycle, and mass shielding (Fig. 10.5) [2, 7, 8].

The deposition of energy by radiation strongly depends on the type of radiation under consideration, both macroscopically and microscopically (Fig. 10.6). Because of the complex mixture of radiations occurring in Space, consisting of sparsely ionizing components (photons, electrons, pions, muons and protons) and densely ionizing components (heavy ions, neutrons and nuclear disintegration stars), different dosimetry systems have been applied that specifically respond to the quality of the radiation under consideration.

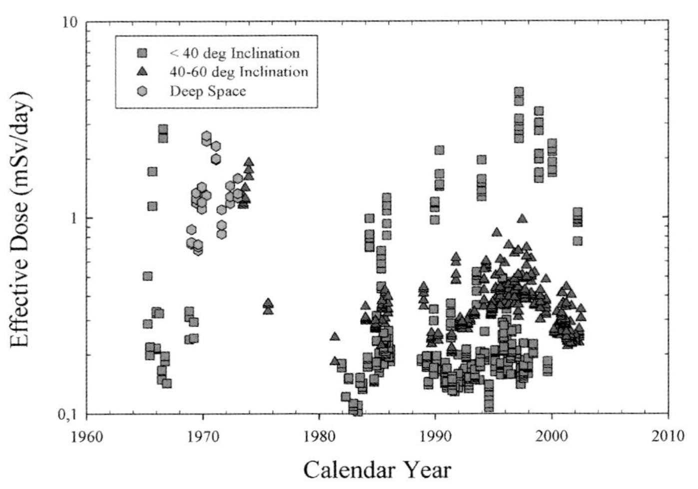

Fig. 10.5 Effective radiation equivalent doses, measured with passive radiation detectors inside a spacecraft in low Earth orbit missions and missions to the Moon. (Credit G. Reitz, DLR.)

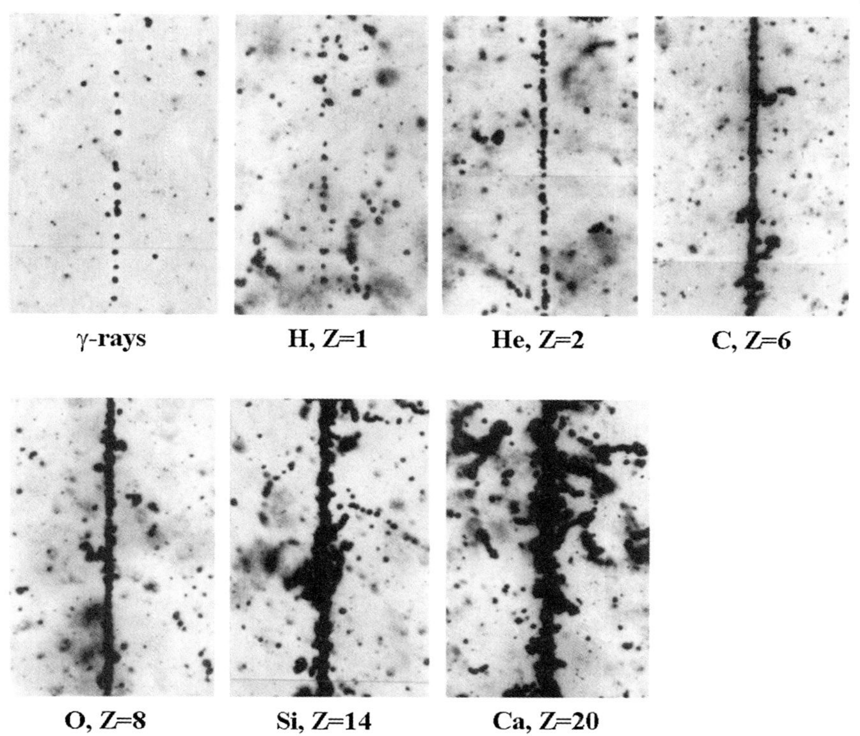

Fig. 10.6 Tracks in photo-emulsions of electrons produced by γ-rays and tracks produced by different nuclei of the primary cosmic radiation (from hydrogen H to calcium Ca). For biological radiation effects the efficiency of a radiation type increases as the ion density along the tracks increases.

The contribution of the sparsely ionizing component of the radiation in Space has been mostly determined by lithium fluoride thermoluminescence dosimeters (TLD). A TLD is a (usually doped) inorganic crystal. It "absorbs" a radiation dose by its valence electrons being excited to a higher energy state. The number of electrons at the higher energy state is directly proportional to the amount of ionizing radiation the crystal is exposed to. When the crystal is heated, these electrons fall back to their resting energy and emit photons, causing the crystal to glow. The emitted light intensity as a function of the temperature is called the glow curve. In a heating cycle the amount of emitted light, i.e. the integral of the resulting glow curve, is proportional to the total dose received by the crystal since the last time it was heated ("annealed"). The sensitivity of TLDs is nearly constant in the energy range of interest [9].

For densely ionizing radiation, the spatial pattern of energy deposition at the microscopic level is important (Fig. 10.6). Lesions in biological sensitive structures, such as biomolecules or chromosomes, are induced with higher efficiency than by, for example, X-rays. Their fluency has been determined mainly by use of plastic track detectors or nuclear emulsions. Plastic detector systems are diallylglycol carbonate (CR39), cellulose nitrate (CN), or polycarbonate (Lexan), covering different ranges of linear energy transfer (LET). LET is an important value that determines the biological effectiveness of a radiation under consideration. The tracks of heavy ions are developed by etching in caustic solutions, e.g. in 6 N NaOH. The track etching rate grows as a function of the LET of the particle. Plastic detectors allow determination of the fluence, charge and LET spectrum of the heavy ions. Generally, different plastic detector systems are arranged in a stack, and the combination of their data is used to generate a LET spectrum adequate for dosimetry calculations (Fig. 10.7). The density of nuclear disintegration stars has been determined by nuclear emulsions. The absorbed dose deposited by neutrons can be estimated from TLDs differing in their relative contents of the iso-

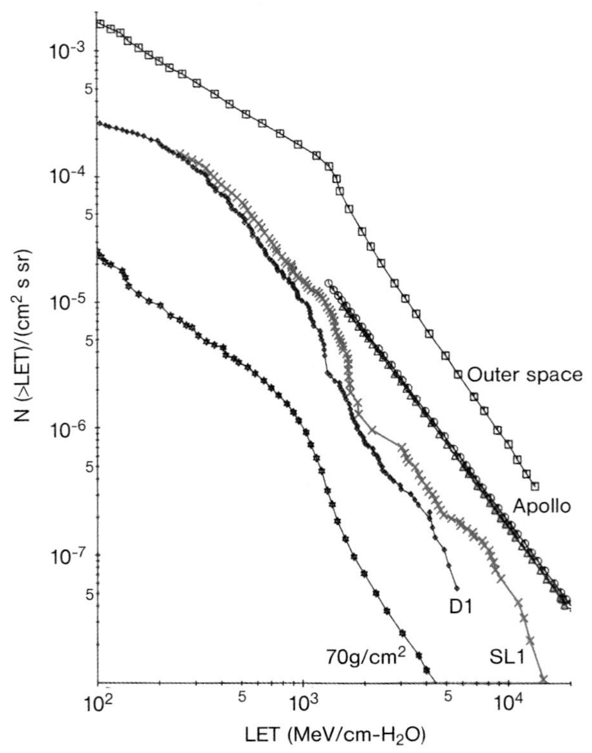

Fig. 10.7 Integral linear energy transfer (LET) spectra of heavy charged particles, measured inside the spacecraft with plastic track detectors outside of the geomagnetic field during the lunar missions Apollo 16 and 17, in low Earth orbit during the missions Spacelab 1 (SL1) and D1, and calculated for outer Space without shielding and behind a shielding of 70 g cm^{-2}. (From Ref. [15].)

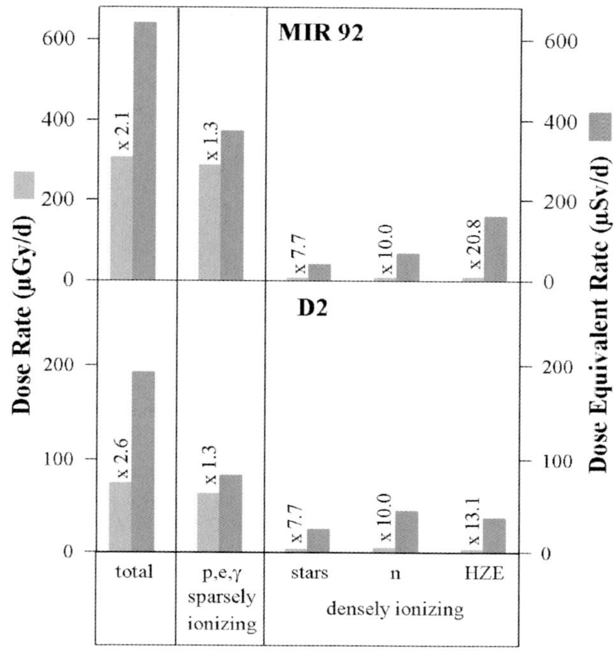

Fig. 10.8 Absorbed dose rate (light grey; $\mu Gy\,d^{-1}$) and dose equivalent rate (dark grey; $\mu Sv\,d^{-1}$) of the sparsely ionizing and the three densely ionizing components of the radiation field measured during the Space missions MIR-92 (51.5°, 400 km), and Spacelab D2 (28.5°, 296 km). The factors (e.g. × 2.1) mean the quality factor Q, which takes into account the different biological effectiveness of the radiations in Space. Multiplication of the absorbed dose (Gy) by Q gives the dose equivalent (Sv). (Credit G. Reitz, R. Facius, DLR.)

topes ^{6}Li and ^{7}Li. Figure 10.8 illustrates the contribution of the different types of radiation measured during two different space flights in LEO. During the Space mission MIR-92 (51.5°, 400 km) a total dose equivalent of $640\,\mu Sv\,d^{-1}$ was measured, and during the Spacelab mission D2 (28.5°, 296 km) a total dose equivalent of $192\,\mu Sv\,d^{-1}$. For comparison, the dose equivalent rates for the ISS (51.5°, 400 km) are about 60 mSv per 3 months ($667\,\mu Sv\,d^{-1}$) and at the surface of the Earth 3 mSv year^{-1} ($8\,\mu Sv\,d^{-1}$).

Notably, these passive dosimetry systems ("passive" in the sense that they do not need power during the mission) integrate over the time of exposure. Their advantages are their independence of power supply, small dimensions, high sensitivity, good stability, wide measuring range, resistance to environmental stressors, and relatively low cost. However, long duration Space missions, such as onboard of the ISS or future interplanetary missions, require time-resolved measurements, especially for radiation protection purposes. This requirement has been met by the "Pille" device, a small, portable and Space-qualified TLD reader suitable for reading out TLD repeatedly on board [9].

In addition to passive dosimeters, active dosimeters have been developed to provide real-time dosimetry data. The measurement principle is based either on ionizations (e.g. ionization chamber, proportional counter, Geiger-Müller Counter, semiconductors, charged coupled devices CCD) or on scintillations (e.g. organic or inorganic crystals). A combination of two silicon detectors, the Dosimetry Telescope (DOSTEL), has been flown onboard of the Space Shuttle, the MIR station and the ISS. Particle count rates, dose rates and LET-spectra were measured separately for GCR, the radiation belt particles in the South Atlantic Anomaly (SAA), and solar particle events (SPE) [10].

During human space flight a personal dosimetry is required for each astronaut and separately for different activities, above all during extravehicular activities (EVA), when the astronauts are shielded only by the material of the Space suit. Several active devices such as small silicon detectors or small ionization chambers may be used, but they need power and are difficult to design in sufficiently small dimensions. In most cases, passive integrating detector systems have been used, such as TLDs, also in combination with the "Pille" device [9]. However, these personal dosimetry systems provide only data of the "surface" or skin dose. To assess the depth dose distribution within the human body and, especially, at the most radiation sensitive organs, such as the brain, the blood forming organs and the gonads, human phantoms are required equipped with different dosimetry systems at the sites of sensitive organs. The anthropomorphic phantom "Matroshka" was exposed for one year to the radiation in Space outside of the ISS (Fig. 10.9) to determine the depth dose distribution of radiations within the human body during EVAs [11].

10.3.1.2 **Biological Effects of Cosmic Radiation**

To study the effects of single particles of cosmic radiation in biological systems, effect–particle correlations were accomplished in different ways:

- by use of visual track detectors that were sandwiched between layers of either biological objects in a resting state, like viruses, bacterial spores, plant seeds or shrimp cysts, or embryonic systems, like insect eggs, realized in the so-called Biostack concept (Fig. 10.10) [12];
- by use of nuclear track detectors in fixed orientation to biological targets of interest, like implantations beneath the scalp of animals or helmet devices for astronauts (see also Chapter 9);
- by correlating the occurrence of radiation effects, such as the light flash phenomenon, with orbital parameters, such as passages through the SAA of the radiation belts.

Results from experiments in Space investigating the radiobiological importance of the HZE particles of cosmic radiation are summarized in Refs. [7, 13–15].

The responses of a single microbial cell to the passage of a single HZE particle of cosmic radiation have been studied on spores of the bacterium *Bacillus subtilis*

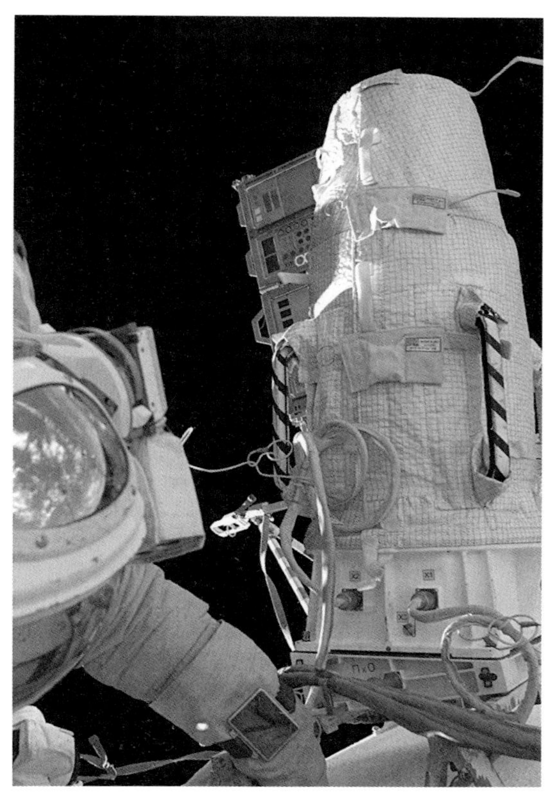

Fig. 10.9 Anthropomorphic phantom "Matroshka", attached to the Russian module Svezda of the ISS, during removal by an Extra-Vehicular Activity (EVA) after exposure for one year to the radiation in Space outside of the ISS. (ESA photograph, 18 August 2005.)

in the Biostack experiments [12] (reviewed in Ref. [16]). Taking the results from several Biostack experiments in Space, starting with Apollo 16 (Apollo 16 and 17, Apollo-Soyuz, Spacelab, EURECA) as well as those obtained at accelerators, one can draw the following general conclusions:

- The inactivation probability for spores, centrally hit, is substantially less than one.
- The effective radial range of inactivation around each HZE particle, the so-called impact parameter, extends far beyond the range where inactivation of spores by δ-rays (secondary electron radiation) can be expected (Fig. 10.11).
- This far-reaching effect is less pronounced for ions of low energies (1.4 MeV per mass unit), a phenomenon that might reflect the "thindown effect" at the end of the ion's path.

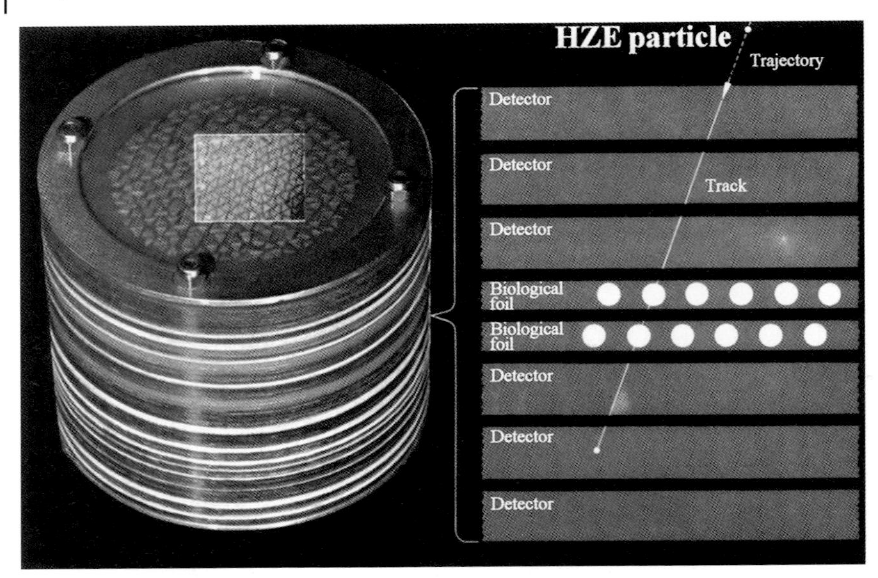

Fig. 10.10 The Biostack concept to localize biological effects produced by single HZE particles of cosmic radiation. Biostack experiments were flown onboard Apollo 16, 17, Apollo-Soyuz, Spacelab 1, D1, IML-1, IML-2, LDEF, Cosmos 1887 and 2004, and EURECA.

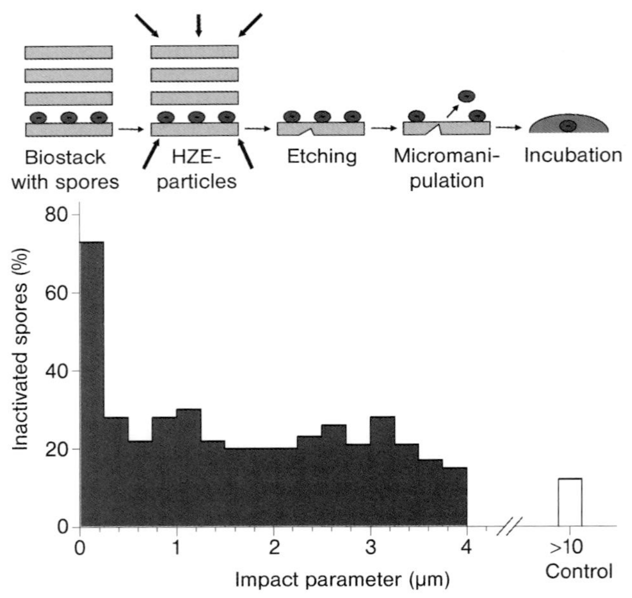

Fig. 10.11 Inactivation (%) of *Bacillus subtilis* spores by single HZE particles ($Z > 12$, LET $> 200\,keV\,\mu m^{-1}$) of cosmic radiation as a function of the impact parameter. Results of the Biostack experiment flown onboard the Apollo-Soyuz Test Project (ASTP). (From Ref. [13].)

The dependence of inactivated spores on impact parameter points to a superposition of two different inactivation mechanisms: a short-range component reaching up to about 1 µm may be traced back to the δ-ray dose (secondary electrons) and a long-range one that extends at least to somewhere between 4 and 5 µm off the particle's trajectory, for which additional mechanisms have been conjectured, such as shock waves, electromagnetic radiation, or thermo-physical events.

With plant seeds that were exposed to cosmic HZE particles in fixed contact with track detectors, methods were developed to determine the impact parameter of the most sensitive target, which is the meristem of root or shoot (Fig. 10.12). In seeds of *Arabidopsis thaliana* or *Nicotiana tabacum*, hit by an HZE particle, development was significantly disturbed, as demonstrated by loss of germination (early lethality) or embryo lethality. Seeds of impact parameters $<120\,\mu m$ in relation to their shoot meristem were severely damaged. Seedling abnormalities, such as hypertrophy or deformation of cotyledons, hypocotyl or root, or chlorophyll deficiency occurred with high frequency as a consequence of the passage of a single HZE particle close to the shoot or root meristem. Evidently, these severe impairments were based on irreparable damage to the genetic apparatus, as demonstrated by the high frequency of multiple chromosomal aberrations developed in *Lactuca sativa* seeds hit by an HZE particle. Among *Zea mays* seeds, flown on Apollo-Soyuz, one seed that received two hits by HZE particles ($Z>20$, LET = 100–150 keV μm^{-1}) in the central region of the embryo developed a somatic mutation

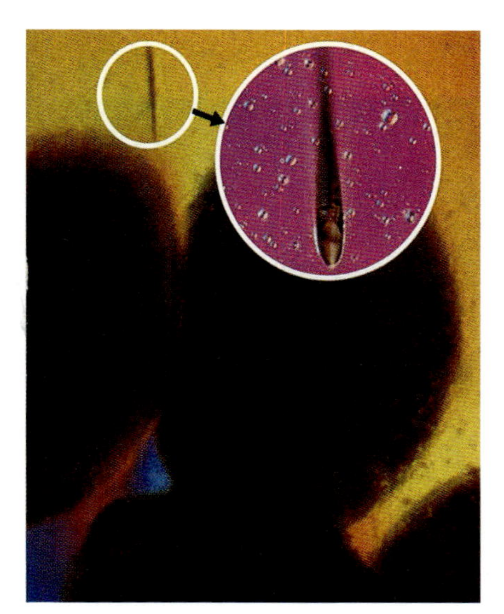

Fig. 10.12 Biostack method used to determine the impact parameter for the most sensitive target in plant seeds (*Arabidopsis thaliana*, large black shadows) after exposure to HZE particles of cosmic radiation; the CN-detector shows the etch cone of the particle track (enlarged circle).

– large yellow strips in all leaves – of an extent that had never been observed before, neither in flight nor in ground experiments.

Among animal resting systems, the mosaic egg of the brine shrimp *Artemia salina*, resting in encysted blastula or gastrula state, represents an investigative system that, during further development, proceeds to the larval state, the free swimming nauplius, without any further cell division. Therefore, injury to single cells of the cyst will be manifested in the larva. From a series of space flight experiments, outside the geomagnetic shielding (Apollo 16 or 17) or in LEO (Biostack on Apollo-Soyuz, Biobloc on Cosmos 782, 1129, 1887 or Salyut 7), a wealth of data has been compiled on the response of this encysted embryonic system to single HZE particle hits. It was clearly demonstrated that the passage of a single HZE particle through a shrimp cyst damages a cellular area large enough to disturb either embryogenesis or further development or integrity of the adult. Emergence, characterized by bursting of the egg shell and appearance of the nauplius larva, still enclosed in a membrane, was slightly disturbed by a HZE particle hit. The subsequent step of hatching, characterized by release of a free swimming nauplius, was severely inhibited by a HZE particle hit (approximately 90% loss of hatching). Additional late effects, due to a hit of a single HZE particle, were delay of growth and of sexual maturity, and reduced fertility. In the Biostack experiments, not a single nauplius larva that developed from a cyst hit was normal in further growth and behaviour. Anomalies of the body or extremities appeared approximately ten times more frequently than in the ground controls.

In summary, evaluation of the effects observed in bacterial spores, in plant seeds, and animal embryos demonstrated that single HZE particles induce significant biological perturbations in all these test organisms, although with varying efficiency. The observed effects consist of gross somatic mutations, severe morphological anomalies, disturbance of development or complete inactivation. From biophysical analysis of some of these results it was concluded that the magnitude of these effects could not be explained in terms of established mechanisms and, in particular, that the lateral extension of effectiveness around the trajectories of single particles exceeds the range where secondary electrons could be considered to be effective. One possible explanation is the so-called bystander effect, which describes the ability of cells affected by an agent to convey manifestations of damage to other cells not directly targeted by the agent. Such bystander effects have been observed after irradiation with a charged-particle microbeam (reviewed in Ref. [17]).

A radiobiological phenomenon of special interest is the so-called light flashes. These were first reported by astronauts of the Apollo 11 mission who, on returning from the Moon, "saw" faint spots and flashes of light at a frequency of one or two per minute after some period of dark adaptation. Investigations on the frequency of visual light flashes in LEO and its dependence on orbital parameters were performed on Skylab 4, on Apollo-Soyuz and recently on MIR and the ISS. The highest light flash rates were recorded when passing through the SAA. These high light flash event rates during the SAA passages can be assigned either to the high proton fluxes or to the occurrence of some particles heavier than protons in the inner belts of trapped radiation. With the SILEY experiments on board of the Space stations MIR and ISS, two separate mechanisms for the induction of light flashes were identified: a direct interaction of heavy ions with the retina, causing excitation or ionization, and proton-induced nuclear interactions in the eye (with a lower interaction probability) producing knock-out particles [18, 19]. Possible mechanisms for stimulating the retina could be electronic excitation, resulting in UV radiation near the retina, ionization in a confined region associated with δ-rays around the track, or shock wave phenomena when HZE particles pass through the tissue matrix.

The light flash phenomenon gives an example of HZE particle hits that are "seen" by the astronaut. The question arises as to what happens to the other organs or tissues of the body exposed to cosmic radiation. Of special concern is the central nervous system (CNS), where damage to a relatively small groups of cells that cannot replace themselves may result in severe physiological effects.

In astronauts, after long-term space flights, an elevation of the frequencies of chromosomal aberrations in peripheral lymphocytes has been reported. Obe et al. [20] have investigated lymphocytes of seven astronauts that had spent several months on board the MIR Space station. They showed that the frequency of dicentric chromosomes increased by a factor of approximately 3.5 compared with preflight control and that the observed frequencies agreed quite well with the expected values based on the absorbed doses and particle fluxes encountered by individual astronauts during the mission. These data suggest the feasibility of using chromosomal aberrations as a biological dosimeter for monitoring radiation exposure of astronauts. However, for cosmonauts involved in multiple Space missions the frequency of chromosomal aberrations was lower than expected, suggesting that the effects of repeated space flights on this particular endpoint are not simply additive events. Changes in the immune system in microgravity and/or adaptive response to Space radiation may explain the apparent increase in radio-resistance after multiple space flights [21].

In addition to chromosomal aberrations, other biomarkers for genetic or metabolic changes may be applicable, such as germ line minisatellite mutation rates or radiation induced apoptosis, metabolic changes in serum, plasma or urine, hair follicle changes and decrease in hair thickness, triacylglycerol-concentration in bone marrow and glycogen concentration in liver. Whereas the first three systems mentioned are non-invasive or require only blood samples for analysis, the latter systems are invasive and therefore appropriate for radiation monitoring in animals only. Dose–response relationships have been described for most of the intrinsic dosimetry systems.

10.3.1.3 **Interaction of Radiation with other Space Flight Factors**

Space flight factors that act over extended periods of time, such as microgravity, radiation, and those that depend on prolonged confinement in a closed environment, are of particular interest with respect to combined influences. Concerning a potential interaction of radiation and microgravity, in most studies cellular systems, botanical material – seeds as well as whole plants –, and animal material

– insect eggs, larvae, pupae and adults, as well as mammals (rats) – have been used (reviewed in Ref. [22]).

To test the influence of microgravity on radiation response during space flight, in addition to the microgravity conditions, an on-board 1×*g* reference centrifuge has been used in parallel, in some cases in addition with methods to localize the heavy ions hits in the biological system, such as described by the Biostack concept (Fig. 10.10). In other experiments, the controlled application of additional radiation during space flight was used. This method was first used on Gemini, when chromosomal aberrations were studied in human blood cells irradiated with β-rays from ^{32}P. Later on, during the Biosatellite II mission, plants and insects were irradiated in-flight with relatively high doses from a ^{85}Sr source. The biosatellite Cosmos 690 mission carried an on-board γ-radiation source (^{137}Cs) to irradiate rats with doses up to 8 Gy. More recently, yeast cells were irradiated in-flight during the Space Shuttle mission STS-84. However, in the past, not many Space missions provided appropriate means for irradiating biological samples on-board. To circumvent this lack of opportunity, biological samples were irradiated before or after space flight. This method was extensively applied during the Cosmos 368, 782 and Salyut mission and more recently in DNA repair studies with cellular systems within the cell cultivation facility Biorack of ESA during the Spacelab missions IML-1, IML-2 and SMM-03.

By use of a combination of the Biostack concept (Fig. 10.10) and an onboard 1×*g* reference centrifuge, the combined effects of microgravity and individual cosmic ray HZE particles were investigated on embryogenesis and organogenesis of the stick insect *Carausius morosus*. It was repeatedly shown that, in radiation sensitive early embryonic stages of development, a combination of HZE particle hit and microgravity acted synergistically on mortality, indicated as reduced hatching rate, and on body anomalies, such as deformities of abdomen and antennae (Fig. 10.13). In experiments with the fruit fly *Drosophila melanogaster*, after exposure of the early development stages to ^{85}Sr γ-rays (up to 14.3 Gy) during space flight, malformations were observed with an increased frequency compared with the concurrent controls. In larvae and adults of *Drosophila*, genetic effects were studied, which included lethal mutations, visible mutations at specific loci, chromosome translocations and chromosome non-disjunctions. Synergism of space flight factors and radiation was observed also in chromosome translocations and thorax deformations. These effects have been suggested to be due to an increase in chromosome breakage followed by a loss or exchange of genetic information. From these results it can be concluded that embryonic systems appear to be especially susceptible to a synergistic interaction of radiation and microgravity.

It has been conjectured that microgravity might interfere with the operation of some cellular repair processes, thereby resulting in an augmentation of the radiation response. Experimental support in favour of this hypothesis has been provided in a Space experiment utilizing a temperature-conditional repair mutant of the yeast *Saccharomyces cerevisiae*, in which the extent of repair of DNA double strand breaks (DSBs) was reduced by approximately a factor of two compared with the 1×*g* ground control; however, this observation could not be confirmed in a

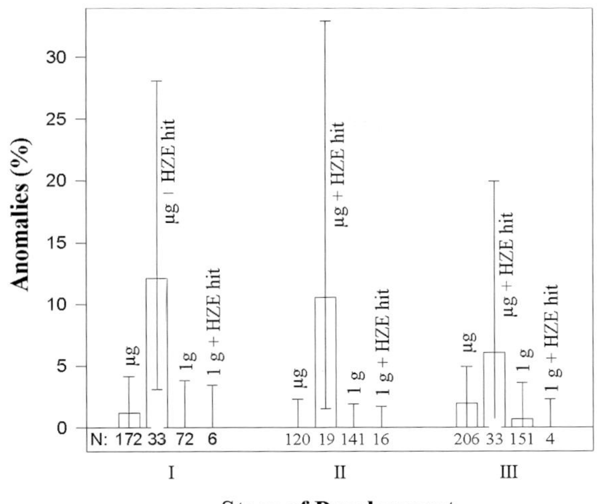

Fig. 10.13 Frequency of developmental anomalies, observed in larvae of *Carausius morosus*, exposed at different embryonic stages to space flight conditions, either at microgravity or on the 1×g reference centrifuge and analysed by use of the Biostack method. Age of eggs during space flight: stage I (16–23 d), stage II (30–37 d), stage III (45–52 d); *N* = number of larvae investigated. (From Ref. [23].)

follow-up experiment. Examining several different repair pathways in different unicellular systems that were irradiated prior to the Space mission, evidence was provided that cells in the microgravity environment possess an almost normal ability to repair radiation-induced DNA damage (Fig. 10.14). In this study, the following repair functions were investigated: (a) the kinetics of rejoining of radiation-induced DNA strand breaks in *Escherichia coli* cells and human fibroblasts; (b) the induction of the SOS response in cells of *E. coli*; and (c) in the efficiency of repair in cells of *Bacillus subtilis* of different repair capacity. Comparison of cells that were allowed to repair in microgravity with those under gravity (1×g reference centrifuge on board or corresponding ground controls) did not show any significant influence of microgravity on the enzymatic repair reactions studied (Fig. 10.14). Although after being irradiated on ground the samples were kept inactive (e.g. frozen, as spores, or at a repair-prohibiting temperature) until incubation in Space, it can not be excluded that the very first steps of repair initiation, such as gene activation, already occurred on ground. Therefore, studies on gene activation related to DNA repair require irradiating of cells directly in Space. This will be achieved with the Biolab facility of ESA on board of the ISS.

If, however, the synergistic effects of microgravity and radiation in biological systems, which has been observed in several instances, can not be explained by a disturbance of DNA repair in microgravity, other mechanisms must be considered:

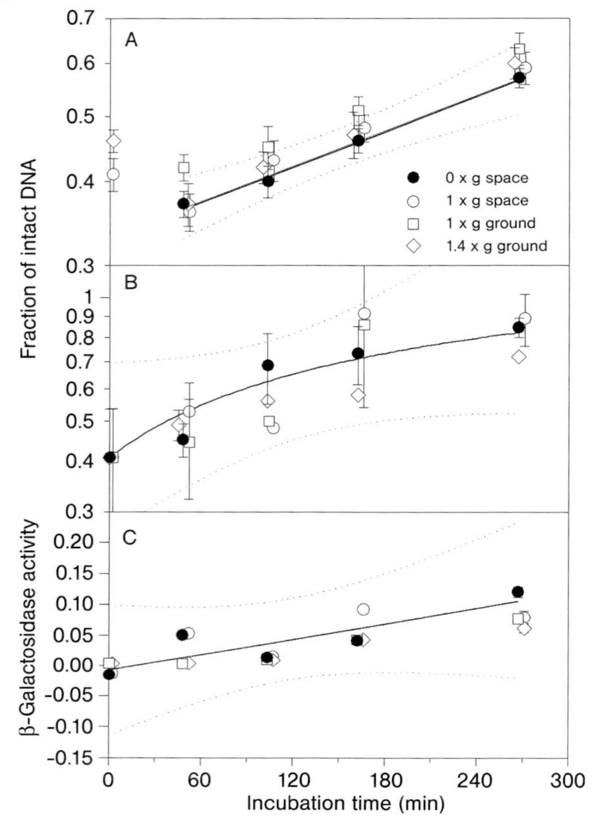

Fig. 10.14 Repair of radiation-induced DNA damage under microgravity conditions. (A) Rejoining of DNA strand breaks in cells of *E. coli* B/r. (B) Rejoining of DNA strand breaks in human fibroblasts. (C) Induction of SOS response in cells of *E. coli* PQ37. (From Ref. [24].)

- at the molecular level as consequences of a convection-free environment;
- at the cellular level as impact on signal transduction, on receptors, on the metabolic/physiological state, on the chromatin, or on the membrane structure;
- at the tissue and organ level as modification of self-assembly, intercellular communication, cell migration, pattern formation or differentiation.

10.3.1.4 **Radiation Protection Guidelines for Astronauts in LEO**

Present guidelines for radiation protection in LEO missions have been derived, starting from a postulated "acceptable" risk for late cancer mortality, which had been justified by comparison with mortality rates from "normal" terrestrial occupations [25, 26]. For Space missions in LEO, a life time excess risk for fatal cancer of 3% due to radiation exposure was judged reasonable, taking into account the fact that Space crews have to cope with other serious risks besides the radiation risk. This risk of 3% is comparable with that in less safe but ordinary industries, such as agriculture and construction. Furthermore, the National Council on Radiation Protection and Measurements (NCRP) recommends age and gender dependent limits. For example, if the career of an astronaut extends over 20 years, the total risk decreases because the susceptibility for radiation-induced cancer decreases with age. Correspondingly, the risk is higher per unit exposure for shorter periods of exposure. The career whole body dose equivalent limit (Sv) for a lifetime excess risk of fatal cancer of 3% as a function of age and gender has been recommended as follows: for males at age of 25, 35, 45, or 55 years the limit has been set at 1.5, 2.5, 3.25 or 4.0 Sv, respectively; for females the limits are lower and at age of 25, 35, 45, or 55 years the limit has been set at 1.0, 1.75, 2.5 or 3.0 Sv, respectively. Under no circumstances should pregnant females be allowed to fly. The special risks for the embryo-foetus are malformations and, particularly, mental retardation, and the risk of cancer is supposed to be greater than for the adults. The dose measurements obtained from previous space flight activities within the geomagnetic shield, in LEO and inside the ISS (Figs. 10.5 and 10.8), have shown that the exposures are sufficiently low that, so, far no special actions were necessary to keep within these NCRP limits.

10.3.2
Life and Solar Electromagnetic Radiation

The Sun has played a vital role as source of energy for prebiotic and biological evolution. The most important role of the Sun for our biosphere is its electromagnetic radiation. However, solar radiation is a Janus face, exerting beneficial as well as hazardous effects on our biosphere (Fig. 10.15). Our biosphere relies on solar infrared (IR) radiation to the keep temperature of the environment clement. The visible range of solar radiation (VIS) is required for vision and it powers photosynthesis. The short wavelength UV radiation range promotes vitamin D synthesis, photo-repair of DNA damage and kills pathogens. However, the UV range of solar radiation is more often damaging to organisms because both proteins and nucleic acids have a maximum absorption in the UV range. Hence, life has developed several protection and repair mechanisms to cope with the hazardous parts of solar electromagnetic radiation [27].

Solar optical radiation is subjected to absorption and scattering as it passes through the Earth's atmosphere, which results in a cut-off at wavelengths below 290 nm. It is especially the stratospheric ozone layer that acts as a protective filter, which effectively cuts off incoming UVC radiation and greatly reduces the amount of UVB radiation reaching the present Earth's surface (Fig. 10.4).

Numerous lines of isotopic and geologic evidence suggest that the Archean atmosphere was essentially anoxic. As a result the amount of ozone in the

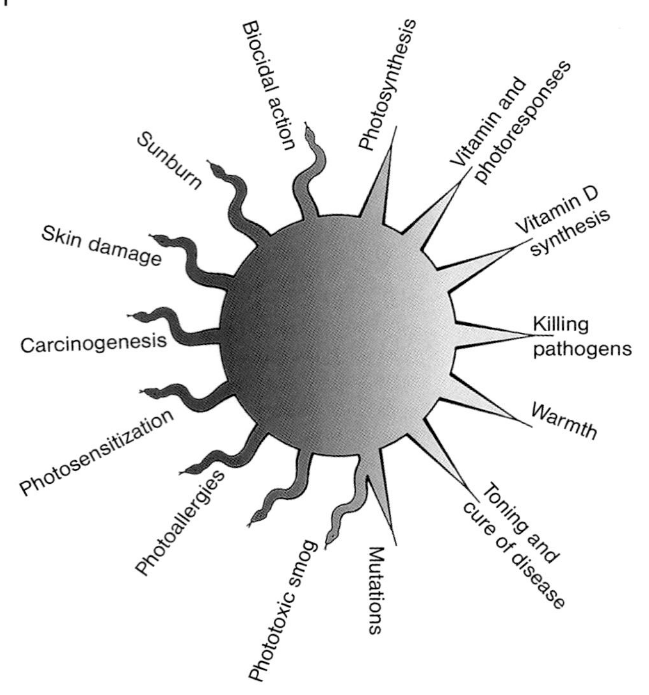

Fig. 10.15 The "Janus face" of the Sun, being beneficial as well as hazardous to life.

stratosphere, if any, would have been insufficient to affect the surface UV radiation environment. Thus, UVB and UVC radiation would have penetrated to the Earth's surface with its associated biological consequences. It took over 2 billion years, until about 2.1 billion years ago, when the Earth's atmosphere was subject to rapid oxidation (as a consequence of oxygenic photosynthesis), before a stratospheric ozone layer was photochemically built up. This UV screen allowed life to spread over the continents and to colonize the surface of the Earth.

10.3.2.1 **Biological effects of Extraterrestrial Solar UV Radiation**

Extraterrestrial solar UV radiation has been found to be the most hazardous factor of Space, as tested with dried preparations of viruses, and of bacterial and fungal spores (summarized in Ref. [28]). The reason for this is the highly-energetic UVC and vacuum UV radiation that is directly absorbed by the DNA, as demonstrated by action spectroscopy in Space (Fig. 10.16) [29]. The full spectrum of extraterrestrial UV radiation kills unprotected spores of *B. subtilis* within seconds [30]. The photobiological and photochemical effects of extraterrestrial UV radiation are based on the production of specific lesions in the DNA that are highly mutagenic and lethal. The most damaging photochemical lesions are thymine-containing

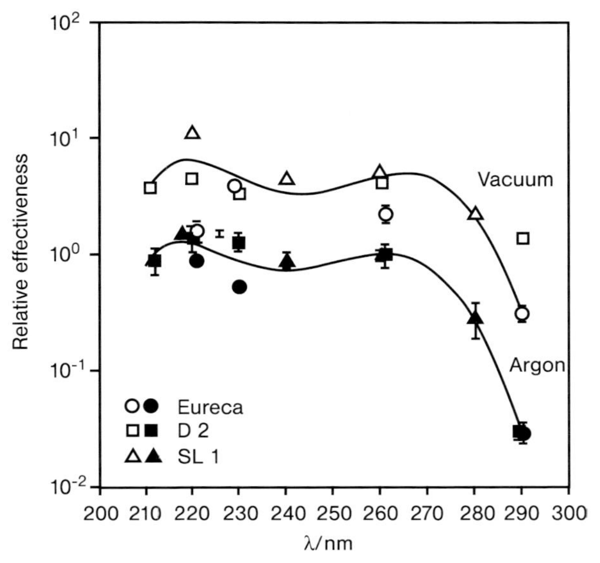

Fig. 10.16 Spectral effectiveness of extraterrestrial solar UV radiation in inactivating spores of *Bacillus subtilis* irradiated in argon at 10^5 Pa (filled symbols) or in Space vacuum (open symbols); cumulative data from three Space experiments normalized to one at 260 nm. (From Ref. [29].)

dimers, cyclobutadipyrimidines, the (6-4) pyrimidine–pyrimidone adducts and their Dewar valence isomers (reviewed in Ref. [31]).

Spores of *B. subtilis* exposed simultaneously to solar UV radiation and Space vacuum responded with an increased sensitivity to a broad spectrum of solar UV (>170 nm) as well as to selected wavelengths. Upon dehydration (e.g. in Space vacuum) DNA undergoes substantial conformational changes. This conversion in the physical structure leads to altered DNA photochemistry, resulting in a tenfold increase in UV-sensitivity of *B. subtilis* spores in Space vacuum as compared with spores irradiated at atmospheric pressure (Fig. 10.16) [29, 30]. In attempts to isolate and identify the photoproducts generated within the DNA of *B. subtilis* spores exposed to UV radiation in vacuum, two thymine decomposition products, namely the *cis-syn* and *trans-syn* cyclobutane thymine dimers, as well as DNA–protein crosslinking were found in addition to the spore photoproduct, the major photoproduct induced by UV in wet spores [30, 32].

The results give some insight into the likelihood of interplanetary transfer of life, the so-called hypothesis of Panspermia [33]. This theory postulates that microscopic forms of life, e.g. spores, can be dispersed in Space by the radiation pressure from the sun, thereby seeding life from one planet to another, or even between solar systems. However, the Space experiments have shown that bacterial spores will be killed efficiently by extraterrestrial solar UV radiation within a few seconds of exposure. Hence, the original theory of Panspermia seems not to be a likely route to interplanetary transfer of life.

However, thin layers of clay, rock or meteorite provided certain UV-shielding to bacterial spores, and rocks a few cm in diameter were sufficient to completely protect bacterial spores against the intense insolation [34]. These data support the hypothesis of Lithopanspermia, as an alternative scenario of potential transport of living matter from one planet of our solar system to another [35–37]: Large impacts on Earth – or any other planet – may eject rocks that could eventually fall on Mars or on other planets of our solar system [38]. Since, on Earth, soil and rocks are colonized by microorganisms and spores, it cannot be excluded that these organisms have been swept along into Space. Recent studies with lichens exposed to Space on the Biopan facility of ESA have demonstrated the high resistance of these symbiotic communities to outer Space [39].

10.3.2.2 **Role of Solar UV Radiation in Evolutionary Processes Related to Life**

Space experiments using the full spectrum of extraterrestrial solar UV radiation as authentic energy source represent new approaches towards our understanding of the role of the UV radiation climate in the evolution of life and potential evolutionary steps on other celestial bodies. So far, the consequences on the biosphere of an increasing environmental UVB irradiation due to a decreasing concentration of the stratospheric ozone were merely based on model calculations [40]. Experimental evidence was obtained in an experiment on board of the German Spacelab D2 mission, where the terrestrial UV radiation climate at different concentrations of stratospheric ozone was simulated using extraterrestrial sunlight and optical cut-off filters (Fig. 10.17a). Ozone concentrations down to very low values were simulated [41]. Biologically effective irradiances as a function of the simulated ozone column thickness were measured with bacterial spores (*Bacillus subtilis*) immobilized in a biofilm [42]. After the mission, the biofilms were incubated, stained and evaluated by image analysis with respect to optical densities, indicating the biological activity for each exposure condition. With decreasing simulated ozone concentrations, the biologically effective solar UV irradiance strongly increased (Fig. 10.17b). At a "hypothetical" complete loss of the ozone layer, the biologically effective irradiance was nearly three orders of magnitude higher than the present value at the surface of the Earth for average total ozone columns. These data demonstrate the importance of the ozone layer in protecting our biosphere against the harmful short wavelength part of solar UV radiation. The experimental data matched very well with model calculations [1].

In experiments using the Biopan facility of ESA on board of a Foton satellite, bacterial endospores were exposed to a simulated UV radiation climate of Mars [43]. The authors tackled the question of the toxicity of the Martian regolith for terrestrial microorganisms exposed to the UV radiation climate of Mars, information pivotal for assessing the habitability of the surface of Mars and for planetary protection efforts.

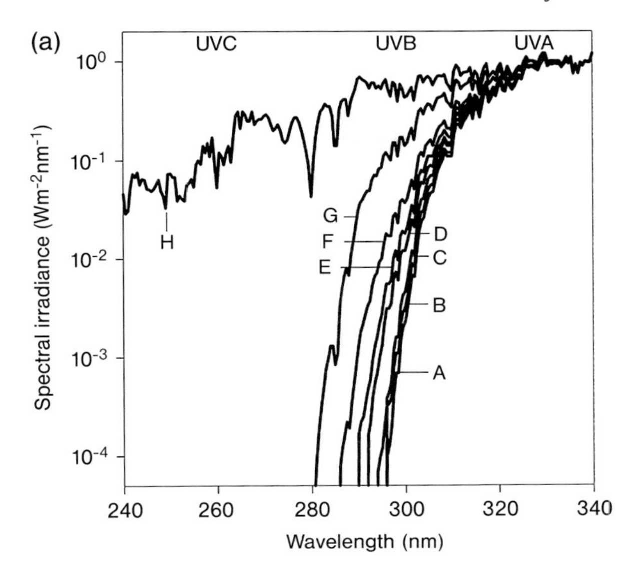

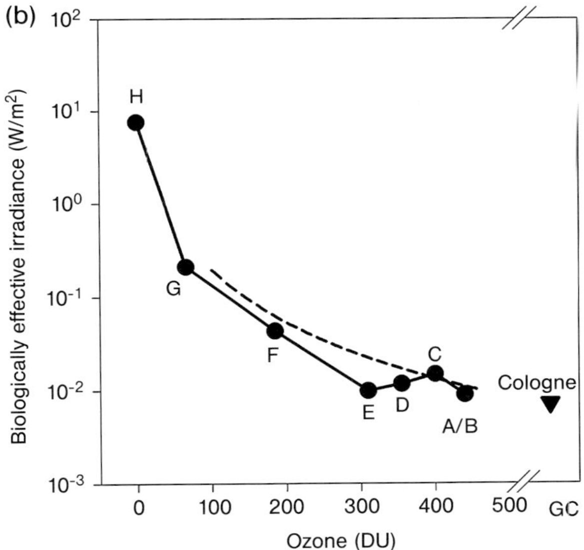

Fig. 10.17 Increase in biologically effective UV irradiance with decreasing simulated ozone-column thickness as measured by the spore biofilm technique in a Space experiment. Extraterrestrial solar spectral irradiances filtered through a quartz (7 mm) plate (H) or additionally through cut-off filters simulating different ozone column thicknesses with a progressive depletion from curve A–G (a); biofilm data of biologically effective solar irradiance for the different ozone-column thicknesses A–G, the extraterrestrial UV spectrum (H) and ground control (Cologne) (b). The dashed line shows the corresponding curve for calculated DNA damage (data modified from Ref. [41]). (DU = Dobson Unit; 1 DU is defined as 0.01 mm thickness of ozone at 0 °C and 10^5 Pa pressure.)

10.4 Outlook: Radiation Biology and Future Exploratory Missions in the Solar System

10.4.1 Habitability of Mars

There is growing evidence that the physical and chemical surface properties of early Earth and early Mars were very similar, and that prior to 3.5 billion years ago and even sporadically in more recent time periods the climate on Mars was wet and more temperate, allowing the presence of large quantities of water in the liquid state on its surface. At that time life had already started on Earth. Under the assumption that life emerges at a certain stage of planetary evolution, and if the right environmental physical and chemical requirements are provided, it is legitimate to assume that conditions on early Mars were as favourable for life to emerge as on early Earth. A closer look at the radiation climate of Mars and its effectiveness on terrestrial model systems may provide clues for estimating the habitability of Mars, in early epochs or even today. The first measurements of the radiation environment (ionizing and UV) on Mars will be performed during the upcoming landing missions to Mars, Mars Science Laboratory of NASA and ExoMars of ESA [44].

The radiation climate on the surface of Mars is governed by the following components:

- solar UV radiation, including the UVC and UVB range ($\lambda > 200\,nm$);
- charged particles of galactic and solar origin as well as neutral secondaries;
- emissions from radionuclides of Mars itself.

Owing to the low shielding by the Martian atmosphere, $5–16\,g\,cm^{-2}$ (depending on altitude), and the lack of a magnetic field, a substantial portion of the cosmic radiation reaches the surface of Mars. In addition, the highly energetic SPEs (up to several GeV) need to be considered. Therefore, the biologically effective annual dose rate amounts to $0.1–0.2\,Sv\,a^{-1}$, depending on altitude [45, 46]. The intensity of the cosmic radiation is also a function of the direction of the incident particles: with increasing zenith angle the amount of protection increases. With increasing depth, the regolith provides a certain radiation protection, although one has to consider the secondary radiation produced.

Although the radiation dose rate on the surface of Mars is about 100× higher than on Earth, it cannot be considered as a limiting factor for microbial life on Mars. A terrestrial example is the most radiation resistant bacterium *Deinococcus radiodurans*. It can tolerate radiation doses up to 5000 Gy without any significant inactivation. Thus, the annual dose rate on Mars, which is 3–4 order of magnitudes lower, would not impair growth of these microorganisms possessing potent DNA repair systems. Likewise, *B. subtilis* spores with a D_{37} of 600 Gy (dose causing a survival rate of 37%) would be able to survive the radiation exposure on the

Martian surface for extended periods. If microbial life once existed on Mars, spores could be suitable candidates to survive dry interim periods between two wet epochs that have been suggested to have occurred during Martian history. In the Martian subsurface, at greater depths, radiation doses are gradually reduced [36].

10.4.2 Radiation Protection Guidelines for Astronauts during Exploratory Missions

In contrast to the situation in LEO, the doses expected during exploratory mission, to the Moon and especially to Mars, are likely to infringe the limits established for LEO missions in certain instances (summarized in Ref. [47]). For a lunar base, these estimates for the total radiation doses conceivably to be incurred at the blood-forming organs (BFO) from GCR and from SPE indicate that some exposure levels, for example for SPEs, exceed the limits set forth in radiation protection guidelines for LEO, e.g. the ISS operations. In general, the exposure levels for GCR are fully compatible with the exposure limits recommended for LEO activities (Table 10.1). The value finally depends on the number of EVAs with much lower shielding.

However, the doses expected when encountering a large SPE could possibly induce acute radiation injuries even behind 5 $g\,cm^{-2}$ Al and even in the so-called shelter with 10 $g\,cm^{-2}$ Al, as shown for the doses deposited by the worst case reference event in deep Space (Table 10.1). If such an SPE event would be encountered with only 1 $g\,cm^{-2}$ Al or even less shielding, e.g. in a Space suit during EVA, severe and even incapacitating acute radiation injuries could ensue with a substantial probability for a fatal outcome unless adequate medical support could be supplied.

Table 10.1 Estimated radiation equivalent doses (Sv) for the blood-forming organs (BFO) during a 6 months long lunar mission (behind different shielding thicknesses and at different times of the solar cycle); values in italics exceed the annual dose limits for Low Earth Orbit (LEO) given in Ref. [25][a]. GCR = Galactic Cosmic Rays; SPE = Solar Particle Event. (Modified from Ref. [47].)

Shielding ($g\,cm^{-2}$)	***0.3***	***1.0***	***5.0***	***10.0***
GCR (Sv)				
At solar minimum		0.195	0.177	0.161
At solar maximum		0.074	0.070	0.066
"Worst case" SPE (Sv)				
Interplanetary travel	*4.21*	*3.52*	*1.93*	*1.26*
Lunar surface	*2.11*	*1.76*	*0.97*	*0.63*

a National Council on Radiation Protection and Measurements (NCRP) limits for LEO: 0.25 Sv for 30 d; 0.50 Sv annual equivalent doses; and 0.4–4.0 Sv for career dose, the latter depending on gender and age [25].

Table 10.2 Estimated equivalent doses (Sv) received during two reference missions to Mars at the blood-forming organs (BFO) as a consequence of exposure to galactic cosmic radiation (GCR) or to one "worst case" solar particle event (SPE). Values in italics exceed the annual dose limits for LEO given in Ref. [25]. (Modified from Ref. [47].)

Mission duration	*Type of radiation*	*Shielding ($g\,cm^{-2}$)[a]*				
		0.3	***1.0***	***5.0***	***10.0***	***20.0***
	GCR					
1000 d	At solar minimum[b]		*0.993*	*0.918*	*0.852*	*0.769*
	At solar maximum[c]		0.402	0.383	0.364	0.339
500 d	At solar minimum[b]		*0.828*	*0.754*	*0.687*	*0.605*
	At solar maximum[c]		0.317	0.299	0.280	0.255
Any	"Worst case" SPE[d]					
	Interplanetary travel	*4.21*	*3.52*	*1.93*	*1.26*	
	Mars surface	0.32	0.31	0.28	0.25	

a Aluminium.
b Based on 1977 solar minimum data.
c Based on 1970 solar maximum data.
d Based on SPE from September 1989 data multiplied by a factor of 10 to mimic in addition the February 1956 SPE that produced, so far, the most intense ground level event [25].

Late effects, such as enhanced morbidity or mortality from malignant cancers occurring up to 20 and more years after exposure have to be considered as well as early effects that may consist of morbidity such as anorexia, fatigue, nausea, diarrhoea, and vomiting (the symptoms of the so-called prodromal syndrome) or cataract formation and erythema and early mortality within days to a few weeks from failures of the hematopoietic, the pulmonary and the gastrointestinal system. Depending on the phase of the mission when such symptoms occur, they may well be associated with performance losses of the affected crew, which in turn might increase the risk for critical failures.

Human visits to Mars will add a new dimension to human space flight, concerning the distance of travel, the radiation environment, the gravity levels, the duration of the mission, and the level of confinement and isolation to which the crew will be exposed. Table 10.2 lists the equivalent doses for the blood-forming organs estimated for the different missions during different phases of the solar cycle. For missions during solar minimum as well as during a "worst case" SPE, the expected radiation dose exceeds the annual equivalent dose limit of 0.5 Sv, which has been established for missions in LEO, such as the ISS, with the aim of keeping the radiation induced lifetime excess of late cancer mortality below 3% [25].

To minimize the risk from Space radiation during exploratory missions, future research and development is required within the following categories:

- an adequate quantitative risk assessment for accurate mission design and planning;
- the surveillance of individual radiation exposure by personal dosimetry during all phases of the mission;
- surrounding crew habitats with sufficient absorbing matter;
- countermeasures to minimize health detriment from radiation actually received by selecting radiation resistant individuals or by increasing resistance, e.g. by radioprotective chemicals. The opposite selection process, whereby individuals with an identifiable genetic disposition for increased susceptibility to spontaneous – and implied – radiogenic cancerogenesis are detected, must in any case be part of the standard crew selection.

References

1 Cockell, C.S., Horneck, G. *Photochem. Photobiol.* **2001**, *73*, 447–451.

2 McCormack, P.D., Swenberg, C.E., Bücker, H. (Eds.), *Terrestrial Space Radiation and its Biological Effects*, Plenum Press, New York-London, **1988**.

3 Reitz, G., Facius, R., Sandler, H. *Acta Astronaut.* **1995**, *35*, 313–338.

4 Cucinotta, F.A., Wu, H., Shavers, M.R., George, K. *Gravitat. Space Biol. Bull.* **2003**, *16*, 11–18.

5 Labs, D., Neckel, H., Simon, P.C., Thullier, G. *Solar Phys.* **1987**, *107*, 203–219.

6 Horneck, G., Rettberg, P. (Eds.), *Complete Course in Astrobiology*, Wiley-VCH, Berlin, **2007**.

7 Swenberg, C.E., Horneck, G., Stassinopoulos, E.G. (Eds.), *Biological Effects and Physics of Solar and Galactic Cosmic Radiation*, Plenum Press, New York, **1993**.

8 Reitz, G. *Acta Astronaut.* **1994**, *32*, 715–722.

9 Apáthy, I., Deme, S., Fehér, I., Akatov, Y.A., Reitz, G., Arkhanguelski, V.V. *Radiat. Meas.* **2002**, *35*, 381–391.

10 Beaujean, R., Kopp, J., Burmeister, S., Petersen, F., Reitz, G. *Radiat. Meas.* **2002**, *35*, 433–438.

11 Reitz, G., Berger, T. *Radiat. Prot. Dosim.* **2006**, *120*, 442–445. Doi:10.1093/rpd/nci558.

12 Bücker, H., Horneck, G., Studies on the Effects of Cosmic HZE Particles on Different Biological Systems in the Biostack Expriments I and II Flown Onboard of Apollo 16 and 17, in: *Radiation Research*, Nygaard, O.F., Adler, H.I., Sinclair, W.K., (Eds.), pp. 1138–1151, Academic Press, New York, **1975**.

13 Horneck, G. *Nucl. Tracks Radiat. Meas.* **1992**, *20*, 185–205.

14 Kiefer, J., Schenk-Meuser, K., Kost, K., Radiation Biology, in: *Biological and Medical Research in Space*, Moore, D., Bie, P., Oser, H. (Eds.), pp. 300–367, Springer, Berlin-Heidelberg-New York, **1996**.

15 Horneck, G., Baumstark-Khan, C., Facius, R., Radiation Biology, in: *Fundamentals of Space Biology*, Clément, G., Slenzka, K. (Eds.), pp. 291–335, Springer, Berlin-Heidelberg-New York, **2006**.

16 Nicholson, W.L., Munakata, N., Horneck, G., Melosh, H.J., Setlow, P. *Microb. Mol. Biol. Rev.* **2000**, *64*, 548–572.

17 Morgan, W.F. *Radiat. Res.* **2003**, *159*, 567–580.

18 Avdeev, S., Bidoli, V., Casolino, M., De Grandis, E., Furano, G. et al. (22 other authors), *Acta Astronaut.* **2002**, *50*, 511–535.

19 Casolino, M., Bidoli, V., Morselli, A., Narici, L., De Pascale, M.P., Picozza, P.,

Reali, E., Sparvoli, R., Mazzenga, G., Ricci, M., Spillantini, P., Boezio, M., Bonvicini, V., Vacchi, A., Zampa, N., Castellini, G., Sannita, W.G., Carlson, P., Galper, A., Korotkov, M., Popov, A., Vavilov, N., Avdeev, S., Fuglesang, C. *Nature* **2003**, *422*, 680.

20 Obe, G., Johannes, I., Johannes, C., Hallmann, K., Reitz, G., Facius, R. *Int. J. Radiat. Biol.* **1997**, *72*, 726–734.

21 Durante, M., Snigiryova, G., Akaeva, E., Bogomazova, A., Druzhinin, S., Fedorenko, B., Greco, O., Novitskaya, N., Rubanovich, A., Shevchenko, V., von Recklinghausen, U., Obe, G. *Cytogenet. Genome Res.* **2003**, *103*, 40–46. DOI: 10.1159/000076288.

22 Horneck, G. *Mutat. Res.* **1999**, *430*, 221–228.

23 Reitz, G., Bücker, H., Rüther, W., Graul, E.H., Beaujean, R., Enge, W., Heinrich, W., Mesland, D.A.M., Alpatov, A.M., Ushakov, I.A., Zachvatkin, Y.A. *Nucl. Tracks Radiat. Meas.* **1990**, *17*, 145–153.

24 Horneck, G., Rettberg, P., Kozubek, S., Baumstark-Khan, C., Rink, H., Schäfer, M., Schmitz, C. *Radiat. Res.* **1997**, *147*, 376–384.

25 NCRP, *Guidance on Radiation Received in Space Activities*, NCRP Report No. 98, National Council on Radiation Protection and Measurements, Bethesda, MD, **1989**.

26 NCRP, *Radiation Protection Guidance for Activities in Low-Earth Orbit*, NCRP Report No. 132, National Council on Radiation Protection and Measurements, Bethesda, MD, **2000**.

27 Rothschild, L.J. The Sun: the Impetus of Life, in: *Evolution on Planet Earth*, Rothschild, L.B., Lister, A.M. (Eds.), pp. 87–107, Academic Press, Elsevier, San Diego, **2003**.

28 Horneck, G. *Origins Life Evolut. Biosphere* **1993**, *23*, 37–52.

29 Horneck, G., Eschweiler, U., Reitz, G., Wehner, J., Willimek, R., Strauch, K. *Adv. Space Res.* **1995**, *16(8)*, 105–118.

30 Horneck, G., Bücker, H., Reitz, G., Requardt, H., Dose, K., Martens, K.D., Mennigmann, H.D., Weber, P. *Science* **1984**, *225*, 226–228.

31 Cadet, J., Weinfeld, M. *Anal. Chem.* **1993**, *65*, 675A–682A.

32 Lindberg, C., Horneck, G. *J. Photochem. Photobiol. B: Biol.* **1991**, *11*, 69–80.

33 Arrhenius, S. *Die Umschau* **1903**, *7*, 481–485.

34 Horneck, G., Rettberg, P., Reitz, G., Wehner, J., Eschweiler, U., Strauch, K., Panitz, C., Starke, V., Baumstark-Khan, C. *Origins Life Evol. Biosphere* **2001**, *31*, 527–547.

35 Clark, B., Baker, A.L., Chen, A.F., Clemett, S.J., McKay, D., McSween, H.Y., Pieters, C.M., Thomas, P., Zolensky, M. *Origins Life Evol. Biosphere* **1999**, *29*, 521–545.

36 Mileikowsky, C., Cucinotta, F.A., Wilson, J.W., Gladman, B., Horneck, G., Lindegren, L., Melosh, H.J., Rickman, H., Valtonen, M., Zheng, J.Q. *Icarus* **2000**, *145*, 391–427.

37 Clark, B., *Origins Life Evol. Biosphere* **2001**, *31*, 185–197.

38 Melosh, H.J. *Nature* **1988**, *332*, 687–688.

39 Sancho, L., personal communication.

40 SCOPE, *Effects of Increased Ultraviolet Radiation on Global Ecosystems*, Scientific Committee on Problems of the Environment, Paris, France, **1992**.

41 Horneck, G., Rettberg, P., Rabbow, E., Strauch, W., Seckmeyer, G., Facius, R., Reitz, G., Strauch, K., Schott, J.U. *J. Photochem. Photobiol., B: Biol.* **1996**, *32*, 189–196.

42 Quintern, L.E., Horneck, G., Eschweiler, U., Bücker, H. *Photochem. Photobiol.* **1992**, *55*, 389–395.

43 Rettberg, P., Panitz, C., Rabbow, E., Horneck, G. *Adv. Space Res.* **2004**, *33*, 1294–1301.

44 Vago, J., Kminek, G., Putting Together an Exobiology Mission: The ExoMars Example, in: *Complete Course in Astrobiology*, Horneck, G., Rettberg, P. (Eds.), pp. 321–351, Wiley-VCH, Berlin, **2007**.

45 Wilson, J.W., Townsend, L.W., Schimmerling, W., Khandelwal, G.S., Khan, F., Nearly, J.E., Cucinotta, F.A., Norbury, J.S., Transport Methods and Interactions for Space Radiations, in: *Biological Effects and Physics of Solar and Galactic Cosmic Radiation, Part A*, Swenberg, C., Horneck, G., Stassinopoulos, E.G. (Eds.), pp. 187–786, Plenum, New York, **1993**.

46 Saganti, P.B., Cucinotta, F.A., Wilson, J.W., Simonsen, L.C., Zeitlin, C. *Space Sci. Rev.* **2004**, *110*, 143.

47 Horneck, G., Facius, R., Reichert, M., Rettberg, P., Seboldt, W., Manzey, D., Comet, B., Maillet, A., Preiss, H., Schauer, L., Dussap, C.G., Poughon, L., Belyavin, A., Heer, M., Reitz, G., Baumstark-Khan, C., Gerzer, R. *HUMEX, a Study on the Survivability and Adaptation of Humans to Long-duration Exploratory Missions*, ESA SP-1264, ESA Publications Division, Noordwijk, **2003**.

Index

Biology in Space and Life on Earth. Effects of Spaceflight on Biological Systems. Edited by Enno Brinckmann

ISBN: 978-3-527-40668-5

t

v

w

Related Titles

Horneck, G., Rettberg, P. (eds.)

Complete Course in Astrobiology

2007
ISBN 978-3-527-40660-9

Shaw, A. M.

Astrochemistry

From Astronomy to Astrobiology

2006
ISBN 978-0-470-09137-1